OLD
FORTY-FOUR

OLD FORTY-FOUR

A Historical and Geological Excursion
Over New Mexico's Old Route 44
(Now Part of US-550)

Dirk Van Hart

SUNSTONE
PRESS

SANTA FE

Sunstone books may be purchased for educational, business, or sales promotional use.
For information please write: Special Markets Department, Sunstone Press,
P.O. Box 2321, Santa Fe, New Mexico 87504-2321.

Book and Cover design › Vicki Ahl
Body typeface › Book Antiqua
Printed on acid-free paper ∞

―――――――――――――――――――――――――――――――――――

Library of Congress Cataloging-in-Publication Data

Van Hart, Dirk, 1941-
   Old Forty-Four : a historical and geological excursion over New Mexico's old
Route 44 (now part of US-550) / by Dirk Van Hart.
      p. cm.
   Includes bibliographical references.
   ISBN 978-0-86534-837-0 (softcover : alk. paper)
  1. New Mexico--History, Local. 2. United States Highway 550 (Colo. and N.M.)
--History, Local. 3. United States Highway 550 (Colo. and N.M.)--History.
4. Roads--New Mexico--History. 5. Geology--New Mexico. I. Title.
   F797.V24 2012
   978.9--dc23
                              2012007597

―――――――――――――――――――――――――――――――――――

WWW.SUNSTONEPRESS.COM
SUNSTONE PRESS / POST OFFICE BOX 2321 / SANTA FE, NM 87504-2321 /USA
(505) 988-4418 / ORDERS ONLY (800) 243-5644 / FAX (505) 988-1025

Contents

List of Figures

Chapter 5. Bernalillo — Select Points of Interest.

Chapter 6. Rio Jémez Corridor.

Chapter 19. San Juan Basin Vegetation.

Chapter 20. Counselor and the "Upper" Canyon Largo.

Chapter 21. Side Journey down the Canyon Largo.

Chapter 22. Lybrook.

Chapter 23. Indian Country I—the Jicarilla Apache Nation (*Tinde*).

This book is about a well-traveled highway in New Mexico. The highway itself provides the organizing element for the book's narrative. The road—the erstwhile New Mexico Route 44 redesignated in 2000 as the southern segment of U.S. Route 550—is the principal highway that connects the heavily populated Rio Grande Valley in central New Mexico to the once remote San Juan River Valley in far northwestern New Mexico. Old timers and regular users of this route will likely always think of it as "NM-44," or "Old Forty Four." Throughout this book, however, I will refer to what was once Old Forty Four as "US-550," but with the understanding that I am referring only to the segment of US-550 that in 2000 replaced NM-44 and that connects Bernalillo on the Rio Grande to Bloomfield on the San Juan River.

This book is not a travelogue. Rather it is a series of stories, explanations, asides, history, geography, and geology—all those things that make an area interesting—linked together by the highway itself. Why this particular highway? Aren't there many others in the state of New Mexico that have equal or more appeal?

By nature I'm a curious guy. I found this route just loaded with visual information that led to a multitude of questions. Reader, follow me on this. Many of New Mexico's roads are more dramatic than NM-44, but often their scenery is remote and always "out there" or "way over there," and therefore beyond the traveler's reach. When I say "information" I mean observable, interesting, and comprehensible details, unlike what we see along many of the mind-numbing interstate highways. It's something like comparing a banana split, garnished with chocolate and whipped cream, crowned with a red cherry and elegantly served on a colorful platter, to a scoop of vanilla ice cream slopped onto a paper plate. The segment of US-550 between Bernalillo and Bloomfield is jam-packed with exquisite geology, bizarre scenery, and intriguing history, all made understandable with a little help.

Accessible and understandable! In this regard I disagree with some who, I suppose, wish to insulate the viewer from the landscape via a veil of mystery. For example, John De-Witt McKee, late professor of English at New Mexico Institute of Mining and Technology in Socorro, wrote the following concerning the New Mexico landscape

> "What is it then that holds us to this curious, raw, new, old and savage land? It is not love, for the land itself is too aloof for love. It is not landscape, for there is no landscape here. There is only the land, which can no more be trapped for taming than can the fleeting watermelon color of the mountains, coming and going between on eye-blink and the next. Landcapes can be whistled in and brought to heel, ordered and arranged in frames. But this! So seemingly inert, impassive, barren, this land will not submit to capture" (McKee 1981).

Well, maybe. The landscape certainly can't be trapped and put in one's pocket, but it certainly can be understood—if perhaps only a little bit at a time—and therefore enjoyed in an entirely new way. The landscape is definitely not inert, at least not when viewed through the prism of time. As an exemplar of the importance of time, I recall an episode from the very successful TV show *Star Trek—the Next Generation* (1987–1994) in which alien intruders appear on the bridge of the starship *Enterprise*. The intruders live in a different time realm that operates very much faster than the scale in which the ship's crew lives. To the intruders the crew members are standing and sitting immobile and frozen. The crew is in fact moving about as usual for them, but at a rate too slow for the intruders to recognize as motion. In the context of our landscape, we're the intruders, the landscape is the crew. Everything is moving and changing, but at vastly different rates. What we see from our car window is much like a single frame extracted from a reel of movie film. In this book I have tried to give the reader a sense of this temporal context, at least in a backward direction deep into the geologic past. I'll leave speculations in the forward, future direction for someone else.

Most of us don't realize that the appreciation of landscape is a rather recent phenomenon. For centuries the only source of history was the Bible. The book of *Genesis* doesn't mention anything like mountains until the story of the Great Flood. Most people, Europeans at least, therefore thought that such dramatic landscape was not part of God's original plan. Such places were imbued with mystery and inhabited by who knows what terrible and dangerous creatures. The very ground and the down direction was associated with dirt, filth, and stuff you don't want to get in your mouth. Those places were to be feared and avoided. Only in the early 19th century, during the so-called Romantic Era, did intrepid folk begin to take note and interest in landscapes.

The same was true in New Mexico. In the early days the residents were struggling just to get by and had precious little spare time to muse (at least in writing) about their surroundings. They ventured out into the hinterlands only when they had to, and when they did it was certainly not for sightseeing. Only when the railroads successfully penetrated these lands and made them easily accessible did people in large numbers begin to fully appreciate the wonders of the New Mexico landscape.

We are very much linked to the landscape. It shapes our institutions, history, and very lives, and we in turn modify it. This book attempts to meld the landscape, i.e., its geography and geology, with the area's human history. The two compliment each other and their combination increases the value of the overall product. For maximum effect I have placed great emphasis on maps. These can synthesize many disparate facts and, if done correctly, can nicely blend them into a comprehensive visual package. For that reason, maps are extremely effective teaching instruments. I have often been frustrated, and annoyed, by books dealing with geography and history because far too many of them lack a sufficient number of maps to compliment the text. That usually leaves me groping backwards through the text struggling to understand what the author is trying to tell me.

As to the writing of this book, I must first interject the bare outlines of my biography. I am a professional geologist and have been so since 1965. In 1986, having just lost my job, my family and I relocated to Albuquerque after a varied career in petroleum and natural-gas exploration. After such disparate stop-gap activities such as high-school teaching, international geological consulting, and field-hand labor, in 1994 I happened upon an opportunity to provide contract geological support for environmental projects at Sandia National Laboratories in Albuquerque. Finally, in 2003, I semi-retired. During the years in our new home state of New Mexico I had silently ranked the state's roads as seen through the eyes of a geologist. I rated our drives as intriguing, so-so, or blah. One route in particular, NM-44 (now of course part of US-550) from Bernalillo northwest to Cuba and on to Bloomfield in the valley of the San Juan River, drew me back again and again. I wanted to learn more about its history and geology, but could not find much information in a readily-accessible form. Therefore I decided to write this book, and in early 2000 began my research. Step by inexorable step the project established a life of its own, and of course it ballooned.

The first decision I faced was how to organize the book. How should I combine the different domains of geography, geology, and history into an interesting whole. How to keep from jumping aimlessly about from place to place and from time to time in the treatment of my material? It came down to recognizing that the central subject of this book, Old NM-44, the southern leg of highway US-550 and its neighborhood, is numerically organized by the New Mexico State Highway Department via mile posts (MPs). They range from a point of origin at Mile Post 0 (MP-0) just east of Bernalillo at the intersection of US-550 with I-25, to an end point a little north of MP-151 in beautiful downtown Bloomfield. This linear, numerical, easily-seen continuum just had to play the dominant role in organizing the narrative's subject matter. However, historical accounts don't necessarily follow the order of the mile posts, so I've blended them with their position along the road as best I could.

I have tried to control my tendency to wander off onto geographical and historical tangents. Such tangents onto related subjects and issues often augment the main subject and are to be expected and welcomed. They're also fun. My choice of tangents into regions adjacent to and reachable from US-550, and of those subjects invoked by the route, is a personal one. They're the ones I like and the ones I think that matter. The final result has been a book organized into 32 chapters, each covering either a major subject or a leg of the highway's course. I have generously cited my sources of information, and the reader will find the full list of references near the book's end.

Finally, I appeal to the reader's sense of curiosity. Those who consider this highway to be on a par with the mile-crunching grind of some of the nation's interstate highways should perhaps instead rent a book-on-tape. Consider this gentle prod by the British writer G.K. Chesterson (1874–1936): "There is no such thing on earth as an uninteresting subject; the only thing that can exist is an uninterested person." And then this fine chestnut by the 20th century American theoretical chemist and biologist, Linus Pauling (1901–1994): "The satisfaction of curiosity is one of the greatest sources of happiness in life."

I really love that last one. It sums up what I'd like to convey to the reader of this book. If one is willing to open one's eyes, recognize and enjoy the landscape's patterns, appreciate the area's past, allow time for an exploratory side trip or two, and ask mental questions, I think this unique and wonderful New Mexico highway will become a favorite.

Acknowledgments

Many have graciously given me their time and knowledge for the writing of this book. I would like to acknowledge the following people and entities, listed by general location along Old-44 from south to north.

Bernalillo: The Sandoval County Historical Society. The Society generously allowed me to use some of their vintage photos. Martha Liebert, the virtual backbone of the Society, read the entire text and gave me her recommendations, encouragement, and the pleasure of her wonderful sense of humor. Thank you Martha.

Cuba and Cuba area: Carlos and Rita Atencio for an interview in their home; Vandora P. Casados, Cuba Water Clerk; Fred Eichwald, Magistrate judge; John Hernández, proprietor of Young's Hotel, for three fascinating interviews at his home; Carla May Johnson, Cuba-area rancher, for two interviews at her ranch west of town; Gail Peters, a wonderfully hospitable Deer Lake resident up in the hills east of Cuba; Barbara Trujillo, Cuba librarian; Victor Velarde, principal of Cuba High School; Louis and Carol Wiese, Cuba ranchers, for two informative interviews at their home; and Mrs. Lovena Wood, Penistaja area rancher, who took the time to show my wife and me around.

Counselor Trading Post: Kelly Aragón, then-proprietor, for an interview at her home behind the post.

Canyon Largo: Patricia Irick, Largo Canon School, for her research tips via E-mail.

Bloomfield-Farmington-Aztec area: Larry Baker, Nancy Espinosa, and Diane Hayden of the San Juan County Salmon Ruins Museum and Research Library; Jim Copeland, Farmington Bureau of Land Management office; Susie Henry, for an interview at her home in Aztec; Elizabeth Kaime, for an interview in an Aztec restaurant; Debbie Mohler, proprietor of the Triangle Café; Cathy Truby of Aztec; Inez Truby, for an interview in her Bloomfield home; and Bob Young, Farmington writer.

Others: Fellow geologists Bob Grant and Dave Love; Mike Coleman, at the New Mexico Environment Department; Laura Glaessner and Julie Dehaven, of the Earth Data Analysis Center (EDAC) at the University of New Mexico; Pat Hester, at the Albuquerque Bureau of Land Management office; Rupert López and Alex Gallegos, ex-CCC boys of camp SCS-8-N; and Laurel Wallace, of the New Mexico Department of Transportation in Santa Fe; James Creager, freelance photographer, for permission to use his aerial photo of Cabezón Peak. Also to Klara B. Kelley and Harris Francis, who generously allowed me to cite information gathered by them in about the trading posts in their unpublished *Navajoland Trading Post Encyclopedia in Progress*.

A great deal of the material incorporated into this book is based on published sources. Although I have taken great pains to cite all of these in the References Cited section, the citations tend to be rather "clinical" and fail to fully convey my humble acknowledgment of the debt I owe their earlier work. I doff my field hat to those authors. To my colleagues in New Mexico geology who read this book, I fervently hope that any errors you might uncover are not overly egregious.

I would like to doubly thank Dr. Paul Catacosinos, retired professor of geology who hard-edited the entire manuscript, not once but twice, and vastly improved it each time. Thank you Paul.

And there is my dear wife Rusty, who somehow bore the role of writer's widow as her husband spent endless hours hunkered down in his office. She did the final read of the manuscript and found a large number of little glitches to which I was utterly blind.

As stated in the preface, "Old Forty Four" (a.k.a. New Mexico state highway 44 or NM-44) is the precursor to the southern segment of federal highway US-550, and is the subject of this book. It serves as the principal connecting link between the populations centers of New Mexico's middle Rio Grande Valley and the San Juan River Valley, and as the gateway to the mountains and canyons of the fabulous San Juan Mountains of Colorado and the canyon country of the "Four Corners" region (Figure 1).

Figure 1. Location of Old NM Route 44 (heavy black line) and US-550.

100 Miles

	Southern leg of US-550 = "Old NM Route 44," i.e., the subject of this book
550	Northern leg of US-550

However, because the subject of this book is the highway, we need to start with a definition of just what that highway is. Its southeast end in particular has undergone some major surgery over the years. Not helpful is the fact that the latest (1978) U.S.Geological Survey (USGS) topographic map (Albuquerque quad, 1:100,000) shows NM-44 continuing east and southeast from Bernalillo to Sedillo on I-40 east of Albuquerque (Figure 2). In about 1988 the New Mexico State Highway and Transportation Department lopped off everything east of Bernalillo from NM-44. It redesignated the part from Bernalillo east to Placitas and south to the Sandia Crest Road as NM-165, that part east to San Antonito on NM-14 near Sandia Park on the east side of the Sandias as NM-536, and that part east and southeast to Sedillo as NM-306 (Riner 2004). Throughout this book the terms "NM-44" and "US-550" refer only to that stretch of highway from Bernalillo northwest to Bloomfield.

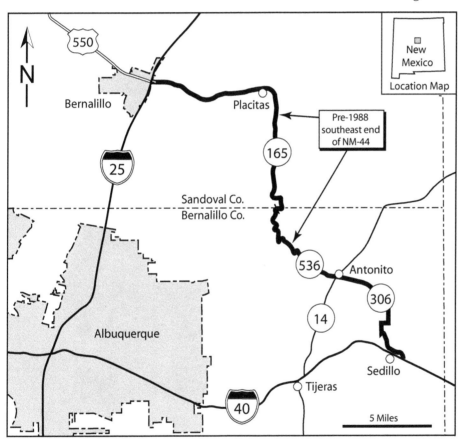

Figure 2. Southeastern pre-1988 terminus end of NM-44.

The"killer road"

In January 2000, state highway NM-44 was overlaid by the southern reach of US-550. The full US-550 now runs 302 miles from Bernalillo, New Mexico, north to Montrose, Colorado (Figure 1). NM-44 no longer exists. By November 2001 the entire 151.7-mile length of the erstwhile NM-44 between Bernalillo and Bloomfield had been widened and upgraded. What had once been a dangerous two-lane road, especially the reach between Cuba and Bloomfield, now became a four-lane thoroughfare and one of the state's finest.

Prior to the reconstruction, NM-44 was known as a killer and was always on the list of the state's top ten deadliest (the most deadly was US-666, now US-491, between Gallup and Shiprock, Figure 1). The two narrow lanes and the lethal mixture of speeding tourists, moping locals, and lumbering semis encouraged foolish attempts to pass, and drunken drivers made the situation worse. Whole families were wiped out. A popular bumper sticker at the time said, "Pray for me, I drive NM-44" (Linthicum 2001).

Beginning during the San Juan basin oil and gas boom of the early 1950s, community leaders had been lobbying for an improved NM-44. In 1973 persistent safety concerns led to the formation of the "Highway 44 Association" in a desperate move to push for improvement. At that time the road, especially that between Cuba and Bloomfield, was characterized by potholes and broken shoulders caused by the steady pounding of traffic (Gantner 1973). Heavy haulage to support the natural-gas and power-generation industries in the San Juan basin and the new Navajo Indian Irrigation Project assured that heavy truck traffic would only increase. (The inadequate narrow gauge railroad that had served the Four Corners from southwestern Colorado since 1923 was shut down in 1968 and never replaced.) Between 1989 and 1993 there were 60 fatalities, and in 1994 alone a dozen people died in five accidents along the 50-mile stretch between Cuba and Nageezi (Figure 1; Linthicum 2000).

In January 1995 the Planning Division of the New Mexico State Highway and Transportation Department (NMSHTD issued its "Long-Range Comprehensive Transportation Plan" for the state. This document proposed a system of four-lane arterial highways that would link all the state's major cities. Reconstruction of NM-44 was part of the plan. In 1999 the NMSHTD had widened the 20.5-mile stretch between Bernalillo and the south edge of San Ysidro. The segment from MP-143 to the south edge of the City of Bloomfield had been widened as well. The big 118-mile gap between MP-24 at the west edge of San Ysidro and MP-143 (with one mile left out within the Village of Cuba) remained.

With the enthusiastic support of then-Governor Gary Johnson, in 1998 the NMSHTD made a bold move by hiring a supercontractor, Mesa PDC (owned by Koch Materials Co.), to construct the entire road in one stage, rather than doing it piecemeal over a period of years. However, New Mexico law prohibited the designer from also constructing the road. Accordingly, Mesa subcontracted an engineering company, CH2M Hill, to design the work, and a construction contractor, Flatiron Structures, to manage the construction activities. Mesa then subcontracted four construction firms to work on four segments, spread across the 118 miles, simultaneously: 1) E.L. Yeager, from mile 24 at San Ysidro to mile 54 near La Ventana; 2) Sundt Companies, mile 54 to mile 63.5 on the south edge of Cuba; 3) FNF Construction, mile 65 on the north edge of Cuba to mile 115 near Nageezi; and 4) Lafarge/Western Mobile, mile 115 to mile 143.

The job was completed in 27 months. If done segment by segment under normal procedures the project would have taken about 27 years. Mesa was paid $46 million for the design and overseeing the construction and $62 million for a warranty. The warranty covers the highway for 20 years or a certain amount of usage—the first such arrangement in the United States. Warranties have been used in Europe and have proven to be very cost effective. This means that the NMSHTD will not have to maintain or repair the road for 20 years.

No one was more appreciative for the widening of the highway than the citizens of the Village of Cuba, situated at the highway's midpoint. Flatiron Structures, the construction manager, had set up its headquarters in the village. While there the company's employees patronized the businesses and occupied every available residence in town. The company also put a large number of local citizens on its payroll. At the end of February, 2002, Flatiron shut down its operations and moved on. It left behind well-maintained vehicles, computers and furniture for the schools, library, and fire department, all at below-market prices. Just as important, it left behind a good taste in the mouths of the Cubanos (*Cuba News* 2002c).

To commemorate the project's on-schedule completion in November 2001, the village hosted a Highway 550 Four-Lane Completion Fiesta the following month. The Cubanos rolled a 7,856-foot-long bean burrito and stretched it across 400 tables up and down the town's main drag, which of course was now US-550. The original idea was to construct the burrito one mile long, but when someone checked the *Guinness Book of World Records* and discovered that the existing record was 7,710 feet, the citizens decided to go for the record. The mega-burrito contained over 8,500 tortillas, 2,500 pounds of beans, and 250 pounds of cheese. Once completed it was measured by a licensed surveyor and certified by the health department as edible. Within a few hours the burrito was gone (*Cuba News* 2001c).

Some Essential Geologic Vocabulary and Concepts

I've been very determined to aim this book at the interested lay person and not to the geologist. In fact the unique character of this book, I believe, is that it attempts to reveal the geologic underpinning of the scenery and history in a way that is comfortable to the observant and curious general traveler. However, like any new field of study, there is an essential vocabulary involved that can be minimized, but not entirely eliminated. Items in this new vocabulary include a few words for rock types, kinds of geologic illustrations, and elements of geologic time.

Geologists over the course of about two centuries have studied and mapped the distribution of rocks exposed on almost all parts of the earth's surface. Such studies eventually wind up in published form in the scientific literature. In order for this information to be conveyable to others, the definition of specific terms thus needs to be standardized. On geological illustrations the different rock types are defined and classified based both on their physical composition (their "lithology") and on their geologic age. Following is a brief look at six essential elements of this relevant vocabulary: 1) types of rocks, 2) bodies of rocks, 3) some geologic structures, 4) geologic illustrations, 5) the concept of base level, and 6) geologic time. These are not something to fret about this early in the game, but I offer them here for later reference.

Rocks

A "rock" is defined as a solid aggregate of one or more minerals. A "mineral," in turn, is a naturally occurring, inorganic, chemical element or compound with an orderly internal atomic structure, and with a characteristic chemical composition and physical properties. Geologists recognize three main categories of rocks: "igneous," "sedimentary," and "metamorphic," based on their mode of origin.

Igneous rocks

These are formed by the cooling and consolidation of a hot, liquid material into solid minerals, that are either extruded onto the earth's surface ("volcanics" such as lavas and ash flows) or intruded into older pre-existing rocks of any type, usually deep below the earth's surface ("intrusives"). Volcanics, in contrast to intrusives, usually occur in discrete layers on the earth's surface, while intrusives typically occur as unlayered, massive bodies. If intrusive igneous rocks crop out today at the surface it follows that they have been uplifted and that the older rocks that once occurred above them have been eroded away. Typically then they are older and therefore harder, and they tend to form mountain ranges. The name of one intrusive igneous rock type, "granite," occurs in the text.

Granite. A massive, intrusive igneous rock, containing easily visible mineral grains of feldspar (gray, white, or pink, which when freshly broken reflect light from cleavage planes) and quartz (gray, glassy, without cleavage planes).

Sedimentary rocks

Most sedimentary rocks are consolidated "sediment." Sediment is loose mineral material of various grain sizes such as clay, silt, sand or larger fragments, all formed by the physical or chemical breakdown of pre-existing rocks, and then transported and deposited in layers by running water, wind, or ice. These layers of sediment later consolidate in varying degrees to form sedimentary rocks. Some sedimentary rock is formed by the precipitation of mineral matter from solution in shallow seas or at fresh-water springs. Both types of sedimentary rocks are formed at or near the earth's surface. Sedimentary rocks are the most widespread category exposed along US-550. The main kinds of sedimentary rocks that we will encounter are given names based on the size of their particles:

Conglomerate. Consolidated (i.e., cemented) gravel, which is a mixture of coarse particles such as pebbles and cobbles.

Sandstone. Consolidated sand-sized particles, which are finer than gravel. Sandstone (and conglomerate) layers tend to be relatively hard and quite durable, and they typically erode to form ridges or cliffs.

Siltstone. Consolidated silt, which is finer than sand. This term is rarely used in this book.

Mudstone. A consolidated mixture of silt and clay, i.e., consolidated mud, which is typically massively-bedded. It is relatively soft and typically erodes rapidly to form valleys.

Shale. Similar to mudstone, but with a characteristic called "fissility"—a tendency to split part along closely-spaced bedding planes. Sometimes mudstones are erroneously called shales.

Gypsum. This type of sedimentary rock is common only along two segments of US-550. It is a hydrated calcium-sulphate salt ($CaSO_4 \cdot 2H_2O$), precipitated from saline solutions, rather than deposited as discrete particles such as sand and gravel. An uncommon rock type, it forms in isolated embayments of the ocean where sea water can be concentrated by evaporation. The rock forms prominent ridges in arid climates.

Travertine. A form of calcium carbonate ($CaCO_3$) formed by fresh-water springs. It typically forms as mounds or sheets sloping down from the water source.

Metamorphic rocks

Metamorphic rocks are former sedimentary or igneous rocks that have been recrystallized or otherwise altered by high temperatures and/or pressures. Along US-550 we see this type only from afar. Like intrusive igneous rocks, these were formed deep below the earth's surface and have since been uplifted and unroofed. They are therefore typically very old and hard, and tend to form mountains. One metamorphic rock type mentioned in this book is quartzite.

Quartzite. Recrystallized and thoroughly welded quartz sandstone. Unlike sandstone though, which when struck breaks around the sand grains, quartzite breaks through the grains, leaving a rather glossy, scalloped surface. Quartzite is one of nature's hardest and most chemically stable rock types, and can be found in many a xeroscaped yard.

Bodies of rocks

Formations

The term "formation" is used a great deal by geologists and is of fundamental importance. A formation is defined as a body of rock, usually layered sedimentary rocks, that has characteristic physical properties such as thickness, grain size, color, bedding types, etc., and that can be mapped in the field at a reasonable scale, typically 1:24,000 (the scale of standard 7.5 minute U.S. Geological Survey topographic maps). Therefore, formations are rather large units. Each formation is given a proper name taken from a geographic locality or prominent geographic feature (a town, river, mountain, etc.) near where the unit was first officially described. The location where the formation is first described in detail becomes the formation's "type section." This reference place is always consulted when questions arise about the nature and definition of the unit. Formations can be split into members and lumped into groups.

Members

If the original formations are too generally defined to map in sufficient detail, a formation may be subdivided into two or more members.

Groups

If the original formations are too finely divided to practically map, a group of formations may be combined together into more inclusive groups.

Simple geologic structures

Several basic geologic structures are referred to in the text. It is important to understand the meaning of these terms in order to appreciate the geology along US-550.

Anticlines.

Folded sequences of layered rocks that are convex upwards, whose limbs dip outward away from the central core, and whose cores contains the oldest rocks.

Synclines

Folded sequence of layered rocks that are concave upwards, whose limbs dip inwards toward the central core, and whose cores contains the youngest rocks. They are the opposite of anticlines.

Faults

Faults are fractures or zones of fractures in rocks along which relative movement has taken place. The relative movement can be lateral (horizontal) or, more commonly, vertical. On geologic maps faults are usually shown as bold lines. The up and down-sides are usually identified with a "U" and "D," respectively.

Basic geological illustrations

In this book I've used two basic types of geologic illustrations: geologic maps and geologic cross sections. With a little practice and exposure the novice can become quite comfortable with both.

Geologic maps

A geologic map shows the distribution of different bodies of rock, both layered and unlayered types, that are exposed on the earth's surface (Figure 3A). The rock units, i.e., formations, members, and groups, are given patterns or colors that differentiate them by age. The distribution of rocks on the surface, as revealed by patterns on geologic maps, is determined by the rocks' structure (i.e., horizontal, tilted, or folded into anticlines and synclines), which in turn controls the nature of the landscape.

Geologic cross sections

Geologic cross sections show a profile of the rocks in a vertical plane along a line drawn on a geologic map, as well as the distribution of the rocks below the surface along that line (Figure 3B). They are essential companions to geologic maps, and generally use the same patterns and colors. A truly accurate cross section is drawn to true scale, i.e., the vertical ("V") and horizontal ("H") scales are the same. However, these sections can be overly wide and skinny because the vertical distance shown (usually 100s or 1,000s of feet) is often so much less than the horizontal distance (usually miles). To alleviate this inherent problem, cross sections are usually drawn with a built-in vertical exaggeration of anywhere from two times (V:H=2) to as much as 50 times (V:H=50) or even more.

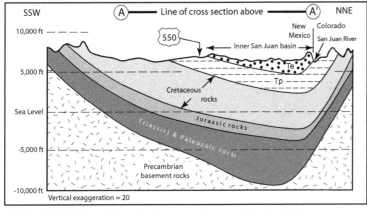

A. Geologic map (see geologic time scale in Figure 4).
Modified from NMBG&MR 2003; McFadden et al. 1983.

B. Schematic geologic cross section across San Juan basin.
Modified from Stone et al. 1983.

Figure 3. Simplified geology of northwestern New Mexico.

Concept of base level

The concept of "base level" is essential to understand the evolution of landscapes, especially that along US-550. As important as it is, the concept is not that difficult to grasp. We know that water runs downhill, and base level is defined as the lowest elevation to which running water can fall. The ultimate base level of course is the sea, which over the short term at least we can assume to be constant. Over the long term though sea level varies considerably. In fact, about 65 million years ago, when much of the land surface of northwestern New Mexico was at or only slightly above sea level, that sea level, compared to today's, was somewhere between about 750 feet (Haq, et al. 1987) to 350 feet (Miller 2006) higher than it is today.

There are any number of local base levels upstream from the sea that affect portions of the streams of a drainage basin. For example, the ultimate base level for the Rio Grande in central New Mexico is the Gulf of Mexico, but because the river's water is impounded at Elephant Butte Reservoir, the local, effective base level is the elevation of that reservoir. Similarly, local base level for the San Juan River in far northwestern New Mexico is the elevation of Lake Powell in southern Utah. In each case the relevant base level is the local base level.

The lower the relevant base level, the greater the vertical fall of running water from upstream sources. The greater the vertical fall, the higher the velocity and the more energetic will be the flow and the erosive power of that water on its downward journey to base level. Given enough time a stream will erode and smoothen its course as it approaches its base level, thus producing a stream "profile" with a minimum slope. As the profile matures and thus stabilizes, the stream velocity and erosive power diminish, and by the late stages the stream flows along just barely fast enough to reach its base level. Because local base levels sometimes change, true equilibrium can be disturbed. An example of this is where a stream's local base level is a lake. If the barrier that dams up the stream (thus creating the lake) were to be breached, the lake would drain and the rejuvenated stream would seek another, lower base level farther downstream and would accordingly carve a new profile.

Geologic, or "deep" time

The sixth essential element in our new vocabulary is that of geologic time (Figure 4). We all know what time is. We live by it. We're obsessed by it. We've all heard the adages, "time is money," "time is of the essence," and other noxious old saws. "Geologic time" differs from "normal time" only in that it is so much more vast than what we are able to experience in our short lifetimes and even in the 5,000 years of written history. Geologic time is "deep" time, and it involves millions and even billions of years.

Common abbreviations for geologic time, used throughout this book, are "ka," "Ma," and "Ga":

ka

Thousand years ago, from the Latin *kilo anni*, literally a thousand years. We're all familiar with the prefix "kilo" from kilogram and kilometer. We're also somewhat familiar with the term "anni," the plural of "anno," used for dates occurring in the Christian era such as 2009 AD–the year 2009 *anno Domini*, literally "in the year of the Lord." Today (perhaps unfortunately), the letters CE, "Christian Era," are often substituted for AD, and the letters BCE, "before Christian Era," are substituted for BC.

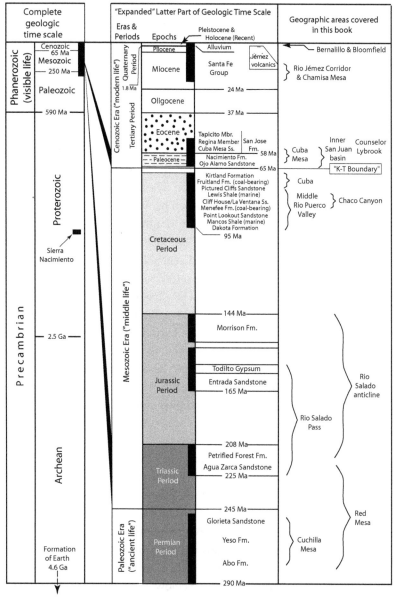

Ga = billion years ago; Ma = million years ago.   Patterns correspond in part to geologic units in Figure 3.

Figure 4. Geologic time scale and geologic units exposed along US-550.

Ma

Million years ago, from the Latin *mega anni*, literally "million years," We use the prefix "mega" often in such terms as megabytes, or megatons. In geology "Ma" is used to assign an absolute age or date, i.e., map "X" shows the landscape that existed at 50 Ma (50 million years in the past), or item "Y" is dated at 30 Ma (30 million years old).

Ga

Billion years ago, from the Latin *giga anni*. For example, the hard drives in our computers have so many gigabytes of storage capacity. But let's be careful here. To avoid confusion geologists sometimes prefer not to use the term "billion" and instead to use the term "thousand million" years (1,000,000,000, or 1×10^9). What confusion you ask? Because in Europe the term "billion" refers to the American term "trillion," or a "1" followed by 12 zeros (1×10^{12}). However, we will use "Ga" to stand for "our" billion years, the number 1×10^9.

31

A brief history of geologic time

It took centuries for educated people (they were the only ones who could read) to accept the concept of geologic time and the notion that the earth might be much older than mankind had up to then contemplated. The reason of course was that for centuries the only existing source of scholarship (and assumed history and prophesy) was the story of Genesis in the Bible. The six days of creation (*Genesis*, 1:31) predicted that the world would exist for "six ages" of indefinite duration. The apostle Peter later provided a possible duration when he wrote, "One day is with the Lord a thousand years and a thousand years are as one day" (Cutler 2003; *New Testament, II Peter*, 3:8). An age of 6,000 years for the world seemed to have ironclad support backed by the ultimate reference book.

In 1654 James Ussher (1581–1656), archbishop of Armagh in northern Ireland and a biblical scholar, set about to determine the age of the earth by adding up the lifetimes of Biblical personages and the reigns of kings. After much excruciating calculation he concluded that the earth was created on Sunday, October 23, 4004 BCE and that it had remained unchanged ever since (Cutler 2003). Again, the idea of an earth that was 6,000 years old had been given even more credibility.

But ideas very slowly began to change. In the late 17th century Nicolaus Steno (1638–1686), a Danish physician living in Italy, was carefully observing the physical world about him. In his 1669 work, the title of which I shorten here to *De Solido*, meaning roughly "on solids," he posited that a sequence of layered rocks becomes younger upwards. In other words, each layer of rock was deposited upon the older, preexisting layer below it. The layers themselves then became a proxy for time. Steno was suggesting that the earth contained information that was intelligible to mere mortals. This was a bold, and quite revolutionary leap, and although he didn't realize it, he had just laid the foundation for the science of geology (Cutler 2003). However, Steno was silent about the absolute age of the earth. He simply had no idea what it might be.

In the mid-18th century, Georges Buffon (1707–1788), a French amateur scientist, believing (correctly) that the earth had originally cooled from a molten lump, calculated how long it would take a given lump—a sphere the size of the earth—to cool to its present temperature. He came up with the exquisitely precise 74,832 years (Goudsmit and Claiborne 1966). That was an enormous increase from Ussher's 6,000 years. In the latter part of the 18th century an English physician/scientist/farmer named James Hutton (1726–1797) began a systematic study of the layered rocks exposed in his native England. His observations compelled him to argue for a vast age of the earth. This was echoed in 1859 by the English biologist Charles Darwin (1809–1882), who believed that the time required to account for the observed sequence of fossils that he saw in the rocks to be in the range of perhaps 400 million years.

The geologic time scale

By the middle of the 19th century scientists had intensely studied many sequences of layered sedimentary rocks, especially those in England, along with their enclosed—and upward-changing—fossil remains, and had assigned names to the different sequences. They had thereby constructed a "relative" time scale, allowing observations such as "this sequence overlies that one and is therefore younger," or "that sequence underlies this one and is therefore older." But the "absolute" age of the earth remained a mystery because of the lack of an effective yardstick.

Just then, when a revolution in thought seemed to be at hand, a powerful dissenting voice made itself heard. In about 1865 the English physicist Lord Kelvin (a.k.a. William Thomson, 1824–1907), harking back to the work of Buffon and the cooling of spheres, made calculations of geothermal heat (the flow of heat from the Earth's interior). He concluded that the earth was not very old at all, and that the cooling of the earth took maybe only a few tens of millions of years. There the issue remained until the end of the 19th century.

In about 1895 the French physicist Henri Becquerel (1852–1908) mistakenly left a piece of a salt of uranium on a photographic plate. The streaks on the plate showed that the uranium salt was throwing off particles. He realized that certain "radioactive" elements decayed over time. Finally, in 1907 the English physicist Lord Rutherford (1871–1937) connected the dots. He saw radioactivity as a potential geologic clock—a mechanism to date rocks that contained radioactive minerals undergoing decay. He also realized—no less significantly—that radioactivity produced heat as a byproduct of the decay (Goudsmit and Claiborne 1966). Kelvin's earlier calculations had not taken into account the continuous addition of heat in the earth's interior by radioactive decay. The impasse was broken. Today the estimated age of the earth is 4.6 billion years. Ussher would be aghast!

The minerals in igneous rocks typically contain slight amounts of radioactive elements. Combining the known rate of radioactive decay of these elements with the measured the ratios of the decayed vs. undecayed "daughter" elements in the rock, it is possible to assign the rock an absolute age, usually in millions of years plus or minus an error factor, e.g., 35 Ma +/- 1 Ma. If an igneous rock, such as a volcanic flow, or a bed of volcanic ash that had settled from the atmosphere following an explosive eruptive event somewhere, is contained within a sequence of layered sedimentary rocks, the sequence too can be dated: the layers below are older than that date, and the layers above are younger.

Via the combination of extensive geologic mapping around the world and the dating of thousands of bodies of igneous rock mixed with sequences of sedimentary rock, we have been able to calibrate the relative geologic time scale, based on fossils, and to construct an absolute time scale, especially for the past 250 Ma of geologic time. Earlier than that the absolute values become progressively fuzzier as the error factors become greater.

Nevertheless, the calibrated geologic time scale is an essential ingredient of the vocabulary threaded throughout this book. It is not to be feared because we already use relative and absolute time scales in our everyday lives. For example, a relative time scale of events in American history might include chapters on the Thirteen Colonies, the Revolutionary War, the rising of the nation up to the Civil War, the Civil War itself, etc. It becomes an absolute time scale by the addition of dates pegging the beginnings and ends of these chapters. The geologic time scale that we use today differs only in that the chapters have strange names and that the dates go way back into the past. Trust me, you can get used to the scale and, eventually, become quite comfortable with it. By far most of the rocks we encounter along US-550 are quite young, i.e., younger than 290 million years, and most of them are younger than 100 million years (Figures 3 and 4). The fact that I refer to these dates as "young" is another one of those things you will get used to.

Finally, it is important to realize that time is continuous but the accumulation of sediment and the preservation of rock are not. It is instructive to compare this concept of time vs. rocks, to a music-recording session. In the sound studio the clock on the wall advances relentlessly without interruption. There is the chatter of conversation preceding and following the recording session and then there is the music itself. All these sounds make up a continuous audio event, but the master tape recorder is turned on and off as needed to isolate the music from the babble before and after. The tape thus records and saves only a selection of the continuous audio event—the part that was occurring when the recorder was turned on.

Now, a rock sequence at a given location is, like our tape, an incomplete physical record of the elapsed time, because the rock sequence consists only of that rock that has been preserved. Much rock or sediment may have been deposited and then eroded away before it could be preserved, or there may have been periods of time during which no deposition of sediment took place at all.

In sum, the complete geologic time scale (Figure 4) is a world-wide master scale that classifies the span of total geologic time via the compilation of many incomplete, local rock sequences that have been correlated together. Note that in the figure the vertical scale expands from the left to the center. On the left the vertical bar represents the 4.6 billion years (Ga) since the Earth was formed, while in the bar in the center shows the more useful, 290 million year (Ma) chunk of time applicable to our story. The black bars on the right side of the center time column represent the ages of the rock that have been preserved to provide our scenery along US-550.

2

History of Highways in Northwestern New Mexico

The modern, four-lane highway US-550 between Bernalillo and Bloomfield didn't just happen. It has a history. The transition from the earliest days of animal-driven transport to today's high-speed conveyances required significance technological and political adjustments to changing needs. Those adjustments are the subject of this chapter.

Earliest days

Prior to the arrival of the Spanish horse in the mid-16th century, human transportation in North America was by foot or boat. In fact no one on the planet had ever moved horizontally on land faster than a galloping horse. For all practical purposes the speed of human transportation was that of a walking horse—about three miles per hour. This was indeed a three-mile-per-hour world. Until the arrival of the railroads, human transportation was powered by draft animals—horses, mules, oxen—and the critical constraint on route selection accordingly was the availability of water and forage for them. Trails therefore followed streams and springs whenever possible and struck out cross country only for short distances to leapfrog from one drainage basin or water hole to another. Heavier loads require teams of animals hauling wagons, and, in addition to water, the animals needed wider, straighter, and more level trails. Routes were again adjusted accordingly.

Before the arrival of the Spanish the only trails were Indian foot trails used for hunting and for trade connecting the pueblos. The first true thoroughfare in what would become New Mexico was the *Camino Real*, the "Royal Road," which followed Indian trails. The Camino Real was not a road, *per se*, but really only a trail. It led north from the trading center of Chihuahua, Mexico, to El Paso del Norte (present El Paso/Juárez), and followed the Rio Grande as best it could to Santa Fe (Figure 5).

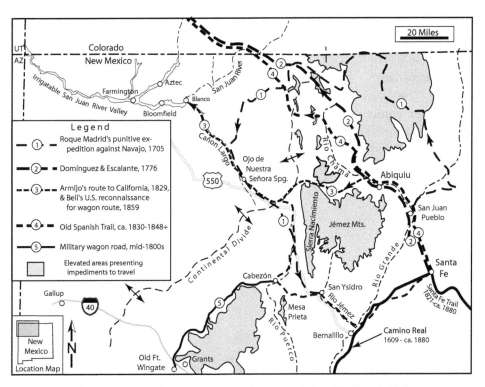

Figure 5. Early transportation routes west and northwest from Rio Grande Valley, 1705–ca. 1880. (Modern geographic features added for reference; modified from Crampton and Madsen 1994; Hendricks and Wilson 1996; Torrez 1988; BLM 1992.)

Long caravans of wagons along the Camino Real supplied the province for more than two centuries. Beginning in the early 1820s Anglo traders from Missouri arrived in Santa Fe along the Santa Fe Trail and brought with them low-priced manufactured goods from "the States" and Europe, and the Chihuahua's trade monopoly suffered. However, over the next two decades Yankee traders continued to carry their goods all the way to Chihuahua along the Camino Real.

The Rio Grande Valley in the mid-19th century experienced a commercial revolution with the arrival of German Jewish merchants who set up shops in numerous small communities. They served as vanguards to open up the area to imported manufactured goods and to provide an Eastern outlet for local commodities.

Neither the Santa Fe Trail nor the Camino Real was graded, paved, or improved in any way, except by the inevitable pounding and compaction by wagon wheel and animal hoof. Each was in fact a series of subparallel strands spread out over a broad front that converged only at favorite campsites or springs (Moorhead 1954).

But a trail is not a road. "Roads," as we know them, simply did not exist. Trails, if not properly maintained, will become drainage ditches. For a trail to become a road it must be graded to a level higher than that of the adjacent terrain, compacted to a consistency to prevent vehicle wheels from penetrating the surface, and be properly drained.

In the middle of the 18th century the Spanish had established a colony northwest of the Rio Grande Valley, named "Abiquiu," on the site of an old pueblo high atop a river terrace overlooking the Chama River (Figure 5). Abiquiu served as a jumping-off place for exploration, trade, and slave-raids via a natural corridor to points beyond to the northwest in the San Juan country (Sánchez 1997).

The Canyon Largo, another natural corridor to the San Juan country, was also well known to the Spanish at this time. Earlier, in 1705, the Spanish soldier Roque Madrid had led a punitive expedition into the heart of Navajo country, the *Dinétah*, to avenge Navajo raids on the Rio Grande settlements. His force consisted of about 400 men and 700 horses, and his choice of route therefore had to allow passage and sustenance for that considerable contingent. After a sharp battle with the Navajo at Tapicito in mid-August (more about this in Chapter 21) he headed south, up-canyon, for home. About 15 miles south of the battle site he stopped at a well-known spring called *Ojo de Nuestra Señora*, the only large spring in a wide area (Hendricks and Wilson 1996). The ease of passage along the Largo, and the existence of its dependable spring, would be long remembered.

By the latter half of the 18th century the Spanish authorities began to consider connecting their domains in New Mexico and California. Accordingly, in 1776 the government charged two Franciscan friars, Francisco Atanasio Domínguez (b. 1740) and Francisco Silvestre Vélez de Escalante (1749–1780), along with eight companions, to pioneer a route to California (Figure 5). They were gone for six months and the expedition never reached California. They circumvented the upper Colorado River drainage basin to avoid both hostile Indians and the formidable canyon country of the upper Colorado River and its tributaries. Just north of the Grand Canyon in north-central Arizona they turned back. Although unsuccessful, their trek highlighted the imposing difficulty of the terrain and the need for additional exploration.

The cartographer on the voyage was a remarkable man of his time named Bernardo Miera y Pachéco. *Don Bernardo* was born in 1713 in Spain and came to El Paso del Norte in 1743. In 1774 he and his family moved to Santa Fe and in 1756 he became the Alcálde Mayor of Galisteo and Pecos. When he departed with Domínguez and Escalante he was a man of some reputation. Two years after his return home, in 1779, he produced his famous and beautiful map covering the route as well as of the geography of New Mexico (more about this map in Chapter 9). He died in Santa Fe in 1785 after a long and distinguished career.

In the early 19th century the Mexican government wished to establish a trail to California in order acquire mules. Potential passageways northwest from Abiquiu were well known by then because Spanish and Mexican traders and raiders had criss-crossed that country for years. Under instructions from the governor, Antonio Armijo with a party of about 60 left Abiquiu in November 1829 (Figure 5). He traveled west, crossed the Continental Divide, descended the Canyon Largo to the San Juan River, cut over to the Animas River near present-day Aztec, and continued northwest to join up with the general trace of the Domínguez and Escalante trail in southwestern Colorado (Torrez 1988). Like the earlier trail, this early version of what would later be called the "Spanish Trail" deliberately arched far to the north. Armijo arrived in Los Angeles after a trek of 86 days. His path was later modified to become a route leading directly northwest from Abiquiu. Both versions of the Spanish Trail were distinct from the Escalante Trail and coincided with it in only a few places. Again, Armijo recognized that the Canyon Largo was a natural corridor to the San Juan country. The 1,120-mile Spanish Trail was used by traders and immigrants until about 1848. Unlike the Santa Fe Trail—a wagon route—the Spanish Trail was a horse and mule route. Freighters led pack trains, strung out for up to a mile, carrying wool and textiles to California and returned with additional horses and mules for sale in New Mexico. The Spanish Trail was then of course Mexican, not Spanish, because Spain had lost control of its colonies in 1821 (Crampton and Madsen 1994).

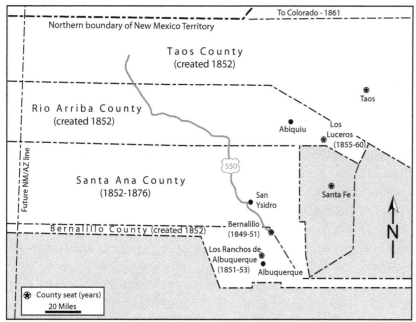

A. 1852 New Mexico Territory (including present-day Arizona. (Modified from Williams 1986.)

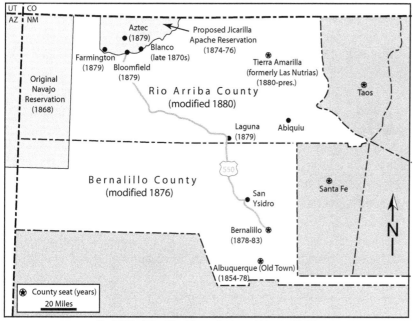

B. 1880. (Modified from Beck and Haase 1969)

Figure 6. Counties and county seats of northwestern New Mexico—I. (Modern US-550 shown for reference.)

Territorial days to 1903

New Mexico became an American Territory in 1850. In 1852 the Territorial Legislature organized the territory into counties, four of which were in northwestern New Mexico (Figure 6A). The meager population and poor geographic knowledge dictated that county lines be simple and rather imprecise. Counties required seats of local government, and the seats required mutual communication and transportation linkage with each other and with the territorial capital of Santa Fe. A loose system of wagon and stagecoach trails therefore slowly became to take shape.

The American "county" has its origin in medieval England where it evolved from the Anglo Saxon "shire." The shire was presided over by a crown-appointed "shire reeve," or sheriff. In the United States the county is the basic unit of local government, and the sheriff is its primary law-enforcement official. Almost all the states are divided into counties, although Louisiana is subdivided into county-like "parishes." In contrast, in New England and the mid-Atlantic states the "township" is the basic unit of local government. (This type of township is not to be confused with the six-mile-on-a-side units of the American Land Survey System, discussed in Chapter 3). The U.S. today has 3,006 counties, Texas the most with 254, Delaware the fewest with three, while New Mexico has 33.

The gold rush of 1859 in the Pikes Peak area of Colorado revealed the need for roads leading north and northwest from the population centers of the Rio Grande Valley. At the same time the U.S. government was taking a keen interest in establishing sites for military roads and posts along the northwestern frontier of the Territory that would become the state of New Mexico. That year Congress authorized several expeditions to explore feasible wagon routes between Santa Fe and southwestern Colorado. One of these tasks was assigned to U.S. Army Lt. William H. Bell. In the summer of 1859 he led his party west from Abiquiu, across the Continental Divide, and down the Canyon Largo to the lower tributaries of the San Juan River (Figure 5). He concluded that a wagon road down the Largo was not only feasible but desirable, and that if it were properly developed it could adequately supply future military posts and settlements along the San Juan River (Torrez 1988).

In 1868 Congress created the Navajo Indian Reservation (Chapter 24). Thousands of Navajo trekked to their new home from the place of their internment at Bosque Redondo near Ft. Sumner in east-central New Mexico. Most of them had no real idea where the reservation was or anything about its boundaries, which had been fixed by lines of latitude and longitude. The Indians simply spread out and occupied their old haunts, many of which were east of and outside of the newly-designated reservation. Many had neither livestock nor farming tools and had to be supplied by the government. In time traders set up trading posts both within and outside the reservation. These posts had to be supplied by wagon from wholesalers in Albuquerque, Farmington, and Gallup, and they in turn by railroads. A crude network of roads thus began to slowly develop (Kelley 1977).

In 1874 a huge swath of land bordered by the new Navajo Indian Reservation on the west and by the San Juan River to the south was set aside as a proposed reservation for the Jicarilla Apaches (Chapter 23). Two years later it was returned to the public domain and opened for settlement. Settlers in the Durango area of southwestern Colorado quickly learned about the available, well-watered bottom lands of the San Juan River and its tributaries. Throughout the late 1870s homesteaders began trickling into the area and established squatters' rights. At the same time Hispanic settlers began to move into the upper part of the San Juan River Valley. One of the first Hispanic communities was Blanco, established at the mouth of Canyon Largo, which even then was the site of much farming and ranching (Figure 6B; Julyan 1996). By the 1880s Mormons were moving into the San Juan country from the west and south. Small communities sprang along the valleys, including Bloomfield (1878), Aztec (1879), and Farmington (1879). In 1876 the Territorial Legislature extinguished Santa Ana County and included it in an expanded Bernalillo County, and in 1880 included the western part of Taos County into an expanded Rio Arriba County (Figure 6B). Northwestern New Mexico at that time consisted of only two counties, Rio Arriba and Bernalillo, indicative of the area's sparse population.

From 1880 to 1882 the Denver and Rio Grande Railroad (D&RG) constructed its narrow-gauge (three-foot) line from Antonito in the San Luís basin of southern Colorado, to the silver mines in southwestern Colorado (Figure 7). In 1905 the D&RG constructed a standard gauge (4 feet 8.5 inches) trunk line from Durango to Farmington, anticipating the conversion of their entire system to standard gauge. This spur was the so-called "Red Apple Flyer," in recognition of its role in hauling agricultural products to the Colorado mining districts. However, it was necessary to trans-ship freight in Durango from standard gauge to narrow-gauge rolling stock, and vice versa (the Farmington spur was converted back to narrow gauge in 1923). Until the 1920s the railroads provided the primary access to the San Juan country.

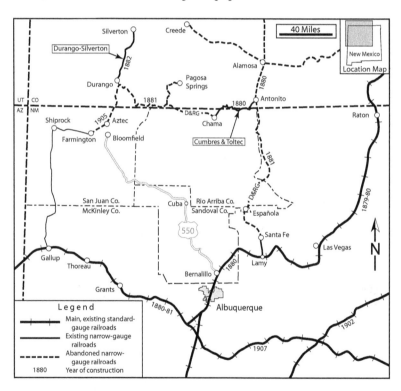

Figure 7. Railroad development in northwestern New Mexico to 1907. (Modern US-550 shown for reference; modified from Williams 1986.)

In 1887 and 1899 the counties were again reorganized. Rio Arriba and Bernalillo counties were reduced in size, and two new counties, McKinley and San Juan, were created (Figure 8A). In 1887 Aztec was designated the country seat of San Juan County after a contentious competition with the citizens of Farmington, and it was reconfirmed as county seat in 1890. For many years the San Juan country was virtually isolated from the rest of the state and was essentially a

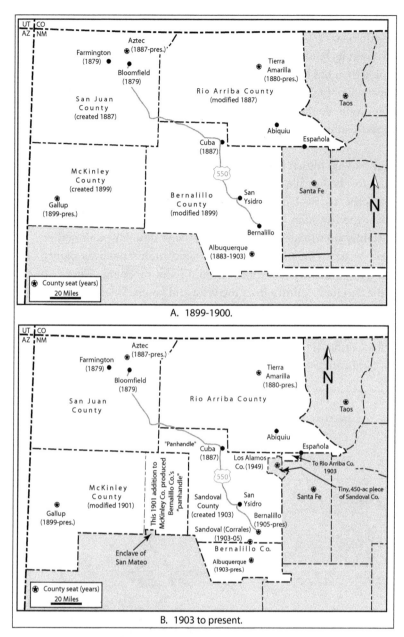

Figure 8. Counties and county seats of northwestern New Mexico-II.
(Modern US550 shown for reference; modified from Williams 1986.)

commercial part of Colorado. To travel from Santa Fe to the San Juan country required a three-day trip by rail (Figure 7). The traveler would take the "Chile Creeper" from Santa Fe north to Antonito, Colorado, where an overnight stay was in order. From there he/she would take the "Cumbres Cumbersome" to Durango, Colorado, where a second overnight stay was made. Finally, the third day would be spent on the "Red Apple Flier" to Farmington (Ricketts 1932).

Winter snows often shut down the narrow gauge Cumbres route, causing much inconvenience among shippers and passengers alike. In the early 1890s it was rumored that the AT&SF Railroad would build a line to connect from the western end of the Colorado Midland (a line in Colorado that the company then controlled) to their main line at Lamy, just south of Santa Fe. This hypothetical line would run south from Grand Junction in west-central Colorado, through the San Juan mining district and Durango, up the Canyon Largo, and down the Chama River Valley and on to Santa Fe (Chappell 1969). Nothing came of this scheme but it does indicate that Canyon Largo was considered a potential major transportation corridor.

Travel in those days was typically by horse-drawn wagon or buggy. Today it is difficult to imagine crossing vast expanses of ground by wagon, lacking roads or even graded trails, forage for the draft animals, communication with the outside world, or even maps. A measure of the difficulty was recorded by Marietta Wetherill, the wife of Richard Wetherill, the amateur archeologist and pioneer of the ruins at Chaco Canyon (Chapter 25). In 1896, her first trip with her new husband by wagon from the San Juan River to Chaco Canyon (Chapter 27)—a distance of only 30 miles—took six arduous days (Gabriel 1992).

1903 to World War II

In the first few years of the 20th century the Territorial Legislature again reorganized the counties of northwestern New Mexico (Figure 8B). McKinley County had been created in 1899 because of the importance of the Gallup coal fields, and in 1901 it was enlarged by annexing a piece of the original Bernalillo County, thereby creating a "panhandle" in the northern part of Bernalillo County. In 1903 the increased importance of the mines in the Jémez Mountains of the northern part of Bernalillo County led to the split-off of that part to form the new Sandoval County (Coen 1925).

Until 1903 little had been done to improve roads and travel in the state was mainly local. That year the Territorial Legislature made the first appropriations to improve roads, using convict labor. In 1909 the Territorial Legislature created the Territorial Road Commission, and in 1912 the first State Legislature renamed the unit the State Highway

Commission. The body consisted of the governor, the Commissioner of Public Lands, and the State Engineer (Roser 1951).

The term "highway" was now being used for the first time. The word denotes a road along which speeds higher than those in the towns were allowed. The U.S. Constitution authorizes Congress to "establish post offices and post roads" (*Article 1, Section VIII*). A "post road" indicated a road along which "speed could be made," at the time meaning "by horse or carriage." This provision initiated the government's system of national highways. However, the invention of the locomotive and the explosive growth of railroads led to the belief that roads were strictly a local concern, and the federal government's interest thus slackened.

A fascinating fact in the history of American roads is that the bicycle provided an important impetus for road improvement. By the 1890s the increased number of bicycle riders began to agitate for improved roads. In 1891 the state of New Jersey became the first to implement a financial-aid program for county-road improvement. In 1893 the federal advisory U.S. Office of Road Inquiry was established, later renamed the U.S. Bureau of Public Roads. By the early 20th century the Bureau began to consider a "federal-aid" road program for the states. Initially the idea of federal aid generated considerable resistance, especially from farmers who did not want to be taxed to provide roads so that "wealthy peacocks" could ride their infernal bicycles. The farmers changed their tune with the introduction of rural mail delivery service, which depended on the existence of good roads. In 1907 the U.S. Supreme Court ruled that Congress had the power and constitutional right to construct interstate highways. The rise of the automobile, especially after Henry Ford's introduction of the affordable Model-T in 1908, created another constituency for good roads. People realized that this new contraption was here to stay and that real roads were absolutely necessary. Finally, the creation in 1914 of the American Association of State Highway Officials (AASHO) gave the states a powerful voice in the national highway movement (Weingroff 2004).

In 1912 the Bureau of Public Roads had made the first federal aid appropriation to the states. The amount of funds allocated to each state was determined by population, area, and mileage of existing public roads. The Bureau placed a resident engineer in each state who served as an intermediary between the state and the federal government. Each state's highway department was required to reach a consensus with the federal government on each project. The state would then be reimbursed by the government after the project was completed and inspected by the highway engineer (*New Mexico Highway Journal* 1931).

Upon statehood in January 1912, New Mexico also qualified for federal highway aid. At the time the state had about 4,000 miles of inter-county roads. Travel on these was mainly between county seats. However, it was definitely horse-and-buggy, with only a few stray automobiles. Road construction emphasized quantity (mileage) vs. quality (Roser 1951). At about this same time in the remote San Juan country, a mesh of wagon roads connecting the network of trading posts on the east side of the Navajo Reservation had become well established. The spread-out Navajo population had a tendency to move about often and these roads were necessary to supply their needs (Brugge 1980). Some of these wagon roads later became graded well enough to admit motor vehicle traffic and later formed the road system of rural San Juan County (Figure 9).

After years of study and political wrangling, Congress enacted the nation's first highway policy, the Federal Aid Road Act of 1916, but America's entry into WWI the next year ensured that funds were scarce. The Act authorized state road projects with federal matching funds. Also in 1916 New Mexico adopted a gasoline tax to supply the funds to match federal-aid grants (Rae et al. 1984). During the war the railroads proved to be unequal to the huge task of military shipping. The incipient trucking industry attempted to step into the breach but was hampered by the deteriorating roads.

At the close of the war the federal government gave the states, including New Mexico, a large quantity of surplus equipment that could be used for road and bridge construction. By 1918 the federal government realized that the 1916 act was insufficient and that something had to be done. A new, modifying piece of legislation, the Federal Highway Act of 1921, attempted to get things back on track. The two acts established the modern highway system in the United States. The first Federal Aid Project (FAP) was kicked off in the northeastern part of the state (Rae et al. 1984). New Mexico's dry climate and furious monsoonal rains provided special challenges for highway engineers. The learning curve during the 1920s was a steep one as floods washed away many of the state's roads and bridges, but valuable lessons were being learned.

Prior to WWI most visitors to the isolated San

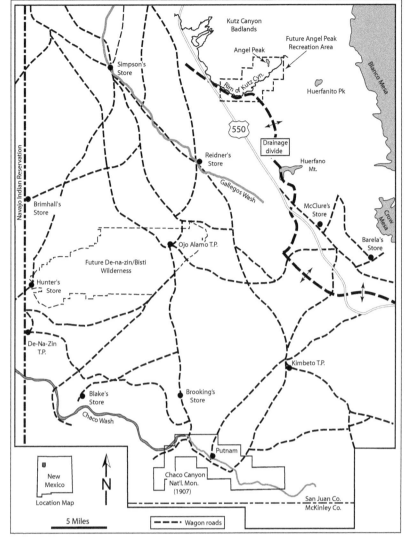

Figure 9. Primitive wagon-road and trading-post network, Chaco Canyon area, 1915–1917. (Modern geographic features shown for reference; modified from Bauer and Reeside 1921; Brugge 1980.)

Juan country typically took the train to Thoreau or Gallup, and left on horseback from there to Shiprock and east up the San Juan River to Farmington (Figure 7; Poling-Kempes 1997). This route became one of New Mexico's first Federal Aid highways. In 1909, Vernon L. Sullivan, New Mexico Territorial Engineer, made a reconnaissance automobile trip from the railroad stop at Gallup north along the old freight trail through Chaco Canyon to Farmington. He proposed this as a "short-cut" route to Farmington (Ricketts 1932). In the spring of 1911, I. Sparks, of the Santa Fe County Commission, with Robert P. Ervein, territorial land commissioner, took a reconnaissance roundabout automobile trip from Santa Fe to the San Juan River. They traveled south to Bernalillo, west up the Jémez River to San Ysidro, north to the Cuba country, and then down the Canyon Largo to the San Juan River. They concluded that an all-weather road down the canyon would be very difficult to maintain due to the shifting sands and flash floods (Ricketts 1932).

In June, 1914 governor William C. McDonald and the state highway engineer, James A. French made a pioneering trip to the San Juan country from the railroad stop at Gallup. They proceeded by automobile over 90 miles of rutted, ungraveled dirt track and arrived in Shiprock two (or possibly three) days later (Ricketts 1932). To reach Shiprock the party had to cross the San Juan River by boat. In 1925 the New Mexico State Highway Department graveled this "million dollar" highway (Ricketts 1930). Because most of the road was on the Navajo Reservation, the federal government footed most of the bill. Instead of two or even three days, freight could now traverse the road between Gallup and the San Juan River in four hours (Ricketts 1930).

By 1923 New Mexico had a loose network of state highways (New Mexico State Highway Department 1923). Only a few miles of the system were paved, and concrete was the pavement of choice at the time. (Personal note: one of my childhood memories was the irritating "thump-thump-thump" the car tires made while passing over the seams in the concrete highways of the time.) Some of the first highways were simply run alongside the railroads and well-established trails. Highways also began to receive official numbers. The original NM-1 generally followed the Camino Real, the Santa Fe Trail, and the railroads between El Paso and the Colorado border. The New Mexico Highway Department assigned other single-digit highway numbers to major connecting routes. The numbers assigned to other, secondary state highways were at odds with geography and seem to have been numbered chronologically as they were laid out.

The mystery (and politics) of highway numbering

It is worthwhile to elaborate on the esoteric and fascinating system of U.S. highway numbers. In 1925 Congress created the Joint Board of Interstate Highways. The board's first task was to propose a system of interstate highways.

First, the board decided to use the official U.S. shield as an interstate highway symbol, and to use numbers instead of names, as had been the practice. East-to-west roads were to be given even numbers, and north-to-south roads odd ones. The main east-west roads were given two-digit numbers ending in zero, and the main north-south roads two digits ending in one. Alternate and branch routes were given three-digit numbers. Accordingly, the board gave the number 60 to the Chicago-to-Los Angeles route that crossed New Mexico. US-60 had five north-to-south branches. The Gallup-to-Shiprock route (mentioned above) was the 5th branch and in 1925 the designation US-560 was proposed. In early 1926 the board added an additional branch, so US-560 became US-660.

At that time a political battle royal broke out in the East. Political leaders in several states thought that a main interstate highway should pass through their states. Accordingly, in mid-1926 US-60 was routed from the Virginia tidewater west to Springfield, Missouri, where it joined up with the Chicago-to-Los Angeles route, which, after much bickering, was redesignated US-66. (Missouri had already printed 600,000 maps showing the original numbers!) The branch line from Gallup to Shiprock in New Mexico thus became US-666 (Weingroff 2003). This story reveals that highway numbering is a systematic process but politics and sheer capriciousness play a role. When the federal revenue-sharing system was implemented in 1926 the federal highways were overlaid on the existing state system. Thus, many of the original state route numbers were retired and replaced by the new federal numbers. Politics entered the fray again much later. In 2003 highway US-666 was redesignated US-491 because of the negative connotation of the number 666 (Weingroff 2003).

During the 1920s various wagon trails connected Bernalillo in the Rio Grande Valley with Cuba. A buggy ride along these trails from Albuquerque to Cuba and on to Farmington in those days was quite an undertaking. The travelers crossed the Rio Grande at Alameda, went north along the Rio Grande, west to the town of Cabezón, and on to Cuba by the end of the second day. Eight additional days were then needed to slog from Cuba to Farmington via the Canyon Largo down to the San Juan River, and finally down the San Juan River to Farmington (James undated).

By 1923 the New Mexico Highway Department planned two maintained highways northwest from Bernalillo (Figure 10). The first one, to be designated NM-55, would follow the buggy trail mentioned above to Cabezón and then north along the Rio Puerco to Cuba. The second, NM-44, would start from Jémez Pueblo, somehow cross the Sierra Nacimiento, and then run north to Cuba. From Cuba the department proposed to continue NM-44 northwest to Haynes Trading Post and on to Simpson's Trading Post. Such a route would follow the topographic high along a series of drainage divides and would require a minimum of bridges and trestles. From Simpson's the road would fork to Bloomfield and Farmington (New Mexico State Highway Department 1923). At this time only a number of dirt tracks meandered from Cuba in the general direction to the San Juan River Valley.

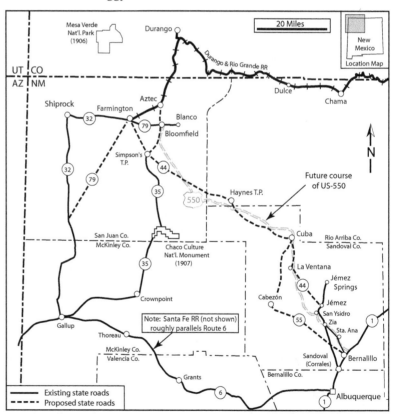

Figure 10. Bernalillo-to-Bloomfield route and select other roads, 1923.
(Modern US550 shown for reference; modified from New Mexico State Highway Department 1923.)

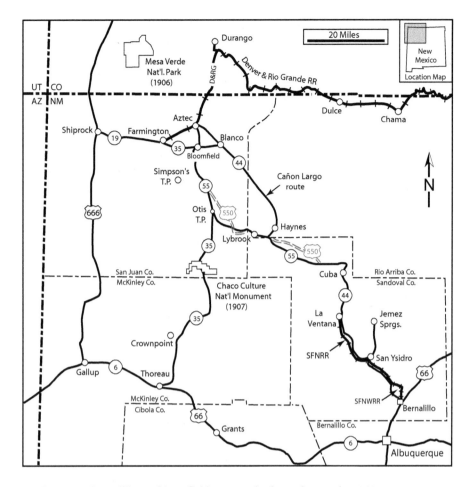

Figure 11. Bernalillo-to-Bloomfield route and select other roads, 1929.
(Modern US550 shown for reference; modified from New Mexico State Highway Commission 1929.)

In 1926 the New Mexico Highway Department began construction of NM-44, a graded dirt road, from Cuba to Bloomfield. It was completed in 1929. That same year NM-44 was reorganized (Figure 11). The segment from Bernalillo to Cuba was still NM-44, but the one from Cuba to Bloomfield was renumbered NM-55. An isolated second part of NM-44 branched off from NM-55 near present-day Counselor, went north past Haynes Trading Post and down the Canyon Largo to Blanco and Aztec (New Mexico State Highway Commission 1929). Prior to this time, as already mentioned, the only dependable connection between the Rio Grande Valley and the San Juan River Valley had been the D&RG railroad. Now for the first time the new highway NM-44/NM-55 combination provided the general driving public with easy automobile access to the fabulous ruins of Chaco Canyon via a turnoff south from NM-55 (Ricketts 1930).

New Mexico Magazine

In the early 1920s the State Highway Engineer was G.D. "Buck" Macy. Through his efforts a publication, *New Mexico Highway Journal*, made its debut in 1923. The monthly magazine featured Highway Department activities and promoted interest in good roads throughout the state. In time the *Journal* began to include articles of a more general nature to appeal to tourists. By 1930 critics were questioning whether highway funds should be used for non-highway activities such as publication of the *Journal*. The State Legislature solved the problem in 1931 by creating a Bureau of Publications. The Game Department's *Conservationist* and the *New Mexico Highway Journal* were combined under a single title, *New Mexico Magazine* (Fitzpatrick 1980). This wildly successful publication has morphed into the state's premier exponent of tourism. New Mexico's sunny climate and colorful scenery attracted a great number of out-of-state tourists with their sleek automobiles and fat wallets. The lucrative tourism business in turn dictated that the state provide a state-wide system of roads rather than one that served mere local needs.

By as late as the mid-1930s, highway grades were not compacted during construction. To advance a section of road grade over a low area such as a ravine, whether a few feet or tens of feet deep, the grade material was simply hauled to the end of the advancing fill and dumped. Compaction of the pile was left to the pounding of traffic and the whims of nature, and the end result was often a hummocky mess. About 1936 the New Mexico Highway Department realized that money could be saved in the long run by spending more during construction to gain a more durable product by watering the fill while emplacement and then compacting the entire fill thickness with a roller. In the early 1930s it was learned that maximum compaction could be attained by adding just the right amount of water prior to rolling. This was determined prior to construction by laboratory analysis (the Proctor compaction test) of the fill material. The fill was constructed piecemeal in six-inch-thick layers, or "lifts," and each lift fully compacted by roller prior to adding the next lift until specifications are reached (Roser 1951). Road construction has clearly been a trial-and-error process.

NM-44 and US-550

In 1944, NM-44 was "removed" from the Canyon Largo and laid over NM-55. NM-55 therefore ceased to exist and the entire route between Bernalillo and Aztec became NM-44. In 1988, in a major reshuffling of state route numbers, the northern terminus of NM-44 was moved south from Aztec to Bloomfield, and the gap between Aztec and Bloomfield was redesignated NM-544. On the southern end, as mentioned in Chapter 1, in 1988 the segment of NM-44 east of Bernalillo was lopped off.

The assignment of the federal highway number US-550 has had a fickle history too. US-550 was originally the 5th branch of the main east-west route US-50 in western Colorado. From 1926 to 1934 US-550 ran south from Montrose to a terminus in Durango, Colorado. In 1934 the southern end was extended into New Mexico through Aztec to a terminus at Shiprock, in 1989 its terminus was yanked back to Farmington, and then in 1999 back to Aztec. Finally, as we have seen, in 2000 US-550 was extended south to Bernalillo where it overlaid both NM-544 and NM-44 (Figure 12).

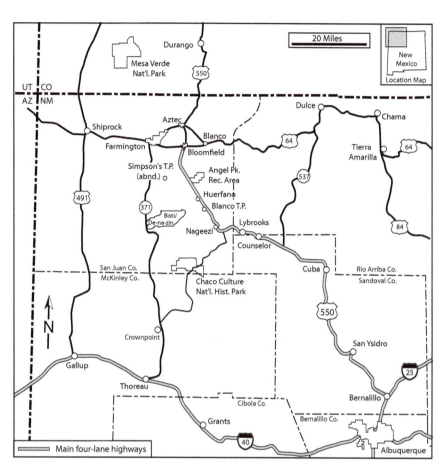

Figure 12. Bernalillo-to-Bloomfield route and select other roads, present. (Modified from current New Mexico Department of Transportation state road map.)

World War II and afterwards

During World War II the deficiencies in the nation's strategic highway system became too obvious to ignore. Many military installations had no access to modern roads, and shortages of manpower and resources limited the amount of road construction that could be done. In 1942 the speed limit was lowered to 35 mph to save gasoline, but in September 1943 the limit was increased back to a hair-whipping 45 mph (*New Mexico Magazine* 1942, 1943). This time of uncertainty gave rise to a highly important piece of legislation, the Federal Aid Road Act of 1944. The act began the process of planning a national system of interstate highways. This act would lead directly to the beginning of the Interstate Highway System in 1956.

In 1945, Southern Union Gas Company discovered the first major natural gas well in the San Juan basin at a place called Barker Dome (north of Farmington astride the New Mexico/Colorado line). This discovery marked a surge in the consumption of natural gas due to the growing needs of an expanding postwar population, to the fact that post-war gasoline and oil fuel shortages mandated the use of alternative fuels, and to the development of new and improved high-pressure pipeline technology. (The perfection of electric welding after 1930 allowed the manufacture of long-distance pipelines, and hydraulic pipe-bending techniques developed during WWII allowed the construction of large-diameter lines.)

For years only the obsolete D&RG narrow gauge railroad served the area from the outside world. Oil men and pipeline contractors begged for an all-weather route from the basin to railheads either in Albuquerque or Gallup. At that time only worn-down, rutted, unpaved passages provided access to the San Juan basin for heavy-truck haulage. In 1950 the Federal Power Commission authorized construction of an interstate transmission pipeline from the Four Corners to southern California. That same year the New Mexico Highway Commission funded a surfaced, two-lane blacktop highway, NM-44, between Farmington and Albuquerque.

Unlike the country north up around Durango, Colorado, northwestern New Mexico lacked the ski slopes and the popular narrow-gauge "little train" to Silverton, but it did have Navajo Dam State Park created in 1963, which was becoming a burgeoning water-sports center. And of course Bloomfield was the natural gateway to the ski resorts in Colorado and the colorful canyon country beyond (Gómez 1994). Good roads were essential to handle the increased demand. In 1950, immediately after blacktopping NM-44, the state highway commission funded the graveling of the 90-mile stretch of dirt road (US-666) between Gallup and Shiprock (Gómez 1994). As the state's ski industry boomed in 1970, New Mexico Governor Campbell authorized construction of NM-64 (later to become US-64), dubbed the "Golden Avenue," to link northern New Mexico's major tourist attractions. This highway was completed in 1972. The future was taking shape.

Topography and population along US-550

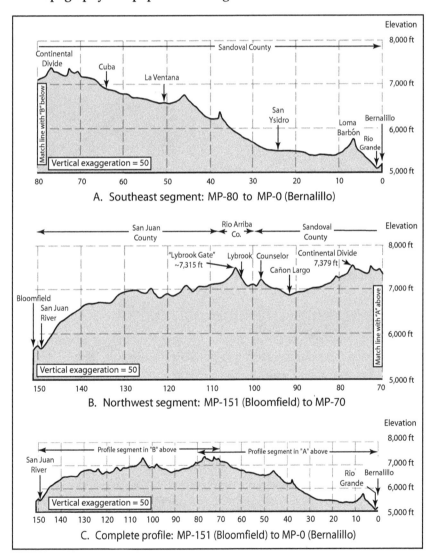

US-550 connects the City of Bernalillo at about 5,050 ft elevation in the Rio Grande Valley, which is part of the Gulf of Mexico drainage basin, to Bloomfield in the San Juan River Valley at about 5,400 feet elevation, which is part of the Colorado River drainage basin. The highway crosses the "hump" of the Continental Divide at an elevation of 7,375 feet (Figure 13). Between these end points, spaced 151.7 miles apart, the population is extremely sparse (Figure 14). In fact, not a single traffic light exists west of MP-3.

Figure 13. Topographic profile along New Mexico portion of US–550. (Data from U.S.G.S. 7.5-minute topographic maps.)

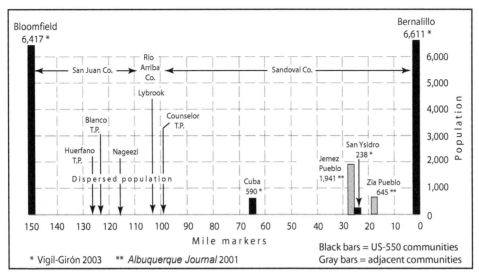

Figure 14. Population profile along and adjacent to US-550.

With the notable exception of the Village of Cuba, the road encounters scenery and little else. Cuba (MP-65), the only population center of any consequence along the route (2000 population about 600), lies almost at the halfway point. The village lies within Sandoval County and therefore faces eastward politically to its county seat at Bernalillo and economically to the metropolis of Albuquerque.

The unincorporated settlement of Counselor (MP-98) is in the far northwest corner of Sandoval County. Politically it faces Bernalillo, 98 miles away, but economically it faces both directions. The tiny unincorporated community of Lybrook (MP-103) awkwardly occupies the far southwest corner of Rio Arriba County (Figure 12). For legal business the inhabitants have to make a 130-mile circuitous haul to their county seat of Tierra Amarilla. All points west of Lybrook lie in San Juan County, which has its county seat in Aztec and its service centers in Bloomfield and Farmington.

Precipitation along US-550 and the Mexican Monsoon

It is no accident that the modern human inhabitation along US-550 is so sparse. The availability of "usable" water is key. The combined heights of the Sierra Nacimiento and the Jémez Mountains act as a giant "water machine" that draws moisture out of the air as snowpack and rain to feed flowing rivers and springs (Figure 15A). The limited human settlement naturally snuggled up against the foothills of these higher regions. However, there is a side to this story not told by the map of annual precipitation. Equally important is the annual distribution of that precipitation. Winter rain and snowfall has a chance to soak into the subsurface and to feed groundwater systems and springs. This precipitation is the most important to settlers.

However, almost half of New Mexico's annual precipitation occurs during the summer months via the "Mexican monsoon" (Figure 15B). Each year a huge mass of moisture surges north up through the Gulf of California and arrives in New Mexico by early July. It brings with it destructive flash floods, crop-devastating hail, fire-causing lightning strikes, and devastating blasts of wind and dust. Water dropped during these torrential downpours runs off and is generally unavailable to plant and animal life.

The variability in strength of the Mexican monsoon over the centuries indeed shoulders considerable responsibility for the evolution of the New Mexican landscape. Segments of our traverse along US-550 reveal landscapes that are characterized by alternating past episodes of erosion of river valleys followed by the backfilling of new sediment into those same valleys. The development of such landscapes therefore operates in exquisite tune with the variable monsoon. During strong monsoonal periods, characterized by relatively wet summers, forests spread over the hillslopes and bind the soil in place. The frequent torrential floods scour out river valleys, incise arroyos, and lower water tables. Two examples of strong monsoonal periods were the "Medieval Warm Period" (ca. 1000–1300 CE) and the period from the mid-19th century to the present. In contrast, during weak monsoonal periods, characterized by relatively dry, arid summers, the forest cover is reduced, the unprotected soil is eroded (albeit slowly) from the hillslopes, and the removed soil/sediment accumulates in the valleys and backfills them due to the reduced flood runoffs. An example of a weak monsoonal period is the "Little Ice Age" from ca. 1300 CE to the mid-19th century (Mann and Meltzer 2007).

It is very important, and extremely satisfying, to be able to generally correlate geological landscapes to climate and historical events. Keep this big picture in mind as we visit landscapes along US-550 in subsequent chapters.

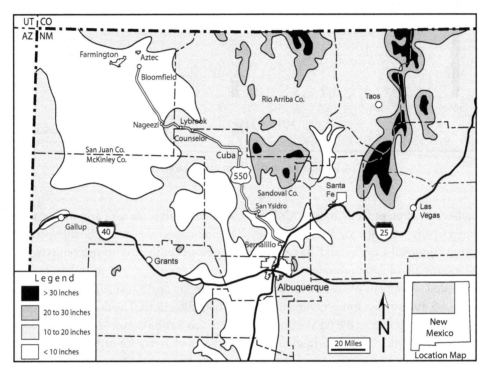

A. Annual precipitation in northern New Mexico. (Modified from Williams 1986.)

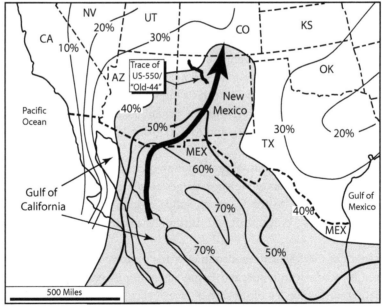

B. Contribution (in percent) to annual precipitation by monsoonal rainfall, July, August and September. (Modified from Douglas et al. 1993.)

Figure 15. Precipitation.

3

The Amazing American Land System

Slicing across US-550 at mile 31.5 is an invisible line running exactly north-south, parallel to a line of longitude called the New Mexico Principal Meridian. No monument or historical marker commemorates the crossing, nor is there a discernable change from one side of the line to the other. The line extends south about 90 miles where it runs through a little lava-capped mesa just east of the town of Acacia in the Rio Grande Valley (Figure 16).

Here the New Mexico Principal Meridian crosses its east-west counterpart, the New Mexico Base Line. Together these two axes, parallel to the cardinal directions, organized New Mexico's land-ownership system and exerted an enormous influence on the state's economic development. This system of subdividing land is part of the greater American Land System, one of America's greatest and enduring innovations. The system has an interesting and long history.

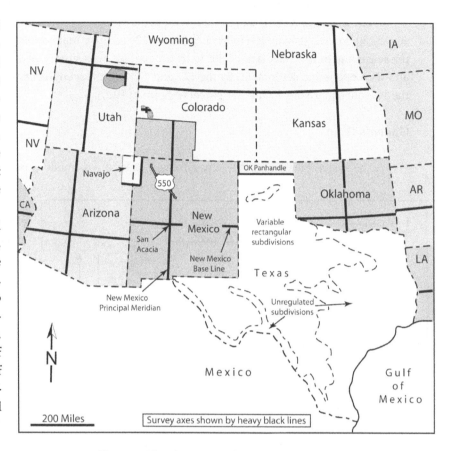

Figure 16. Regular rectangular survey systems, southwestern U.S.

Surveying in ancient times

The rectangular system of dividing and settling land has ancient origins and is largely based on Mankind's quest for order. The four cardinal directions, keyed off the North Star, *Polaris*, symbolized a link of everyday life to the cosmos. It is not surprising that Mankind's early attempt to organize his space was based on the "big four." In the Old Testament of the Bible, God tells Moses to lay out new cities and their suburbs in their newly-occupied promised land of Canaan. The suburbs were to occupy a square with sides measuring 2,000 cubits (about 3,000 feet) on a side and extending 1,000 cubits (about 1,500 feet) along each of the cardinal directions outward from the city walls (*Numbers*, 35:4, 5). The size of the enclosed city is unspecified, but the suburbs would measure about 90 acres—not much elbow room—"for their cattle, and for their goods, and for all their beasts" (*Numbers*, 35:3).

A more rigorous rectangular system was probably invented by an ancient Greek named Hippodamus. He lived in the mid-5th century BCE and was from Miletus, a Greek city in what is now western Turkey. After laying out the city of Miletus in a checkerboard pattern around 466 BCE he was persuaded to do the same in Piraeaus, Athen's military port on the Aegean. Later he sailed with Athenian colonists to southern Italy and took his system along with him. Orderly city planning soon characterized the Greek south of Italy in contrast to the haphazardly-organized cities of the Roman north. By the 3rd century BCE the ever-logical Romans adopted the Greek Hippodamian system and applied it to their new colonies. By the 1st century BCE the Greek system, modified by Roman military needs, became the norm for layout and organization of new Roman cities and expansion of old ones (Cunliffe 1978).

The Romans realized that a stable government required a stable people, and a stable people required a land system that provided simplicity and certainty of ownership. Property boundaries were as fixed and as durable as the political system that installed them. If land could easily be legally defined it could be sold, passed along intact to heirs, and—most importantly—taxed. In addition to leaving Western Civilization with a durable legal system the Romans bestowed onto us the seeds of a comprehensive land system (Sherman 1925).

The practical rectangular system withered away, along with the Roman Empire, in the mid-5th century CE, but somehow it made its way to Holland. Since the 12th century that country had been reclaiming land (the "polder") from the sea. Some of this new land, prior to its occupation, was subdivided into rectangular parcels of various dimensions and then opened to settlement. By the 17th and 18th centuries engineers frequently divided land newly reclaimed from the sea into square or rectangular parcels (Sherman 1925).

Gunter's chain

The systematic, large-scale measurement of land was made possible by the invention by an English mathematician named Edmund Gunter (1581–1626). In about 1620 he devised a chain consisting of 100 elongated links, connected by small rings at their ends, that when extended spanned exactly 66 feet (Figure 17). It is perhaps no coincidence that 66 feet was a length with which Englishmen were quite familiar and comfortable because it was the length of the cricket field (a pitch), a highly popular game since the 16th century. The chain could be conveniently slung over the shoulder and carried from point to point. Prior to Gunter's chain, length was measured by a rod or pole, called a "perch." In the 16th century the perch was standardized at 16.5 feet.

This strange length has its basis in human experience. A plot of land, two perches (33 feet) on a side, was the area that a man could work in one day. A team of oxen, on the other hand, could plow one long furrow, or one "furlong," before needing a rest. A furlong was ten chains in length, or 660 feet. A plot of ground measuring ten chains long by one chain wide was the area that could be worked by the man and his team of oxen in one day, and this area became known as an acre. A square ten chains on a side contains ten acres. A length of 80 chains (eight furlongs or 5,280 feet) became the mile, and a square mile contained 640 acres. The use of Gunter's chain to measure land that was so easily multiplied and divided by the number four became the standard for linear and areal measure for more than 250 years until the introduction of the steel tape in the 1870s. Gunter and his chain have left an indelible impression on almost all the U.S. lands west of the Appalachian Mountains (Linklater 2002).

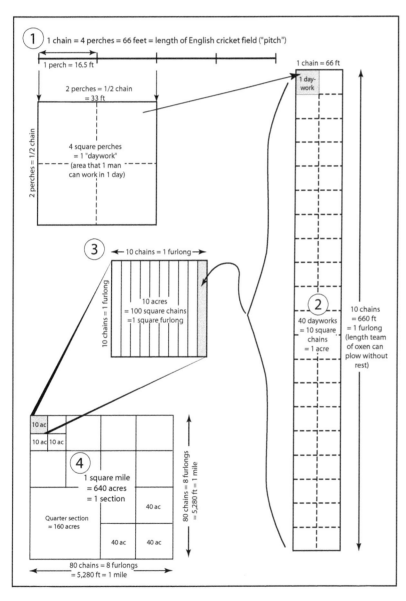

Figure 17. Legacy of Edmond Gunter's chain.

The rectangular system comes to America

Most cities sprang up as river-market towns or as major-road intersections. Roads converged on the market places or river front as best the topography and the river courses permitted. Property boundaries were based on customary acceptance and the act of possession. If the land was surveyed at all it was by the system of "metes and bounds." The land boundary, or bound, was walked from a definitive point, such as a prominent tree or rock, until another definitive point on the boundary was encountered, or met, thus the word "mete." Because the markers tended to burn down or wash away, this traditional system of defining land ownership led to endless controversy, litigation, and strife. There was a better way and it took root in the Northwest Territories of the United States west beyond Appalachian Mountains.

However, the first city in what would become the United States that was affected by city planning and the rectangular grid system was Charleston (Charles Town), South Carolina in 1672. Savanna, Georgia was laid out in a similar manner, and in 1682 William Penn's surveyor platted some 2,000+ acres for the city of Philadelphia on a precise grid. He based his plan on one that the English architect Sir Christopher Wren had drawn up for the rebuilding of London after the Great Fire of 1666, which burned two thirds of the city to the ground. Unfortunately, existing London property was not gridlike and the merchants were not about to have one superimposed on their city (Goodwin 2003). These experimental American cities were harbingers for the laying out of the new cities and vast new territories.

Probably the most influential proponent of a rectangular land development system for the western territories of the growing United States was a Swiss army officer, Henry Bouquet. Born in Switzerland about 1719, he served in Holland under the Prince of Orange and must have noticed some of the rectangular land systems there. In 1754 he came to America as a colonel under the British. He served in the French and Indian War until about 1763 and saw much duty in the western frontier of what is now western Pennsylvania and eastern Ohio. In 1764 he wrote a book about his Indian war experiences and the book was published the following year. In it he proposed a remarkably modern land-development plan for the Northwest Territory (today's Ohio, Indiana, Illinois, Michigan, Wisconsin, and part of Minnesota). Frontier settlements were to be located in rectangular "townships," oriented parallel to the cardinal directions

and subdivided into 640-acre parcels (Sherman 1925).

At the time, seven of the Thirteen Colonies claimed lands under colonial charters and grants beyond the Alleghenies to the Mississippi River, which was then the border with Spanish territory. During the Revolution, in 1777, the suggestion was made that those colonies cede their western claims into a public fund for the benefit of the anticipated new nation. Between 1780 and 1802 the cessions were complete and about 222 million acres became public domain from which Congress could sell to raise funds and to pay Revolutionary War veterans. The men were to be paid in increments of 100 acres, ranging from 100 acres for a soldier or sailor to 15,000 acres for a Major General (Sherman 1925). The fledgling nation anticipated a system of land ownership that was stable, legally defendable, and, of course, taxable.

Bouquet's book came to the attention of the scholarly Thomas Jefferson (1743-1826). Jefferson was certainly aware of the facility that the grid system used by Charleston and Philadelphia gave to economic development. He was also certainly aware of the endless litigation involved in transferring title to property in almost the entire Thirteen Colonies where the old traditional metes-and-bounds land-division method was used. Jefferson believed that the creation of a class of independent owners and tillers of the land was a prerequisite of true democracy. He wrote, "Cultivators of the earth are the most valuable citizens. They are the most vigorous, the most independent, the most virtuous, and they are tied to their country, and wedded to its liberty and interests, by the most lasting bonds" (Watkins 1969). These cultivators required a certainty of ownership of their lands.

Jefferson was fascinated by astronomy and by the decimal system. He therefore proposed subdividing the public domain of the Northwest Territories into 14 new states bounded by cardinal lines based on astronomical observations. These states would in turn be subdivided decimally into sellable parcels. He invented whimsical names for his new states, including "Cherronesus," "Assenesipia," and "Polypotamia," and tweaked some Indian names to create "Illinoia" and "Michigania" (*American Heritage* 1966). Thankfully only the latter two names survived, although in revised form. As chairman of a committee to develop the country's first public land law, Jefferson's report entitled "Report of Government for Western Lands," submitted in 1784, led to the Land Ordinance of 1785. He staunchly believed that the new nation's land should be surveyed before being sold to avoid the chaos experienced by the colonies where the land was surveyed only after being granted and occupied. This brilliant piece of legislation prescribed the subdivision of unoccupied land into square blocks, or townships.

Although there was general consensus about the shape of the townships, considerable controversy existed about their ideal size. One camp argued for dimensions of five miles on a side. Such a block would enclose 16,000 acres in 16, 1,000-acre lots (Figure 18). These blocks lent themselves to easy subdivision into 100-acre units for payment to soldiers, as prescribed by law. Jefferson espoused this system. The other camp preferred blocks six miles on a side, which would enclose 23,040 acres in 36, 640-acre lots. This system exploited the beauty of the number four, which can be multiplied and divided into many whole-number values. The area of 640 acres facilitated quartering and then quartering again and again. Precedents for six-mile square townships existed in the state of Vermont dating from the mid-18th century. In that case the notion was that a six-mile township provided a tract of optimum size for convenient access from the farthest part to the town meeting place usually located in the center.

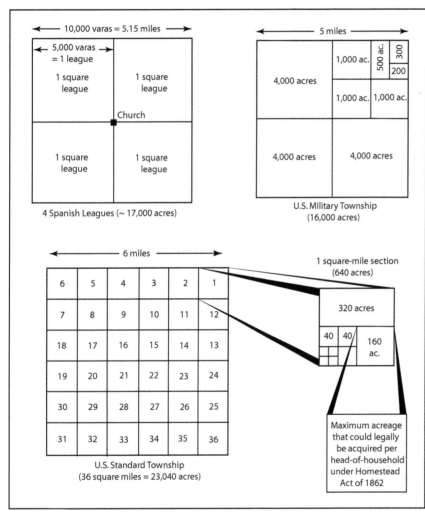

Figure 18. Various rectangular survey systems.

The proponents of the six-mile system eventually gained the upper hand. The 1785 Ordinance for Ascertaining the Mode of Disposing of Lands in the Western Territory decreed that the public domain in the new territories would be surveyed in six-mile-square blocks consisting of 36 one-square-mile, 640-acre sections. Sections, half sections, and quarter sections would provide the basis for today's ranches and farms. A block of land was identified by its section number, by its township number measured in a north-south direction from a base line, and then by its range number measured in a west-east direction from a principal meridian of origin, thus the "township and range system." (Oddly enough, the term "township" performs double duty by referring to the specific number of an east-west row of 36-section blocks as well as any one of those blocks, e.g., township T12N R4W.) Jefferson however was successful in applying his coveted decimal system to the new U.S. currency, which resulted in the American dollar being divided into ten dimes and 100 cents.

The manifestly simple land system, the American Land System, is arguably the wisest and most forward-thinking innovation of the Revolutionary Period. The point of origin for the system was where the Ohio River left Pennsylvania, and Ohio thus became the system's experimental ground. Surveying proceeded piecemeal, and both five-mile and six-mile rectangular systems were tried (which give Ohio a strange quilt work of township patterns). By the time the surveyors reached northwestern Ohio in 1819, the kinks had been worked out and a national, six-mile system was in force. The system was eventually extended to the Pacific, with the notable exception of the state of Texas, which maintained an assortment of metes and bounds as well as irregular rectangular systems (Figure 16).

The Ordinance of 1787, also known as the Northwest Ordinance, was the second brilliant piece of legislation enacted by the Continental Congress and the most important. It was aimed at the vast lands of the Northwest Territory acquired from England in the aftermath of the Revolution, and established the ground rules for settlement. The Northwest Ordinance allowed people to settle the land, form their own territorial government, and later create their own states that would exist on the same footing as the original thirteen. The Northwest Ordinance built on the Land Ordinance and established the means and the precedent by which the United States would expand to the west. Daniel Webster described the ordinance this way: "We are accustomed to praise lawgivers of antiquity . . . but I doubt whether one single law of any lawgiver, ancient or modern, has produced the effects of more distinct, marked, and lasting character than the Ordinance of 1787."

In sum, the American Land System is extremely stable. Municipal and county boundaries denote political jurisdictions and can—and do—change with time. However, all investigations of title revert back to the original surveys. These original surveys are permanent and fundamental to the ownership, sale, transfer, and taxation of the land.

Territorial and state meridional boundaries

The three most common boundaries between territories and states are 1) bodies of water, 2) drainage divides, 3) parallels of latitude, and 4) meridians of longitude. The latter two played important roles in creation of many of the sparsely populated western states and depend on the astronomical observations of surveyors. Lines of latitude are straightforward and are measured in degrees north of south of the equator. Meridians, including the western boundary of New Mexico, are more complicated and give rise to an interesting story. A close look at a map shows that the New Mexico-Arizona border is close to the 109th meridian (109 degrees west of Greenwich, England, today's generally accepted point of origin), but not exactly. In fact, the line is about 109 degrees 3 minutes of longitude west of Greenwich. Why the discrepancy? The answer lies in the fact that the accepted meridional point of origin had changed.

In 1850 Congress declared that the meridian which passes through the center of the dome of the old Naval Observatory in Washington, DC would serve as the meridian for all astronomic purposes (Van Zandt 1976). The New Mexico/Arizona line falls exactly on the 32nd meridian west of the "Washington" meridian. For 62 years, until 1912 when New Mexico was admitted to the union, the meridional boundaries of Arizona, Colorado, the Dakotas, Idaho, Montana, Nebraska, Nevada, New Mexico, Utah,

and Wyoming were referred to as "west of Washington." This is why the meridian at the Four Corners state monument at the intersection of Utah, Colorado, Arizona, and New Mexico is not a nice even increment of degrees of longitude.

The American Land System in New Mexico

The United States acquired New Mexico from Mexico via the Treaty of Guadalupe in 1848. The U.S. agreed to honor those land claims recognized by the Mexican government at the time of acquisition. A welter of Pueblo Indian, Spanish, and Mexican land grants had to be sorted out, surveyed, and adjudicated by the U.S. Government. New Mexico Territory was organized in 1850, and in 1854 Congress established the office of the Surveyor General of New Mexico. The next year the first Surveyor General arrived in Santa Fe. His first responsibility was to extend the rectangular American Land System into the territory. The land grants would be surveyed later and only after they had been confirmed by Congress (Figure 19).

A rectangular system of sorts had preceded the Surveyor General to New Mexico by a century and a half. In 1684 the Spanish governor was given the authority to grant land to the pueblos from the crown. Between 1689 and 1704 the governor supposedly issued grants to 17 of the 19 existing Indian Pueblos. These grants were based on the Spanish *vara* (between 32 and 33 inches, about one pace of the typical man of the 16th and 17th centuries), and 5,000 varas made up a Spanish league. The approximately square grants consisted of four square leagues, or two leagues (about 5.15 statute miles) on a side (Figure 18). The limits of the grant were determined by measuring one league in the four cardinal directions from each corner of the Pueblo church. Spanish law required that sufficient space exist between the Pueblo grants to preclude mutual interference between neighboring pueblos. These little gaps would play an interesting and contentious role later in New Mexico history.

In 1856 the U.S. surveyor-general approved the pueblo grants and Congress approved them in 1858. In the 1890s an investigator for the court of private land claims made an intense study of the 17 Pueblo grants and determined that they were invalid for three reasons. First, the documents were countersigned with the name of a person who evidently didn't exist. Second, signatures at the ends of the grants didn't match those signatures on official documents stored elsewhere and were therefore indicated to be forgeries. Third, the grant to Laguna Pueblo was made ten years before the pueblo was founded. Despite these determinations the grants were allowed to stand (Brayer 1939). Many of the pueblos have increased their territories such that the original Spanish league grant makes up only the core of their reservation. However, Picurís, Pojoaque, San Juan, and Tesuque Pueblos, all in the upper Rio Grande Valley, to this day consist solely of their original Spanish league grants.

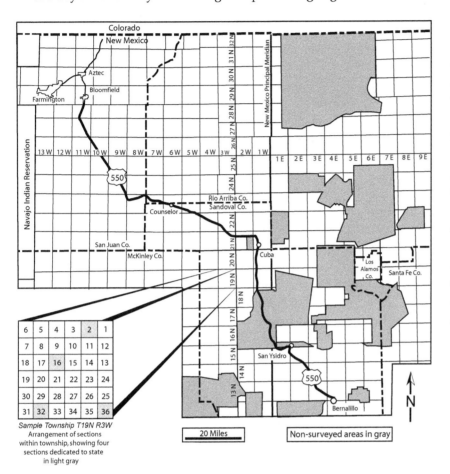

As mentioned earlier, the Surveyor General established his point of origin for the state's survey system on a little hill east of San Acacia in the Rio Grande Valley, about 12 miles north of Socorro. From there he surveyed a line, the New Mexico Principal Meridian, north into Colorado and south to the Mexican border, and a second line, the New Mexico Base Line, west and east between Arizona and Texas (Figure 16). Over the next several decades eight successive Surveyor Generals extended the system in the public domain across the entire state and southwestern Colorado. The system stops at the borders of validated land grants, which were never considered part of the public domain (Figure 19).

Figure 19. The American Land System in northwestern New Mexico.

Unfortunately the Federal Land Office brought with it a philosophy ill-suited for the largely arid west. Since the Land Ordinance of 1785 the Federal Government embarked on a policy of putting the immense Public Domain into the hands of its citizens. The Homestead Act of 1862 granted 160 acres of the Public Domain in the West to any person who was the head of a family, at least 21 years old, a U.S. citizen or an intended U.S. citizen, and had lived on the plot for five years. The venerable 160 acres resulted from the convenient quartering of a standard section of land and was the amount of land deemed sufficient to support a family in a fertile region (Figure 18). It was adequate in the humid East and Midwest, and in the intensely irrigated river valleys of New Mexico. Beyond the valleys, however, water was scarce and the land was suited only for grazing. The ten head of cattle that could eke out a living on 160 acres was a number of livestock woefully insufficient to support a family. In fact, none other than the intrepid John Wesley Powell, the explorer of the Grand Canyon, had recognized this and suggested that settlers be allowed to homestead 2,560 acres in the arid West.

But no one listened to him. The Federal Government held tenaciously to the policy that 160 acres was the maximum that a head-of-household could legally acquire and fervently subscribed to the doctrine that rain would "follow the plow." The government imposed an impossible system which led inevitably to abuses and official corruption. Worse, it led to a total disregard and a general disrespect for the law and led to improvisation. During the 1880s the cattle industry became highly profitable. To acquire an extensive spread of grazing land required only to secure title to an available water supply. Thus the surrounding land of the public domain would be controlled as if it were actually owned, but no taxes would be paid on it (Westphall 1958). The system favored the early and the strong and it encouraged evasion of the law in order to survive. Done once, it was easy to do again. In time the large, aggressive cattle empires would acquire control of the water and much of the land.

Checkerboard land tenure in northwestern New Mexico

The land-ownership situation in northwestern New Mexico today is messy. The pattern of tenure is often referred to as "checkerboarded," and its story is long and tortured. As mentioned earlier, when the Territory of New Mexico was created in 1850 the entire area was considered public domain, with the exception of the land grants. The checkerboarding is partly the legacy of the federal government's land grant to the Atlantic and Pacific Railroad (A&P). During and shortly after the Civil War the U.S. issued four grants to railroads who were planning to construct lines to the Pacific Coast. These were the Northern Pacific, the Union Pacific/Central Pacific, the Southern Pacific, and the A&P. In order to provide the necessary incentive, the companies received a grant of all even-numbered one-square-mile sections for 20 miles on either side of the completed line in states, and for 40 miles on either side in territories such as New Mexico (Figure 20). Existing claims and mineral lands within the zone were exempt, and the railroad could chose "indemnity lands" from a ten-mile-wide strip on either side of the primary grant, expanding the zone of partial railroad control to a belt 100 miles wide (Mosk 1944).

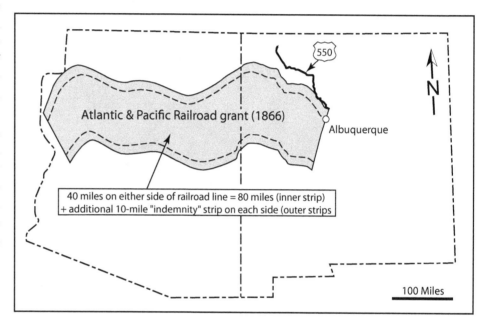

Figure 20. Atlantic and Pacific Railroad land grant. (Modified from Mosk 1944.)

The A&P received their grant in 1866. Most of the grant was forfeited in 1886 but it remained in force for the section of rail line west of Albuquerque to the California border. In 1876 the A&P was taken over by the St. Louis and San Francisco Railroad Co., and in 1880 a one-half interest in the grant was acquired by the Atchinson, Topeka and Santa Fe (AT&SF). In 1884 about 10% of the grant, mainly in Arizona, was sold to the Aztec Land and Cattle Co., and in 1894 the remaining portion of the grant that was owned by the AT&SF was given over to an affiliate called the Santa Fe Pacific Railroad Co. The Santa Fe's main business was to lease the lands for grazing (Mosk 1944).

In 1868 Congress created the Navajo Reservation (see Chapter 24). The Indians moved from their internment camp in Bosque Redondo, where they had been incarcerated since 1864, to their new reservation as well as to their former haunts east of the reservation. Their population grew and grew, and the tribe soon needed more land to support a subsistence economy. Whites too moved in to the area and soon the Navajos and whites were head to head in competition for the scarce water resources. By the beginning of the 20th century the situation worsened. Ranchers from overgrazed and crowded ranges, especially from Texas, had moved in and leased the alternating railroad sections, and over the years many ranches were built up in this way. The leases were renewable, adding greatly to their value. With the leases of odd-numbered sections in place, the ranchers were able to control the intervening even-numbered sections. To overcome the disadvantages of the checkerboard tenure and increase their control over the range the ranchers would homestead and purchase some of the alternate sections. Even so, the scattered and sometimes uncertain tenure in the checkerboard area prevented effective use. The desperate Navajo petitioned the government for more land. In 1907 a presidential executive order enlarged the reservation to the east (Chapter 24). An outcry by the ranchers resulted in the extension being rescinded in 1908 and 1911.(Mosk 1944).

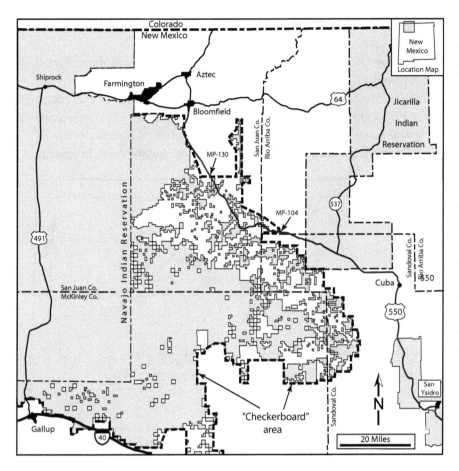

Today the Indian nations are collectively a major land owner (Figure 21). A large swath of land southwest of US-550 is owned by individual Navajos, interspersed with blocks belonging to others. This is the so-called "Checkerboard Area" that lies entirely outside of the Navajo Indian Reservation to the west, and is that land occupied by many Navajo before being relocated to the reservation in 1866. A large block mainly northeast of US-550 is the Jicarilla Apache Nation. This land was originally set aside for the Jicarillas because of its "low value" and remoteness astride the Continental Divide. Today this block includes significant oil and gas fields and the Jicarilla Nation is doing quite well financially (Chapter 23).

Figure 21. Indian lands (gray) in northwestern New Mexico. (Modified from BLM 1994.)

The state is also a major stakeholder in the area. In 1898 the U.S. granted sections 16 and 36 in every 36-square-mile township to the Territory of New Mexico to be used to support education, and in 1910 the Enabling Act for statehood added sections 2 and 32 (Morgan 1958). The U.S. Continental Congress in 1785 had stipulated that only one section, No.16, be set aside for education, so this was a generous increase. Today these lands are managed by the New Mexico State Land Office (SLO), and income is administered by the SLO for a list of beneficiaries, most of whom are educational institutions. In time, via land swaps and other transactions, the state acquired additional land, but in the San Juan basin the "checkerboard" pattern of state ownership is striking (Figure 22).

Much of the land remains part of the public domain. This land is administered by the Bureau of Land Management (BLM), including most of the huge Blanco Gas Field, and by the Forest Service and the National Park Service (Figure 23).

Individuals early on filed claims on the irrigatable land along the San Juan River and its tributary valleys. In contrast, the vast expanse between the San Juan River and the Village of Cuba is almost devoid of private ownership. Private ownership is severely constrained by the absence of reliable water sources (Figure 24).

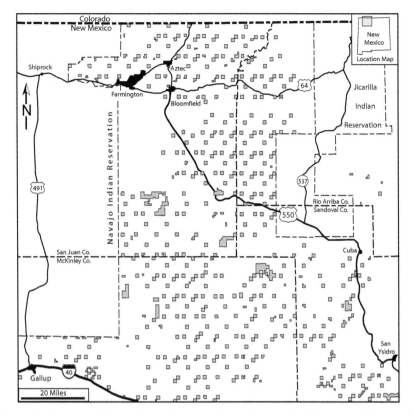

Figure 22. State lands (gray) in northwestern New Mexico. (Modified from BLM 1994.)

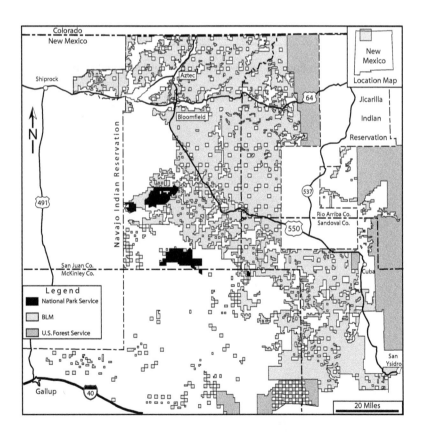

Figure 23. Federal lands in northwestern New Mexico. (Modified from BLM 1994.)

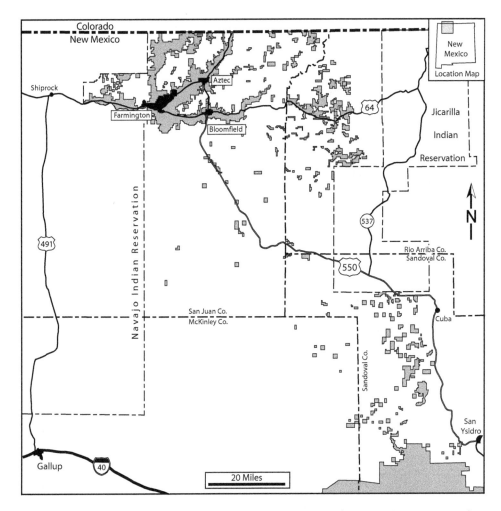

Figure 24. Private lands (gray) in northwestern New Mexico. (Modified from BLM 1994.)

4

Bernalillo—Our Point of Origin

The first leg on our northwesterly excursion along US-550, the Old NM-44, begins at the Rio Grande Valley community of Bernalillo. This hallowed old place rests almost completely within the flood plain of the Rio Grande, and its history has been very much forged by the capricious behavior of the river (Figure 25). The place is sandwiched between the pueblos of Santa Ana to the north and Sandia to the south (Figure 26).

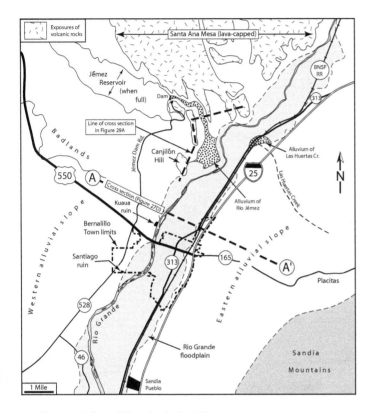

Figure 25. Bernalillo: physical setting

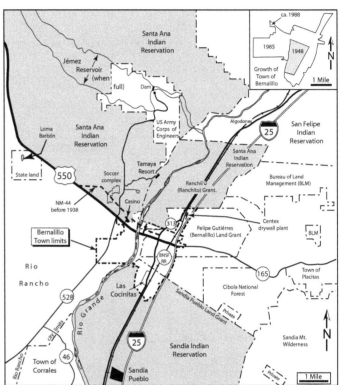

Figure 26. Bernalillo: political and economic setting

Until recently this sleepy hamlet existed totally in the shadow of the metropolis of Albuquerque to the south. However it should be recalled that in the 19th and early 20th centuries the town was an important mercantile center, lumber town, and railroad town. Today Bernalillo is experiencing a renaissance attendant to major economic developments on the adjacent Santa Ana Pueblo, including the Santa Ana Star Casino, the Hyatt Regency Tamaya Resort, the Prairie Star Restaurant, and the Lovelace New Mexico Soccer Tournament Complex. These have spurred the establishment of supporting services along US-550 such as new stores, motels, and fast-food joints. And then there is the hugely successful New Mexico Wine Festival held annually over the Labor Day weekend. The place has been, and still is, an important transportation hub. Since 1905 it has been the seat of Sandoval County (Figure 8B). In 2000, the town had about 6,600 residents.

Geologic setting

The geology of Bernalillo is essentially the geology of the Rio Grande and its precursors. Yes, the present river and its flood plain are only the latest versions, and puny ones at that. Over a period of a million years or so the river has, step-by-step, cut downward hundreds of feet into its valley. During each step the river cut a channel into an older flood plain, and then partially backfilled the channel to form a new, lower and narrower flood plain nested into the preceding one. This step was repeated several times to produce the modern valley. Let's go back in time and take a closer look.

About 2,500,000 years ago (2.5 Ma), a volume of runny, basaltic lavas was extruded out from several vents from an area just north of Bernalillo. The lava flowed out onto the top of a sandy landscape (Figures 27A and 28). Since that time the land around these lavas had been eroded down several hundred feet such that the surviving lava now forms the elevated cap that holds up Santa Ana Mesa (Figure 29A).

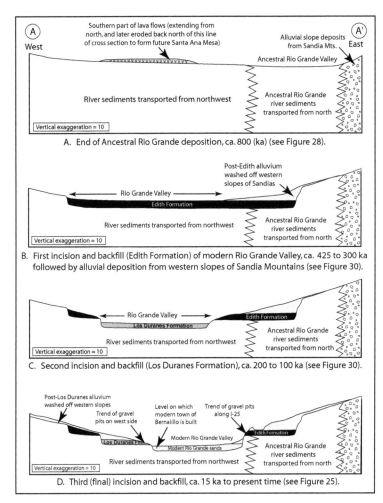

Figure 27. Evolution of Rio Grande Valley near Bernalillo. (Modified from Connell 1998.)

At the south end of the lava field is a strange volcanic feature sometimes called the Bernalillo Volcano, more correctly called "Canjilón Hill," or "deer antler" in northern New Mexican Spanish (Julyan 1996). The lavas of Santa Ana Mesa once extended over the top of Canjilón Hill (Kelley and Kudo 1978). As the column of hot, molten lava was on its rise toward the surface under what would become Canjilón Hill it encountered water-saturated sediment, creating steam. The resulting explosions shattered the lavas and distorted its bedding. The wavy lava beds are a trademark characteristic of Canjilón Hill (Figure 29B).

These steam-blasted volcanics tell us something important about the ancient history of the Rio Grande. The message is that water-saturated river sands, and therefore the water table (i.e., the top of water-saturated sediments) was located just barely below the ancient land surface, some 400 feet higher than it is today. Because a river generally flows at or very near the level of the water table, it follows that the ancient river that deposited these pre-lava sands flowed nearly at that same elevation, about 400 feet above our heads, and that about 400 feet of old river sediment has since been eroded away. This ancient river is sometimes referred to as the" Ancestral Rio Grande" (Figures 27A and 28). For several million years the Ancestral Rio Grande, and its tributaries from the northwest, had been building up, or aggrading, the floor of the valley. At the same time the valley floor gradually subsided, providing accommodation space for the continually-added river sediments.

Sometime after the extrusion of the Santa Ana lava flows at 2.5 Ma, and perhaps beginning as recently as about 800,000 years ago (800 ka), everything changed. From then on the river began to cut downward into its old deposits. This great incision took place in at least three big steps. The first scoured out a valley into the earlier deposits and then backfilled it with cobbly gravel (Figures 27B and 30). The pebbles and cobbles of the channel, now called the Edith Formation, consist largely of extremely old and hard material called "quartzite" (Chapter 1) eroded from mountains to the north. The chunks of quartzite have been beautifully rounded by the vigorous stream action.

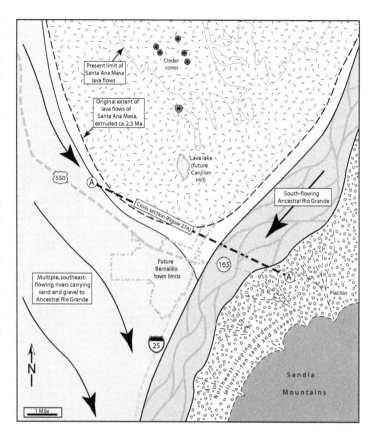

Figure 28. Ancestral Rio Grande, ca. 2.5 Ma to 800 ka (see Figure 27A). (Modern geographic features for reference; modified from Connell 1998; Connell et al. 1995.)

A second incision into, and through, the Edith Formation and the subsequent backfill of this new valley produced a second sequence of cobbly gravel, called the Los Duranes Formation. This unit lies at a lower elevation than the Edith and is inset into it (Figure 27C). The abandoned Edith channel now became what is called a "terrace." The third, most recent incision has excavated a deeper valley and backfilled it with sands of the modern Rio Grande (Figures 25 and 27D). The abandoned channel of the Los Duranes Formation is now, like the higher and older Edith, a second terrace that overlooks the modern flood plain from the west side. The Los Duranes on the east side has been completely eroded away.

The cobbly gravels of both the Edith and Los Duranes Formations are a valuable but relatively low-value natural resource that is extensively quarried and used for construction and landscaping. Size-screened volumes of this material can be found in many a xeroscaped Albuquerque yard. An under-appreciated fact is that the production of sand and gravel is an indicator of a nation's economic well-being. Modern civilization requires ample supplies for use as aggregate in concrete and as road-base material in highway construction. Based on tonnage, the production of construction sand and gravel is the second largest non-fuel mineral industry in the country. The land containing the Edith gravel terrace on the east side of I-25 (opposite the Centex wallboard plant) is located on Santa Ana Pueblo land. The pueblo leases the land to Western Mobil, which in turn operates a large gravel and sand quarry. The company removes about a million tons of material annually. The cobbly gravels of the Los Duranes terrace, located on the west bank of the Rio Grande on pueblo land just south of the Tamaya Hyatt Regency Resort are also quarried.

The third episode of river incision and partial backfill had produced the modern flood plain of the Rio Grande. For at least several centuries until the late 19th century the Rio Grande was a wild river, meaning it flooded when it wanted to. Gradually and furtively it shifted its active channel to the west, leaving behind it, on the east side, a slightly higher, and less flood-prone, alluvial surface that was ideally suited for the positioning of an agricultural community, such as Bernalillo.

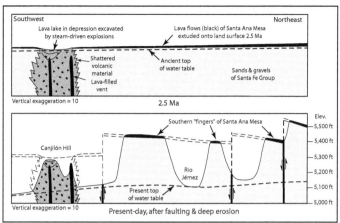

A. Schematic cross sections of Canjilón Hill and southern part of Santa Ana Mesa (see Figure 25 for location).

B. North, low-oblique aerial view of Canjilón Hill. (Photo by author 2001.)

Figure 29. Canjilón Hill

Reduced water flows due to the development of up-stream irrigation projects, and to increased sediment loads caused by overgrazing and logging in the late 19th century, initiated a period of river-bed aggradation (see "Bernalillo vs. the Rio Grande" below). The active channel accordingly built up its bed and its flanking levies such that the river eventually flowed at an elevation higher than that of the surrounding flood plain. For example, by 1960 the channel of the Rio Grande in Albuquerque was six to eight feet above the flood plain outside the levees (Lagasse 1981). Construction of the Jémez Canyon Dam and Reservoir in 1953, and especially the Cochití Dam and Reservoir in 1973, eliminated much of the sediment supply and the river is now beginning to cut down into its most recent channel deposits.

Figure 30. Ancestral Rio Grande, 425 to 100 ka. (Modern geographic features shown for reference; modified from Connell 1998; Connell et al. 1995.)

Early settlement

Tiguex Province

When Coronado arrived in this area in 1540 he encountered 12 Tiwa-speaking villages. (*Tiwa*, of the Tanoan language family, is today also spoken by Taos and Picurís Pueblos located well to the north.) He called the area the *Tiguex* (pronounced TEE-wesh) Province. The northern two of these villages were "Santiago" and "Kuaua" (pronounced koo-AH-wah). Of the original 12, only two, Isleta and Sandia, remain today. The Tiwas built Santiago and Kuaua on the west side of the Rio Grande and on the eastern rim of the Los Duranes terrace, at a safe height above the flood-ravaged modern flood plain at their feet (Figure 25). At that time the river was flowing on the eastern side of its flood plain, affording the pueblos easy access to their irrigated fields on their side of the river.

The Spanish

In 1598 *Don* Juan de Oñate led the first settlers up the Rio Grande to northern New Mexico. He established his headquarters near the confluence of the Rio Chama and Rio Grande, just north of present-day Española. After a few years some settlers moved down the river into the *Rio Abajo*, that territory south of the *La Bajada* (Spanish for "the slope" or "the descent") escarpment. The La Bajada escarpment is a south-facing, lava-capped cliff that effectively subdivided the valley of the Rio Grande into the *Rio Arriba* (Spanish for "upper river") north of the escarpment, and the *Rio Abajo* ("lower river") to the south. The Rio Arriba is less continuous, less wide, and less fertile than the lower valley. The escarpment presented a formidable barrier to travelers as late as the 1920s, but today the La Bajada is best known as the sharp, sometimes icy change in elevation along Interstate Highway 25 between Bernalillo and Santa Fe.

In the Rio Abajo the settlers established farms and ranches between the Rio Grande pueblos of San Felipe and Sandia (Santa Ana acquired its Rio Grande properties later). The first mention of Bernalillo as a settled place, the *Real de Bernalillo*, was in 1696 (Stanley 1964). The name means "little Bernal" or "little Bernardo."

Just who was Bernal, a.k.a. Bernardo? In 1691 a certain Francisco Bernal was a 60-year-old soldier in El Paso. As his pension he received a parcel of land in the area of today's Bernalillo, with the stipulation that he maintain a garrison to protect the area (Stanley 1964). Could the name have come from the old soldier? There is a second possibility. One of the clans living in the area before the Pueblo Revolt of 1680 was the Gonzáles-Bernal family. Could Bernalillo have been named after a junior, or small-statured member of that family? And there's a third. The most prominent resident in the area in 1680 was *Don* Fernando de Chávez, whose oldest son at the time, Bernardo, was six or seven years old. Could the place name have come from him? Regardless, it seems that the namesake of Bernalillo was a little Bernal/Bernardo person (Chávez 1957).

The settlement occupied an ideal strategic location in the fertile Rio Abajo. To add to the strategic value of the site, the nearby Rio Jémez provided a convenient corridor to the excellent grazing lands to the northwest in the Rio Puerco Valley and the Valles Grandes in the Jémez Mountains. It was a perfect spot (Figure 25).

The early community probably consisted of scattered ranches. In 1766 an observer noted that "on both sides of the river there are several small ranches called Bernalillo." One area of settlement was on the west side of the Rio Grande and on the south side of the Rio Jémez. In 1776 this "small settlement of ranchos" was described as "Upper Bernalillo," in contrast to another Bernalillo to the south near the present town (Snow 1976). In the early and mid-18th century, Santa Ana Pueblo acquired land between the two areas, thus separating the upper and lower communities. An additional wedge between the two was the Bernalillo Grant. In 1701 this parcel, containing about 18,000 acres, was issued to Felipe Gutiérres, a soldier at the garrison. A few years later, in 1708, it was reissued, most likely because the governor had been replaced. The exact location of this large grant is difficult to determine, but the final grant, approved only in 1897, consisted of a tad more than 3,400 acres, a far cry from the original 18,000 (U.S. Secretary of the Interior 1874; Figure 26). In time the upper settlement was washed away by repeated floods.

Bernalillo vs. the Rio Grande

The modern flood plain is generally about 1.5 to two miles in width, and the modern Rio Grande itself occupies only a small part of that width. It is important to remember that in the past the river has meandered back and forth across every inch of the flood plain. It is the river's nature to not only transport sand and silt downstream, but also to dump sediment into its channel and therefore to raise, or aggrade its bed. Eventually the situation becomes unstable, gravity prevails, the river shifts, or "avulses" laterally to a lower location and abandons its former channel.

Aerial photographs of the Rio Grande flood plain in the Bernalillo area show many of these abandoned channels. The abandoned, generally eastern part of the flood plain was slightly (a few feet) higher and drier, and therefore more desirable for settlement (Figure 31).

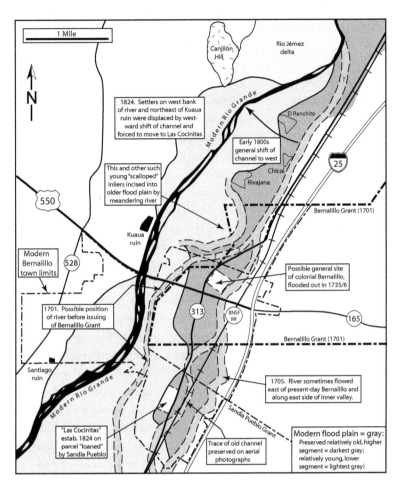

In the area of present-day Bernalillo the river generally flowed along the east side of the flood plain. The unusual shape of the western margin of the Bernalillo Land Grant, issued in 1701, suggests that it might have been determined by the position of the river at that time. Before 1709 the Rio Grande flowed east of the area known today as "Las Cocinitas" (U.S. Secretary of the Interior 1874). A big flood in 1735 washed away several homes and a church in colonial Bernalillo (then located on the west side of the river), as the channel moved several hundred yards to the west and many settlers were forced to relocate to the east bank (Scurlock 1998, p. 186). Between 1709 and 1739 the river began its inexorable shift to the west as it sought an area of lower elevation. The river flooded almost every year between 1753 and 1760. In the early 19th century the continued shift of the river westward washed out settlers north of the Kuaua ruin, and some of the refugees settled on the new east side of the river on the northern edge of Sandia Pueblo land. Then in 1828 came a mega-flood (Scurlock 1998, p. 187). Floods were a rude part of life on the Rio Grande flood plain.

The severity and frequency of flooding began to increase after about 1850 due to more rapid erosion in the river's upland drainage basin caused by excessive grazing and logging. The increased sediment load slowed down the river's flow and the river bed therefore began to aggrade. Another

Figure 31. Bernalillo vs. the Rio Grande. (Modern geographic features shown for reference; based on 1935 Soil Conservation Service aerial photographs #39A, #40A, #41A, #66A, and #67A; Scurlock 1998.)

mega-flood in 1874 marked the beginning of even more serious problems. In the late 1870s and 1880s large-scale development of irrigation in the San Luis Valley of southern Colorado further diminished the river's flow near Bernalillo. The decreased flows were incapable of moving the huge volumes of available sediment. By 1900 this new regime resulted in a rising water table and waterlogging of the soil. Evaporation from the elevated water table precipitated salts in the soil. This deterioration caused the amount of irrigated acreage in the valley to plummet (Scurlock 1998).

Floods continued into the 1920s. The established settlement was the older, slightly higher part of the flood plain on the east side of the valley. The meandering river modified the western edge of this old surface and cut a number of scalloped re-entrants (Figure 31). The main road through town (formerly State Route 1, then US-85/US-66, and now the Camino del Pueblo), as well as the Santa Fe Railroad, wisely kept to this higher, less flood-prone surface. One such scallop produced a "waist" in the growing town and caused the community to form with two-lobes (Figure 31). Not until the flooding was controlled did the town fill in the gap between the two lobes.

In 1925 the Middle Rio Grande Conservancy District was formed. Its mission was 1) to stabilize the river's channel via the use of jetty jacks (12-foot, cross-barred devices interlinked by metal coils), 2) to drain the swamplands via the construction of riverside levees, riverside and internal drains and via a system of diversion canals, dams, and laterals, and 3) to provide irrigation water for farmers (Figure 31). Most of the work was completed by 1935. The water table indeed fell and irrigated agriculture began to recover. During the 1950s many more jetty jacks were installed, and between that time and the completion of Cochití Dam in 1973 the channel of the Rio Grande became more narrow, more restricted, and less braided. The Rio Grande is now a controlled river.

Santa Ana Pueblo (*Tamaya*)

Santa Ana Pueblo, known to its inhabitants as *Tamaya*, is one of five Keresan-speaking pueblos (others include Zia, San Felipe, Santo Domingo, and Cochití). When the Spanish came to stay in 1598 there were about 15,000 Keres people in the area. In 1630, the Keres region was described as rich, with fertile soil and a river filled with fish (Bayer 1994). The old pueblo of Tamaya was probably located high up on Santa Ana Mesa on the north side of the Rio Jémez (Figure 32). The Santanas crossed the river on boats to reach their fields on the opposite side. When the river level dropped they built a bridge. This is the only pueblo known to use boats before the arrival of the Spanish in 1640. Coronado's party saw Tamaya during their wintering in Tiguex 1641–1642. The Santanas joined in the Pueblo Revolt of 1680. The old pueblo was destroyed by the Spanish in 1687 in a foray prior to De Vargas's reconquest in 1692. De Vargas persuaded the Santanas to return (Hume 1971), and they rebuilt the pueblo at its present site on the north bank of the Rio Jémez.

The agricultural land near Tamaya was poor and undependable. In the past the people used lands in a wide area around the pueblo. By 1698 the Village of Bernalillo had become an important center and much of this land had been preempted, but the pueblo needed more land to survive. From 1709 to 1763 the pueblo purchased back land taken by the Spanish settlers along both sides of the Rio Grande north of Bernalillo. The new tracts were called the "Ranchii'u Grant," or "Ranchito Grant." By the early 19th century the Santanas began moving from Tamaya to

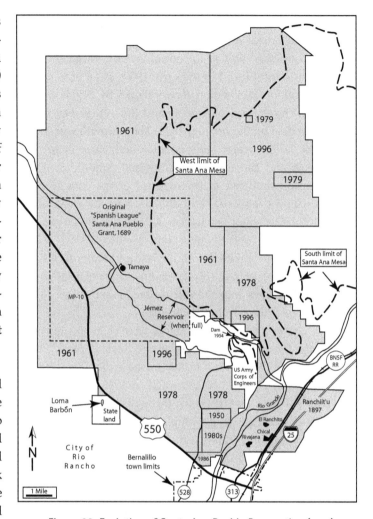

Figure 32. Evolution of Santa Ana Pueblo Reservation (gray).

Ranchito. Today the approximately 475 pueblo members live in three villages, Chical, Rivajana, and Ranchito, all on the old flood plain surface in the Ranchito Grant (Hume 1971; Figure 31). The grant was confirmed by the U.S. Court of Private Land Claims in 1897. Since 1961 the pueblo, via a protracted and intelligent program of purchase and land swaps, has gradually added it to its land base from the original four square Spanish leagues (about 15,400 acres), to more than about 78,000 acres at present (Figure 32).

The bosque

Before the creation of the Middle Rio Grande Conservancy District (MRGCD) in 1925, the river contained scattered stands of cottonwoods that were periodically inundated and sometimes washed away by floods (Shah 2001). Upon completion of the works of the MRGCD in the 1930s and 1940s the floods came to an end. Cottonwoods require occasional flooding to germinate, so new cottonwood growth was prevented. The cottonwood stands therefore grew to maturity and were not replaced by new growth.

Much of the bosque today consists of aging and dying cottonwoods with thickets of salt cedar and Russian olive. Salt cedar (or tamarisk) and Russian olive are two exotic species native to Euro-Asia. Salt cedars were introduced to Albuquerque as early as 1908. In 1919 the Albuquerque Chamber of Commerce purchased several hundred exotic species, including salt cedar, for planting in residential areas and in public parks. By about 1926–1927 they were used widely to control erosion. Russian olives were also introduced in the early 1900s as an ornamental plant and for stabilization of river banks. By 1936 both species had spread extensively up and down the river, aided a great deal by the flood of 1929. Their growth was aided by the falling water table because both have deep root systems and can tolerate alkaline-soil conditions. Their expansion occurred at the expense of the cottonwoods and willows. Salt cedar and Russian olive are water hogs. By the 1950s water use by salt cedar alone was twice the amount used by cultivated crops (Scurlock 1998). Their proliferation combined with the reduced water flows has in some places actually obliterated the channel itself where the water flows into the salt cedar/Russian olive forest and simply disappears (Watts 2000).

In recent years drastic steps have been taken to control the infestation by salt cedar and Russian olive. Santa Ana Pueblo has been a leader in the effort. The Bosque Restoration Project was launched in 1996 to restore the river along its six-mile reach through the pueblo's lands north of Bernalillo, with supplemental funding provided by the Bureau of Indian Affairs, the U.S. Environmental Protection Agency, and the U.S. Fish and Wildlife Service. Work began in 1999. The pueblo has been using heavy machinery to yank out thousands of salt cedars and Russian olives. The former are shredded to mulch and the latter are cut into firewood and distributed to tribal elders. The bare alkaline soil is treated with gypsum and irrigation water and seeded with native, salt-resistant grasses. In places with low alkalinity they have been planting cottonwoods and black willows. The pueblo hopes it is setting a precedent for its downstream neighbors (Ikenson 2001).

Sheep and mercantile empires of the Rio Abajo

As mentioned earlier, the valley of the Rio Grande are divided into a lower (*Rio Abajo*) and upper valley (*Rio Arriba*). The wider and more fertile valley of the Rio Abajo was therefore more prone to the development of large landholdings and concentration of wealth and influence. The Rio Abajo was destined to become a sheep-raising territory. When Coronado came north to New Mexico in 1540 he brought with him, in addition to his men and horses, 500 head of cattle and 5,000 *churro* sheep. Churros were a small, tough, hairy breed of sheep introduced to the New World from southern Spain. Their purpose on this voyage was to serve as meat on the hoof—a mobile commissary. Most were consumed on the way. When Coronado's men headed south for home in 1542, with their surviving livestock, a few sheep were left behind, but these likely soon died off.

In 1581 a small Spanish party blazed a trail north from El Paso up the Rio Grande and partly across the desert. This route would become the *Camino Real*, or "royal road." When *Don* Juan de Oñate brought his settlers north in 1598 along the newly blazed trail he brought with him more than 6,000 head of livestock, including 4,000 sheep. After a long period of internecine strife between Spanish representatives of the church and state, the Pueblo Revolt of 1680 flared up and the Spanish were ignominiously driven from the province. At the time of the revolt sheep numbered only about 5,000. Many of these—not to mention large numbers of horses—fell into the hands of the Indians. In 1692 Diego De Vargas brought the settlers back north. A few years after this reconquest De Vargas distributed livestock to the 1,000 settlers, including 4,000 sheep (Baxter 1987). The Spanish ascendancy in the Rio Abajo had begun.

It took several decades for the flocks to significantly increase in size largely due to Navajo raids that plucked off large numbers of livestock. Sometime in the late 17th century, probably after the Pueblo Revolt, the Navajo had learned from their Pueblo neighbors how to weave and wool had become part of their culture. Despite the raids, Spanish sheep numbers grew rapidly in the 50 years following 1720. The growth was due in part to a system of sharing the risk of sheep ranching called the *partido* system—an economic arrangement something akin to agricultural sharecropping. The owner of the sheep assigned the *partidario* a certain number of animals. The partidario was required to return a certain amount of lambs and wool to the owner after a period of three to five years, but he took all the hits from nature and Indian raids and quite often descended into debt. The partido system thus spawned a peon class while at the same time it freed up the owners of the flocks to pursue other, more mercantile interests (Baxter 1987).

During the second half of the 18th century the flocks increased further in size. Surpluses developed in excess of what was needed to feed and clothe the colony. Sheepmen sought additional pasture and export outlets for the surplus animals. By the 1790s some 200,000 to 500,000 sheep were driven down the Camino Real to Chihuahua, Mexico. At that time fully about one third of heads of households were involved in the woolen textile industry. Most clothing was made from wool in the forms of shawls and blankets. Courser types were used for floor covering. Handmade woolen textiles, in addition to sheep, were major items of trade. Sheep raising and trading became one of the very few export industries until imported manufactured goods began to make their way to New Mexico along the Santa Fe Trail beginning in the 1820s (Baxter 1987). Bernalillo was the center of the sheep-raising industry.

During the first half of the 19th century a small group of rancher-merchant families had gradually became ascendant and had begun to control the livestock market and local political affairs. These men broadened their grasp via an intricate web of intermarriage ties. These *ricos* tended to be European Spaniards or native-born from Mexico City (Baxter 1987). There was very little coinage available and trade was usually by the barter system, often involving sheep. Sheep provided food and clothing—the stuff of life. Sheep were valued even more than real estate because land in New Mexico seemed to exist in unlimited quantities. Sheep therefore became the principal currency for the exchange of wealth. Powerful *dons* (Spanish for "gentlemen," a honorific used before the Christian name and with a connotation of power) acquired great herds, and people to run them via the partido system. The urge to accumulate wealth could not be satisfied through the normal trappings of expensive consumer goods because few such items were available. The ultimate expression of wealth became power, both economic and political. The owners of the great sheep herds became the arbiters of the fates of the less powerful around them. They became *patróns*.

As the herds grew the mutton-sheep drives south to Mexico continued, but wool production and the associated textile industry expanded as well. When the Santa Fe Trail opened in the early 1820s, bringing manufactured goods from "the States" to the colony, the amount of wool production exceeded the needs of the local weavers and raw wool was backhauled over the trail to Missouri. The trail gradually became a two-way street and by the 1830s the powerful Hispanic families were heavily engaged in the Santa Fe commerce.

By the late 1830s members of these families were traveling east along the trail to St. Louis and beyond to Philadelphia and New York to purchase goods directly from the major wholesalers. In time the big sheepmen dominated the foreign merchandise trade as well. With a keen ear to the ground, some of the families realized that to effectively compete with the Anglo traders they would have to learn to speak English. For example, in 1843 *Don* José Leandro Perea (more about him below) took his nephew, Francisco, to St. Louis and enrolled him in a Jesuit college. In 1845 young Francisco returned to Bernalillo and began a long and distinguished career. He also became official translator for his powerful uncle. In 1859 the Pereas shipped 35 tons of merchandise down the Santa Fe Trail to New Mexico, and by 1867 Leandro began outfitting annual shipments (Boyle 1997).

The family name of Perea was one that would become synonymous with the sheep industry and with regional power. The first Perea in the Rio Abajo was Pedro Acencio Perea (b. 1760s), who moved from Mexico City to Corrales in 1780 (Sandoval County Historical Society, undated document). His son, Pedro José Perea (1787–1870), born that same year, in time became the head of a vast mercantile-livestock dynasty headquartered in Bernalillo.

Don José Leandro Perea, the youngest of Pedro José's nine children, arguably became the most powerful of the patróns during the mid-19th century. Born in Bernalillo in 1822, he married María Dolores Chávez of the prominent Chávez family and fathered eight children. Not to be outdone, Leandro's nephew Francisco (1830–1913), only eight years younger, had 18 children by wife number one and 18 more by wife number two. Although many of these children died in infancy, many survived to make the Perea clan a sizable crew.

Don Leandro was now one of the wealthiest men in the territory. To have risen so far he must have been one very tough character. His portrait reveals a steely demeanor and a thoroughly determined face (Figure 33). He was not one to cross. In 1860 the 38-year-old *don* listed assets of $225,000; a decade later they were $408,000. That was indeed a very great deal of money for the time (Boyle 1997). By the 1870s he exerted life and death power over area residents, and some called him the "dictator." This power was shared by other members of his extended family (Bernalillo High School 1976).

After the American victory in the Mexican War of 1846–1848 New Mexico saw a major influx of American military presence. From that time onward, especially after the Civil War, the influence of the Hispanic traders slowly waned. These traders failed to vigorously pursue the military trade and very gradually were replaced by Anglo traders.

Jewish merchants also played a major role in this transition. A wave of Jewish migration to America occurred after a period of economic oppression in the German states of Europe during the Napoleonic era. After 1815 large numbers of German Jews fled to the new world. Many settled in the St. Louis area. St. Louis became the financial hub for the Santa Fe Trail, and many Jewish men followed the trail to begin mercantile ventures in what would become New Mexico.

The U.S. military, with its Forts and troops, required a steady stream of food and supplies. The Jewish merchants became middlemen in this movement of goods. By 1860 the men began to bring German Jewish women to New Mexico to start families. Important family mercantile firms rose up, particularly involving brothers. Most of the families lived in Santa Fe and Las Vegas, but some ventured out. Nathan Bibo was one of the latter (Tobias 1990). He arrived in Santa Fe in 1866, a short time after his brother Simon. The Bibo brothers moved to Bernalillo in 1871 and, at the urging of Francisco Perea, built a store in 1873. By the late 1870s the Bibos were prominent merchants in Bernalillo.

In 1878 *Don* Leandro made a fateful decision that sealed the fate of Bernalillo for the next century. One evening in the winter of that year five representatives of the Santa Fe Railroad arrived at his doorstep. At the time the railroad was advancing south toward Raton Pass on its way to New Mexico from Colorado. The men wished to acquire a parcel of land, owned by the old don, to construct an office, shops, yards, and

Figure 33. Don José Leandro Perea, 1822–1883. (Photo courtesy of Sandoval County Historical Society, cat. #78.005P/007.)

roundhouse as the main division point for a transcontinental route to extend west from Bernalillo. This prospective line would have coursed west up the Valley of the Rio Jémez, past the community of Cabezón, on toward Ft. Wingate to the west and points beyond (Figure 5).

Don Leandro knowingly insisted on a price that was way above what the railroad men were willing to pay. Bernalillo merchant Nathan Bibo (see discussion below about the Bibo-Seligman home) described what he saw: " . . . they left my place, and while Mr. Robinson [Santa Fe's chief engineer] never spoke a word, I could see on his pale face and also by the expressions of contempt uttered by some of the party, that they must have met a most unexpected and unpleasant disappointment." Immediately after their return, their coach was ordered to proceed to Albuquerque. What had happened that memorable February afternoon was the setback for the town of Bernalillo and the making of the new town of Albuquerque" (Bernalillo High School 1976).

The don's motive is still debated today. It may have been to limit the Anglo influx into "his" town and to therefore maintain his status as *patrón*. He may simply have thought, "Let them go somewhere else." And so they did! The Santa Fe Railroad men proceeded 17 miles south to the little hamlet of Albuquerque and the rest is history. Bernalillo was to remain a small town under the don's control—exactly as he had intended. When *Don* José Leandro, *patrón extra ordinaire*, died in 1883 he was buried with some pomp inside the Our Lady of Sorrows Church, in front of the sanctuary (Chávez 1957).

Nathan Bibo was not at all amused by Leandro's decision. He was convinced, correctly, that without the railroad facilities the town of Bernalillo would wither. Disillusioned, in 1884 he moved to San Francisco, where he married. In 1898 a brother followed in his footsteps. Tellingly the two retained their business holdings in New Mexico. However they met financial ruin in the wake of the San Francisco earthquake of 1906. Nathan eventually returned to Bernalillo around 1920 and died there in 1927. In 1922 he wrote: "The appearance of Bernalillo today represents about the scenes which I would have seen in 1878. Forty-four years have elapsed and one of the best located towns in the picturesque valley below the Sandia Mountain Range was kept to decay by the selfish and iron will of a powerful potentate of those days" (Bernalillo High School 1976).

In 1892 another German Jew, Siegfried Seligman, immigrated from Germany to San Francisco, where he met Nathan Bibo. He also met Bibo's niece, a girl from the Block family, whom he soon married. In about 1899 the couple moved to Bernalillo and Siegfried began working for Joseph Bibo at the store. Siegfried's brothers, Julias and Carl, soon afterwards immigrated to the U.S. Each married a Block sister (that made three Seligman-Block unions!) and moved to Bernalillo (Arango 1980a). A short time after 1903 the old Bibo store burned down and the business was moved to an existing dry goods store down and across the street. The name of the new store was changed to the "Bernalillo Merchantile," or simply "the Merc." In 1922 the Seligman brothers bought out the Bibo family's mercantile business.

Bernalillo wine

In the 17th century grape cuttings were brought to the New World from Spain and hauled up the Camino Real by settlers and Franciscan priests. The sandy, alluvial, well-watered soils of the Rio Grande flood plain provided excellent substrate for the vines. The priests produced wine for both sacramental and table use. Bernalillo soon became an important wine-growing center (Liebert 1993). In 1846, J.W. Abert, an American lieutenant with General Kearney, wrote:

"Along the roadside were beautiful vineyards surrounded by high walls of adobes. We rode up to one of them and looking over saw some *Doñacellas* plucking the fruit. They had round flat baskets placed on their heads. These were filled with thick clustered bunches of purple grapes from beneath which the bright eyes of the *Doñacellas* were sparkling" (Liebert 1993).

Travelers along the *Camino Real* would comment on the magnificent home of the Pereas, and described Bernalillo as a village surrounded by peach orchards and vineyards with adobe walls covered with prickly cactus at the top to protect the grapes from raiders (Bayer 1994). In 1854, W.W.H. Davis, Territorial Attorney General observed: "No climate in the world is better suited to the vine than the middle and southern portions of New Mexico . . . At Bernalillo we enter the vine growing region of New Mexico which extends down the valley of the Del Norte [Rio Grande] to some distance below El Paso. Throughout this extent, grapes of a superior quality are cultivated. When fresh from the vine the flavor is very fine and thought to equal imports from Spain and the Mediterranean for table use . . . Several thousands of gallons of wine are made yearly for home consumption . . . little gets to market" (Liebert 1993).

In 1872 the order of the Christian Brothers arrived in Bernalillo to open a school for boys. They planted several thousand grape cuttings in the vicinity of the Our Lady of Sorrows Church (more in Chapter 5). Their operation was managed by the French winemaker, Louis Gros, Sr. In the 1920s Gros left the Brothers to start up his own vineyard and winery. In 1882 the Mallett Brothers came to Bernalillo from France, and in about 1919 the Rinaldi family arrived and brought the northern Italian wine-making tradition with them. The powerful Perea family also had a fine vineyard (Liebert 1993). The region flourished.

By the late 1880s things started to gradually go downhill. As mentioned above, settlers in the San Luis Valley of southern Colorado at this time began to exploit the headwaters of the Rio Grande and establish an extensive irrigation system. Average downstream flows were accordingly reduced to the point that sediment was deposited in the riverbed rather than transported downstream. Water levels rose and the flood plain became waterlogged and alkaline. Finally, the combination of a series of devastating floods in the late 19th century, grapevine root rot in 1900, Prohibition (1920–1933) and pervasive drought in the 1920s and 1930s ruined the industry. Bernalillo is no longer a wine-growing center, but its annual New Mexico Wine Festival is leading a renaissance in the New Mexico's wine industry. The popular three-day event over the Labor Day weekend began in 1988. It has reestablished Bernalillo as the vibrant symbol of the state's very-much-recovering wine industry.

Santa Fe Northwestern Railroad and the lumber business, 1924–1973

The Santa Fe Northwestern Railroad was the brainchild of the irrepressible Sidney M. Weil (more about him in Chapter 12). He came to Albuquerque from back east in 1916 and soon became heavily involved in timber and coal ventures. In 1917 he proposed the construction of a 55-mile, standard-gauge (4-foot 8.5-inch) line from Bernalillo, around the southern nose of the Sierra Nacimiento and north along the west side of the Sierra to the undeveloped coal deposits at La Ventana between San Ysidro and Cuba (Figure 11), and thus gaining access to the huge timber resources of the Jémez Mountains. The Santa Fe Northwestern Railway Company was incorporated in New Mexico in 1920.

The following year a certain Guy Porter from West Virginia, president of the White Pine Lumber Co., acquired control of the timber rights of the huge Cañon de San Diego Land Grant in the southwestern part of the Jémez Mountains (more in Chapter 6). He and his family announced plans to construct a lumber mill in either Bernalillo or San Ysidro, and pushed for the rail line to run up the upper Rio Jémez Valley from San Ysidro into the mountains. In a preemptive move, in 1922 the town of Bernalillo purchased 100 acres of land and presented it to the lumber company for use as a mill site. Needless to say, Porter opted to build in Bernalillo.

During all this action Sidney Weil gradually became marginalized in the timber scheme but continued to promote his rail line to La Ventana. He eventually succeeded and promoted the construction of the Santa Fe Northern (later called the Cuba Extension) Railroad between San Ysidro and La Ventana (Glover 1990).

The line was constructed between 1922 and 1924, and the steam-powered mill, *el molino*, in 1924 with its towering 110,000-gal water tank and a 12-acre log pond (Chapter 5). The finished lumber products were shipped via the main Santa Fe rail line to the Midwest. Many of the local men worked at the mill. Logs were transported on the railroad from the Jémez Mountains, down the Rio Jémez Valley, over a wooden bridge spanning the Rio Grande, and south along the grade of what is now *Camino Don Tomás* to the mill. The wooden-pile trestle bridge once crossed the river immediately north of the Kuaua ruins. This was before the days of upstream dams, and floods were a constant threat. The bridge was so low that it often collected debris that had to be carefully removed (Glover 1990). Not a trace of the bridge remains today.

A declining lumber market in late 1928 and a *coup de grace* to the home construction industry inflicted by the stock market crash in late 1929 shut the operation down from 1928 to 1930. In 1931 the White Pine Lumber Company was reorganized as the New Mexico Lumber and Timber Company and rail-haulage resumed. During the difficult days of the Great Depression the mill experienced considerable, and very serious, labor strife. Major floods in the spring of 1941, following a record-setting wet winter, washed out much the track and many of the bridges. Afterwards logs were hauled to the mill by truck, until 1973 when the mill was shut down for good after a run of almost 40 years. Today only the water tower and sawdust burner remain (see Figure 43B in Chapter 5). For a time in the early 1970s the old water tower was used to hold the community's municipal water supply (less than a five-hour supply) until a new facility could be constructed (*Sandoval County Times-Independent* 1972a).

Roads west

For many years a network of trails led out from Bernalillo west along the Rio Jémez to grazing areas in the Jémez Mountains and the valley of the Rio Puerco of the East, and beyond to the Navajo country. After the mid-19th century the U.S. military constructed a wagon road along the Rio Jémez to carry food and supplies to Ft. Wingate (Figure 5). In time, travelers, traders, and settlers, in addition to the military, agitated for improved roads.

After the peace with the Navajos and their return in 1866 from their internment at Bosque Redondo and resettlement in their homeland, the road west from Bernalillo gradually became a thoroughfare. Just west of San Ysidro the main road forked into two parts—one west, around the north end of Mesa Prieta to the middle Rio Puerco Valley, and one north to the upper Rio Puerco Valley. Bernalillo became a supply center for settlers in these areas and—after the arrival of the Santa Fe Railroad in Bernalillo—provided a railhead for livestock and wool. Freight wagons, passenger and mail stages, and stockmen used these roads in both directions (Bayer 1994).

By the advent of motorized transport a straighter road was needed for safety and higher speeds. By the 1920s State Highway 44 was a dirt road that left Bernalillo and crossed the Rio Grande over a wooden bridge located about 1,500 feet north of the present bridge (Figure 34). The road went directly through the present Santa Ana Star Casino, crossed the barren west mesa, and kept to the higher ground on the south flank of the Rio Jémez Valley (Figure 26). In 1938 the present bridge over the Rio Grande at Bernalillo was constructed and a new, re-aligned and surfaced NM-44 was constructed close to the present alignment.

The incorporated Town of Bernalillo

The old core of the town of Bernalillo is *Las Cocinitas* (Spanish for "little kitchens"). A glance at the map (Figure 26) reveals that the southern end of the town, including Las Cocinitas, is located within the Sandia Pueblo Land Grant, and herein lies an interesting story. In 1814 the Sandia Indians "loaned" some strips of land to a group of poor, landless Spaniards, who had been displaced by floods, for a period of five years. By 1816 the Pueblo became alarmed by the permanent nature of the dwellings and asked the Spanish governor to oust the settlers. However, it seems likely that Sandia was ordered to honor their five-year contract.

Figure 34. Southeast view of old Highway 44 bridge over Rio Grande, ca. 1935. (Note absence of bosque; photo courtesy of Palace of the Governors Photo Archives (NMHM/DCA), negative #51466; photographer T.H. Parkhurst.)

Under Spanish jurisdiction, which ended in 1821, the Indians were considered wards of the Crown and were not allowed to deed their land. During the Mexican period, 1821–1846, these restrictions were relaxed and some sales were allowed. In 1824 two Sandian Indians deeded a large tract of land to a group of Bernalillo settlers "on account of their not having land on which to support their families." In 1858 the U.S. government approved the Sandia Land Grant, but did not take up the issue of the claims of land by non-Indians. In 1924 the Public Lands Board was formed to deal with such matters. In 1927 the board, at hearings held in Bernalillo, ruled that the 1824 deed of land to the settlers was null and void, in keeping with the Spanish practice disallowing the deeding of land by Indians. The board filed suit in federal court to quiet title to the land in the name of Sandia Pueblo (Brayer 1939). This would have returned most of downtown Bernalillo to the Pueblo!

However, before the case could be heard a group of Bernalillo citizens took up negotiations with the Pueblo to purchase the land in question, and a deal was struck. Siegfried Seligman, as trustee for the Bernalillo citizens, acquired a quit-claim from the Pueblo for the land. In 1929 the U.S. government approved the sale and the town secured valid title to its downtown, which included Las Cocinitas (Brayer 1939).

In 1948, after almost two and a half centuries of existence, Bernalillo finally incorporated as a town. The town's limits were located entirely on the east side of the Rio Grande and included most of the inhabited area between the Sandia Acequia on the west and the Santa Fe Railroad tracks on the east (inset in Figure 26).

In 1985 the town annexed a western panhandle that extended across to the west side of the Rio Grande and bumped head on to the expansion efforts of the burgeoning community of Rio Rancho (Jones 1988). In 1987 the town also annexed a skinny northern panhandle along the Santa Fe Railroad tracks to allow construction of the new Centex gypsum drywall plant.

5

Bernalillo — Select Places of Interest

With its recent splurge of economic development, the Town of Bernalillo no longer lies in the shadow of its big neighbor to the south, Albuquerque. Bernalillo is certainly experiencing a renaissance. Of equal importance of course is the town's rich history. This chapter identifies some of the points of interest (Figure 35).

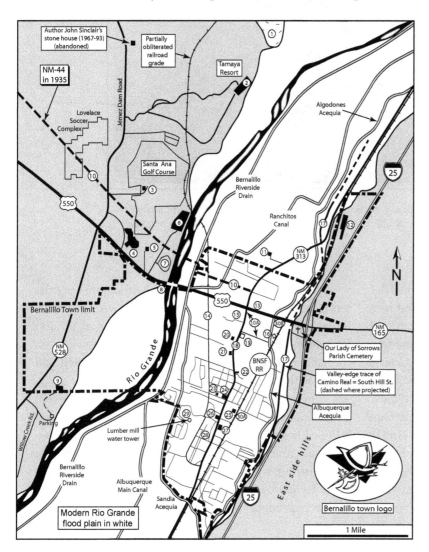

Figure 35. Select points of interest in Bernalillo area

No. 1. Canjilón Hill

This is a volcanic center sometimes called the Bernalillo Volcano (see Figure 29A). For years the feature served the Geology Department of the University of New Mexico as a mapping venue for its field geology classes (Kelley and Kudo 1978). It is located on Santa Ana Pueblo land and is off limits to the general public, but is a quite imposing sight from the Hyatt Regency (see No. 2 below).

No. 2. Hyatt Regency Tamaya Resort and Spa

This first-class resort is a Santa Ana Pueblo enterprise. The 350-room, 500-acre, $80 million operation was opened for business in January 2001. The 18-hole Twin Warriors golf course gracefully wraps around the western flank of Canjilón Hill (see No. 1 above). The resort has been recognized on Condé Nast Travelers gold List, which ranks places with the best spa programs (Rayburn 2003, Cole 2003).

No. 3. Prairie Star Restaurant and Santa Ana Golf Club

The upscale Prairie Star Restaurant is also located on land owned by Santa Ana Pueblo. It is housed in an adobe home, also known as the Brooks' home. The original structure, now the southern foundation of the house, was built by Eva Wade Duke in the 1920s. Harold Brooks was a turkey farmer who in 1948 added on to the earlier home. Brooks planned to use the surrounding land for an orchard, but was forced to drop the idea when the State Engineer informed him that he had limited water rights. Brooks handpicked the 156 vigas for the ceiling from forests in the Jémez Mountains, at a cost of $10! He and his large family lived in the home until about 1959. In 1985 Santa Ana Pueblo acquired the house and ten acres of land, invested some $250,000 in renovation (Hartranft 1986), and added some 10,000 square feet of space. The restaurant opened in 1986 (Pinel 1987). The adjacent 27-hole golf club has hosted several PGA championships.

No. 4. Santa Ana Star Casino and tribal offices

This huge adobe-style structure opened in 1994. Just west of the casino, and virtually incorporated into the tribal office complex at about mile 2.2 on the north side of the highway, is the small campus of the Pilgrim Indian Mission School, built in 1950. The facility was abandoned in the late 1980s and the land was purchased by Santa Ana Pueblo in 1994. The Pueblo remodeled the old school into offices and renovated the chapel. The 40-acre parcel contains a small cemetery (Keystone Environmental and Planning, Inc. 1997).

No. 5. DeLavy House, home of Sandoval County Historical Society

The building was formerly the home and studio of artist Edmond DeLavy. It serves as an archive, museum, and meeting facility for the Sandoval County Historical Society. The Society owes its success mainly to a remarkable woman named Martha Liebert and the generous donation of property by artist Edmond DeLavy.

In the early 1950s Martha *nee* Barr was a graduate student, working on a masters degree at the University of New Mexico. She and her brother had earlier taken a trip to New Mexico and were astounded at the contrast between the Land of Enchantment and their native North Dakota. She was intent on returning. While at UNM she met Joe Liebert, a World War II veteran who was attending the university on the GI Bill. They married in 1957. Upon graduation in 1955 Joe went on to Valley High School in Albuquerque, where he taught for the next 28 years. While Joe taught school, Martha envisioned a public library for Bernalillo, and in 1965 this vision became a reality. This facility would, in 1985, be dubbed the "Martha Liebert Library" (Snider 2007).

The Sandoval County Historical Society was incorporated in 1977. The following year the society took on as its major project the augmentation of a photo collection supported by a grant from the New Mexico Humanities Council. In 1980 Martha became the Society's president and presided over it for the next five years. During this time the photo collection was housed in the Bernalillo Library and displayed in the Town Hall. When Martha retired in 1989 the collection was moved to the DeLavy House. The house plus 2.5 acres of homestead land had been bestowed to the Society upon the artist's death that year (Snider 2007). By 2002 the Society had outgrown its facility and added a wing to the main house. The Society now is a vibrant and viable organization with a significant membership, and Martha is, and has been, the driving force behind the Society's success.

Edmond J. DeLavy (1916–1989) was born in Maine of Swiss and German parents. Early in life he was influenced by the work of artist H.C. Wyeth and decided to become an illustrator. In the 1930s he attended the Pratt Institute in Brooklyn, New York, and became an illustrator for publishers of pulp westerns. He interrupted his career to serve 4.5 years in World War II as an Army surgical technician. After the war he continued his studies and worked again as an illustrator. In 1947, on a tour of the West, he landed in Albuquerque's Old Town and remained there for three years while he worked as a portrait artist and carpenter. He returned east but then came back in 1959 and became acquainted with writer John Sinclair, who was then curator of Coronado Monument (see No. 6 below). Sinclair suggested that DeLavy file for a Homestead piece of land located next to the Monument and build a home there. In 1961 DeLavy began construction of his home, and made the design and did much of the work himself. He returned back east again and in 1963 married a portrait painter. They later divorced and in 1970 he returned to New Mexico, this time for good. DeLavy provided illustrations for the books and articles of his neighbor Sinclair, as well as for University of New Mexico Press and the periodical *New Mexico Magazine*.

No. 6. Coronado State Monument, Kuaua Pueblo Ruin

Kuaua is the Tiwa word meaning "evergreen." The old pueblo was the northernmost of 12 Tiwa-speaking villages first settled in the 14th century (Figure 36). The conventional wisdom is that Coronado spent the years 1540–1542 here or very near here. This thinking has since been revised (see Santiago Pueblo ruins, No. 9 below). Kuaua consists of two plazas surrounded by about 1,200 small rooms constructed of adobe. The pueblo's population must have been about 1,500 (Sinclair 1980a). When the Spanish under *Don* Juan de Oñate returned in 1598 the pueblo was found abandoned.

From 1934 to 1940, major excavation and restoration took place, as well as construction of the museum and curator's house. In 1944, writer John L. Sinclair went to Kuaua as custodian and moved into the curator's house with his furniture, typewriter, 500 books, a dog and three cats (Sinclair 1947a).

John L. Sinclair (1902–1993) is a fascinating character. He was the son of a Scottish sea captain, born in New

York City but raised in Scotland and educated in England. His family intended him to be an aristocrat, but he was less than enthusiastic about that. Instead he was fascinated by farming and livestock breeding. After apprenticeship on various farms in Scotland from 1919 to 1923 he was ready for his new career. In 1923 his family packed him off to America, with a final destination of Vancouver, Canada, where he was to set up a family ranch operation. In the summer of 1923 he embarked on the Chicago-Santa Fe-San Francisco train. Fatefully, on the trip he met a New Mexican cattleman returning home from Chicago, who told the young Sinclair about the Southwest. When the train reached Clovis, New Mexico, he saw saddle horses and cowboys, and was enthralled. He got off the train and watched it disappear on its way to San Francisco. His new cattleman friend suggested he take the train down to Roswell to see what "real" cattle country looked like. And so he did (Sinclair 1996).

Figure 36. East, high-oblique aerial view of Coronado State Monument, Kuaua ruins.
(Curator's house with white roof; photo by author 2001.)

Roswell became Sinclair's first home in New Mexico. He decided to convince his family to start their business there. He made his way back east and in the fall of 1923 sailed for Scotland. However, his aristocratic family would have none of this talk about setting up business in America, a country and culture they despised. If he wanted to be an American he'd have to do it on his own dime! However the family did agree to make him a "remittance man," meaning he would receive an allowance twice a year from the family's estate, with a sizable inheritance to come later. (Of an estate worth hundreds of thousands of dollars he eventually wound up with $800. His family squandered the rest.) He sailed for America and took the train to his new home in New Mexico (Sinclair 1996).

From 1923 to 1937 he worked as a cowboy and sheepman on ranches in the El Capitan area west of Roswell. However, beginning in 1927 he grew increasingly inspired to write about the things he saw and the people he met, especially the rugged homesteaders ("nesters"). In 1933 he made a visit to Santa Fe and the inspiration he received from writers there convinced him to become a professional writer himself. In 1936 he wrote a short article and sold it for $1, and in 1937 he moved to Santa Fe. Ironically, the "cowboy" Sinclair never rode a horse again. He felt that horses were solely for transportation (Sinclair 1996).

That same year he sold his first piece to *New Mexico Magazine*, entitled *Shepherds on Horseback*, for $13.50. He landed a job as a research assistant with the Museum of New Mexico and was charged with writing articles for the *Santa Fe New Mexican* to promote the museum's work. From 1940 to 1942 he worked as curator at the Lincoln County Museum and wrote his first novel, *Time of Harvest* (Sinclair 1943).

In 1942 he moved back to Santa Fe and in 1944 the Museum made him curator of the Coronado State Monument. There he wrote his second novel, *Death in the Claimshack* (Sinclair 1947b). In 1947 he met Ed DeLavy, and eventually they would become good friends. Also in 1947 he married an art teacher, Evelyn Fox, and the pair spent the next 15 years at the monument. In 1962 Sinclair retired from the museum, and after a few years in southwestern New Mexico he and Evelyn moved back to the Bernalillo area for good in 1967. They rented a small stone house on Jémez Dam Road on the Santa Ana Reservation and spent the rest of their lives there. (The house had been built in 1938 as a retreat by Dr. Sophie Aberlie, America's first practicing anthropologist. Another of her former homes became the Visitors Center at Petroglyph National Monument on Albuquerque's west side; Sinclair 1996). While there he wrote a collection of essays entitled *New Mexico, the Shining Land* (Sinclair 1980a), and his third and final novel, *Cousin Drewey and the Holy Twister* (Sinclair 1980b). With illustrations provided by his friend and neighbor Ed DeLavy he wrote *Cowboy Riding Country* (Sinclair1982) about his 14 years as a ranch hand. Finally, in failing health, at the age of 90 he dictated his memoirs (Sinclair 1996).

Evelyn died in a nursing home in 1993, and John followed her in death a month later. The stone house was never again occupied. Santa Ana Pueblo has inexplicably allowed it to decay. The house is only partially visible from Jémez Dam Road and is today closed to the public, a sad loss to local history (Figure 35). Sinclair's papers are stored in the Special Collections of Zimmerman Library on the campus of the University of New Mexico. His portrait, painted by his longtime friend Ed DeLavy, hangs in the library.

No. 7. Coronado State Park

This popular campground is perched atop the old river terrace of the Los Duranes Formation, on the west side of the Rio Grande.

No. 8. New Rio Grande bridge

This 955-foot steel-girder bridge was built in 1938 when highway NM-44 was realigned. The new bridge replaced the old wooden structure that once crossed the river about 1,500 feet north of this point (Figure 34).

No. 9. Santiago Pueblo Ruin

Like Kuaua Pueblo, the ruins of Santiago Pueblo are situated on the terrace of the Los Duranes Formation overlooking the Rio Grande from the west side. The pueblo was excavated in 1934 to 1936 by the same group and at the same time as Kuaua. The site was occupied from the 15th century to the 17th century. It consisted of four room-block wings surrounding a central plaza. About 450 ground-floor rooms were excavated, although it is likely that part of the pueblo had two stories (Vierra 1987). The excavations recovered six crossbow bolt heads. One was found in the chest of a skeleton in the ruin. Coronado's expedition was the only one to carry crossbows. Then in 1986, roadwork along NM-528 revealed metal fragments, nails and a riveted plate from a flexible armored vest. Many scholars now believe that Coronado's 1540–1542 camp was at Santiago Pueblo and not Kuaua (Bauer et al. 2003). Access to the perimeter of the site, from where very little can be recognized, is via NM-528, 1.8 miles south of the US-550 intersection, east on Willow Creek Rd. for 0.2 miles, and left on the bosque access road for 0.2 miles to a parking area. From there it is an easy 0.25 mile walk north on the terrace top to a fence. The site is just on the other side of the fence and is posted as private property. There is little to see.

No. 10. Segment of old Highway 44

A wonderful series of aerial photographs, shot by the Soil Conservation Service in 1935, clearly shows the original route of the old highway NM-44 as it then passed through Bernalillo. From the west the road crossed the Rio Grande on the old wooden bridge (Figure 34), took a 500-ft zig to the south at the Bernalillo Riverside Drain, and then a zag east along a quiet 3,000-foot-long residential lane. At the lane's intersection with today's Camino del Pueblo a street sign flags it as "Old Hwy 44." From there the route ran south along Camino del Pueblo and then east along the Calle de Escuela, named for the Bernalillo elementary school (No. 10A on Figure 35). In the 1920s and 1930s this was a dirt track called the El Callejón de la Máchina, or "the big machine street," because in the 1920s there was a flour mill at the end of this street run by water power near the Bernalillo acequia (*Sandoval County Times-Independent* 1978). The road crossed the acequia, just south of the old Perea family cemetery (see No. 17 below) and wended its way east up an arroyo toward Las Placitas. (The area north of Calle de Escuela is today occupied by a housing subdivision.) In 1938 the present bridge over the Rio Grande was constructed and NM-44 was relocated to its present alignment (see No. 9 above).

No. 11. La Hacienda Grande Bed and Breakfast Inn

Formerly the Gallegos estancia, this house was constructed ca. 1700. The Gallegos family had arrived in the area in 1695. They were granted a tract of land—the Gallegos (a.k.a. Bernalillo) Land Grant—to be used as a farm and ranch. The original building was constructed on the site just off the Camino Real, then the only road to the outside world. The ranch became the center of the growing community of Bernalillo. In the 19th century the house served as a stagecoach stop on the Camino Real. The Gallegos and Montoya families intermarried and became an economic and political power in the community. The home served as a chapel for the town before the construction of Our Lady of Sorrows Church.

During the Civil War, Confederate soldiers raided churches in New Mexico. The church rushed its valuables to the Gallegos/Montoya chapel for protection and supposedly buried them under one of the floors. The rebels never made it to Bernalillo but for some reason the "treasure" was lost. In 1986 or 1987 a man rented the house for several months. The Montoyas stopped by when the rent was due and discovered an eight-foot-deep hole dug in the floor of one of the rooms! The man quickly disappeared. The descendants of the Gallegos and Montoya families sold the house to the present innkeepers, and since 1993 it has been a Bed and Breakfast. The old chapel area is now the kitchen, the estate's winery and granary are now the dining room, and the stable area is now the living room.

No. 12. Centex (American Gypsum Company) drywall plant

Before construction of the plant in Bernalillo there were proposals to build it near Algodones, and next on Santa Ana Pueblo land, but in each case local objections killed the project. After considerable and sometimes bitter controversy—mainly dealing with concerns about air pollution—the plant was built within the newly annexed, 63-acre sliver of land now making up Bernalillo's northern "panhandle." Operations began in 1989. The plant operates 24 hours a day, five days a week, and produces about 30 million square feet of drywall a month (Armijo 2001a).

Employing 80, it is the largest private-sector employer in the town. The raw gypsum is quarried from White Mesa, just southwest of the Village of San Ysidro, about 20 miles west of Bernalillo on Zia Pueblo land. A steady flow of ore trucks rumbles down US-550 to the plant.

No. 13. Louis Gros winery, vineyards, and orchards (site)

In 2004 the site became the location of the new Bernalillo Marketplace, anchored by a Walgreen's drug store.

No. 14. Camino Don Tomás

This street was built atop the old bed of the Santa Fe Northwestern Railroad that linked the timber areas in the Jémez Mountains with the sawmill in Bernalillo.

No. 15. New Mexico Wine Festival venue

The extremely popular New Mexico Wine Festival was initiated in 1988, and has been held every Labor Day weekend at least through 2008.

No. 16. Perea family burial plot

This forlorn, overgrown, and virtually forgotten burial site occupies a sliver of land sandwiched between the tracks of the Burlington Northern and Santa Fe Railroad on the east and the Bernalillo acequia/levee on the west (Figure 37A). The plot was abandoned sometime after 1918. It can be easily accessed by parking at the eastern end of Calle de Escuela where the road abuts the levee of the Bernalillo acequia, walking to the top of the levee and following it north about 100 yards, crossing the acequia at the gate to the cemetery on the other side of the acequia. Ironically, despite the name, the only Perea grave in the cemetery is that of *Don* Leandro's son Pedro (1852–1906) and grandson Abel (1878–1918) (Figure 37B). Both of these men were prominent citizens and served as Territorial delegates to Congress. Their joint headstone was vandalized sometime before 1976 (Sinclair 1976) and today merely the base of the monument remains. Only ten readable grave markers can be found in the cemetery today (2004) and most are in poor shape.

No. 17. "By-pass" route of the *Camino Real*

This old wagon road hugged the base of the old river terrace on the east side of the Rio Grande. Portions of the old track are still intact as South Hill Street on the south and as Llano Road on the north (Jackson 2006).

No. 18. Camino del Pueblo

Bernalillo's main drag (Spanish for "road of the town"), was once called simply "Old Main Street." This had been the "business route" of the old *Camino Real*. By the early 1920s it was a graded road designated State Route 1, and in 1926 it was redesignated US-85. At the time US-85 was the only through-route connecting Mexico with Canada. From 1926 until 1937, a portion of US-66, the main east-west cross-country route, overlapped US-85 between Romeroville (south of Las Vegas, New Mexico) and Los Lunas (south of Albuquerque), and ran through Bernalillo. Thus two of the nation's principal long-distance highways ran down Bernalillo's main street. In 1937 US-66 was realigned to its present east-west course through Albuquerque, and the main drag became simply US-85.

Sometime shortly after 1939 the main street was paved for the first time. In 1964 the road was widened to four lanes and it achieved its modern look. In 1967, as part of the town's radical makeover, many of the streets were renamed to recognize their historical relevance. Accordingly, what had been "Main Street," was rechristened Camino del Pueblo. The next big street, Central, became San Lorenzo after the town's patron saint. Others were named after some of Coronado's soldiers, such as Calle Don Juan and Calle Don Diego (lanes of John and James, respectively). Calle Montoya and Avenida Perea got their names from two of the town's first prominent families (*Albuquerque Tribune*,1967).

A. Derelict Pedro and Abel Perea grave. (Photo by author 2004.)

B. Tombstone of Pedro (son of *Don* José Leandro) and Abel Perea, before vandalism. (Photo courtesy Sandoval County Historical Society.)

Figure 37. Perea family burial plot.

In 1958 Interstate Highway 25 bypassed the town and dealt a nasty blow to the local economy. I-25 merged with and replaced US-85 north of Algodones. In 1988, NM-313 replaced the remaining section of US-85 between Algodones to the north and Albuquerque to the south, and US-85 was no more.

Route US-66 officially went out of existence in 1985, when it was decertified. In 1994 national interest in the legacy of the old road led to the formation of the New Mexico Route 66 Scenic Byway to celebrate sections of the surviving route. It is sometimes forgotten that part of the so-called "Mother Road" passed through little Bernalillo from 1926 to 1937.

In the 1920s the northern part of the Camino del Pueblo was lined with two-story adobe houses were surrounded by orchards, vineyards, and fields of alfalfa, corn, and chile. With the sole exception of the old high school in *El Zócolo* (see No. 19 below) all the buildings were destroyed by fire in the 1930s. Rumor has it that arsonists torched the structures for insurance purposes (Mraz 1982). Most of the houses had been built by the master *adobero* Abenicio Sálazar. He was born in 1858. As a boy he was taught to read, write, and draw by a blue-eyed hermit, probably one of the church builders brought to the Territory by the Archbishop Lamy. Sálazar's first building in Bernalillo was probably the stone jailhouse behind the courthouse (see No. 22 below), built in 1896. In 1922 he built the Sisters of Loretto School, a two-story, 12,000 square foot edifice that is still in use today. He also built the large two-story adobe schoolhouse in Peña Blanca (which has since burned) in 1905, and the entire town of Hagan (now a ghost town in eastern Sandoval County) in 1925 and 1926. Benicio died of a stroke in 1941 while he was instructing his grandson the skills of the *adobero* (Jersig 2001).

No. 19. El Zócalo

The term *El Zócolo* is Spanish for "public square." This 3.5-acre area, on the southeast corner of Camino del Pueblo and Calle del Escuela (Spanish for "street of the school"), includes five historic structures, including the Convent of Sisters of Loretto (formerly the elementary school for girls). In 1874 the town's patrón, José Leandro Perea, granted the property and an adobe building to the Sisters for an elementary school for girls. The school opened in 1878. In 1885 the Sisters added a government-contract school for Indian girls. From 1922 to 1923 the Sisters had a two-story, upper-grade school built just to the south of the earlier structures built by Abenicio Sálazar and his *adoberos* (Jersig 2001). By 1930 the new school had become a standard four-year high school—the old Bernalillo High School.

In 1949 the New Mexico Supreme Court decided the Zellers vs. Huff Case (better known as the infamous Dixon Case), that mandated the separation of Church and State in state-supported schools (more in Chapter 16). As a result, both the elementary and high schools became parochial. Fortunately both schools were the outright property of the Sisters (Chávez 1957). The high school continued on, but in 1966 it finally closed its doors and became essentially abandoned. In 1976, Bernalillo resident Terry Lamm purchased the property at a distressed tax sale and gave it its new name shortly after the area in 1980 became designated the "Abenicio Sálazar Historical District." In 2003 the federal government granted funds to purchase the property. By 2008, after five years of scrambling for funds, Sandoval County has a gem of historical preservation. Restoration of the convent church preserved the original brick floors, tile and tin work. Extensive landscaping converted a weed-infested courtyard to a gracious garden. The two-story school building was converted to office suites (Rayburn 2008). The complex is the pride of Bernalillo.

No. 20. Our Lady of Sorrows Catholic Church

The community's first church was washed away by a flood in 1735. During the next 122 years the people attended services at the church at nearby Sandia Pueblo. In 1857 Bishop Lamy built the new, Our Lady of Sorrows Church. By the 1970s the old church had fallen into disrepair and was being threatened with destruction. An article in the *Albuquerque Journal* during this time reported: "The old church, with 4.5-foot-thick adobe walls, viga roof and gothic windows, is listed with the State Register of Cultural Properties as an outstanding example of New Mexico church architecture during the French-oriented period of Bishop—later Archbishop—Jean Baptiste Lamy" (Bernalillo High School 1976). A new church has been built just to the south. The old structure is now the *Sanctuario de San Lorenzo*.

No. 21. Baca family home (Figure 38)

Located at 413 Camino del Pueblo, this was a coach and horse-change stop on the Santa Fe-El Paso stage line. It is now the Baca Antique Shop. The house was originally built and owned by the Perea family. The Cabeza de Baca family purchased the house in 1901. During the Great Depression another dwelling was built in the adjoining lot and the old house became vacant. It remained so until 1967, when the new house next door burned down. There was once a second floor, but it was removed because it became safety hazard (Brennan undated).

Figure 38. Baca House.

No. 22. Home of José León Castillo

This rambling house was built in the 1890s. It has fallen in disrepair and one of these days will likely go up in flames.

No. 23. Sandoval County stone jailhouse.

This sturdy structure was built in 1896 by *adobero* Abenecio Sálazar (Figure 39).

No. 24. Sandoval County court house

The original structure was built in 1905, and burned down in 1926. All the records were lost and therefore the local history from the creation of the county in 1903 to 1926 is somewhat obscure. The present three-story yellow brick court house was erected in 1927–1928 (Figure 40A). In the early 1970s the county commissioners unsuccessfully attempted to pass a bond issue to build a new courthouse. Having failed that, federal loans were acquired in 1974 to add a pair of stucco, Territorial-style annexes to the front and back sides. Construction was completed the following year. The annexes sandwich the brick court house such that only the very top of the classic structure can be seen from the street (Figure 40B; Tessler 1992). Sandoval County is unusual because of its almost 2.5 million acres, only about 28% is privately-owned, taxable land. The remainder is federal, state, and Indian lands, and their administrators do not answer to the county government. However, the exploding needs of the burgeoning City of Rio Rancho threaten to overwhelm the capacity of the tiny courthouse.

The question of why Bernalillo is the county seat of Sandoval County and Albuquerque the seat of Bernalillo County is a source of some confusion—especially to outsiders (Figures 6 and 8). Sandoval County was carved out of a much larger Bernalillo County and in 1903 the town of Sandoval (present-day Corrales) was designated its seat. The Sandoval county seat was moved to Bernalillo in 1905, where it has remained. The southern part of the original Bernalillo County became the modern county of that name and Albuquerque became its seat. Anticipating the confusion, the *Albuquerque Journal-Democrat* (1903) once suggested that Bernalillo County be named Albuquerque County, but, perhaps unfortunately, the idea did not catch on.

Figure 39. Stone jailhouse built by Abenicio Sálazar. (Photo by author 2004.)

A. Court House in 1964. (Photo courtesy Sandoval County Historical Society, cat. #04.004A/013.)

B. Present Court House with its front and rear annexes. (Photo by author 2004.)

Figure 40. Sandoval County court house.

No. 25. L.B. Putney flour mill ruin (Figure 41)

The mill was built about 1908 for the L.B. Putney Company of Bernalillo by Abenicio Sálazar. Lyman Beecher Putney came down the Santa Fe Trail in the late 19th century and ended up in Albuquerque, where he became a wholesale grocer. He established retail stores in several towns including Bernalillo, Jémez Pueblo, and Cuba (see Chapter 17). The mill processed flour, coffee, salt, and sugar. After L.B.'s death in the early 1900s his descendants ran the company for many years (Wiese 1978) before going bankrupt in 1939.

The mill burned down in June 1972. At the time of the fire some members of the Bernalillo Fire Department were attending the Miss Nude New Mexico contest in Algodones. The unwelcome phone call came for them to rush to Bernalillo, just as the first contestant stepped on stage (*Sandoval County Times-Independent* 1972b). The ruins are sometimes referred to as the "Spooky Mill" because of a hanging that supposedly occurred here (Sinclair 1980). In 2004 the Town of Bernalillo purchased the adjacent 1.5-acre property for $118,000. This parcel is now the site of a station for a commuter-rail line, the "Rail Runner," that runs on the Burlington Northern Santa Fe track between Belen and Bernalillo (see No. 30 below; Diven 2004).

A. Northeast view of ruined mill. (Photo by author 2004.)

B. Whimsical watercolor of northwest view of mill in its heyday. (Artist Ken Kloeppel 2005.)

Figure 41. L.B. Putney flour mill.

No. 26. Bibo-Seligman home

Built in the 1890s, 853 Camino del Pueblo. The names Bibo and Seligman are renowned in northern New Mexico, and particularly in Bernalillo. They were two of the prominent Jewish families that drove much of the commerce of Territorial New Mexico in the 19th century.

No. 27. Bernalillo Mercantile Company (the "Merc"), now True Value Hardware

This thriving business was founded by Nathan Bibo. At the urging of Francisco Perea, the influential nephew of *patrón Don* José Leandro Perea, Bibo in 1873 built a store on property next to Perea's house and vineyard, on the west side of the main street (Sinclair 1976). In the early 1900s the store burned down and in 1919 Bibo purchased an existing dry goods store, about a block to the south on the east side of the street, from a Lebanese family. The name of new store was changed to the "Bernalillo Mercantile," or simply "the Merc" (Figure 42A). This iconic structure became the center of commercial activity for a huge hinterland and drew customers from the surrounding pueblos, from the Navajo in the Checkerboard Area near the Cabezón area, ranchers in the Cuba area, and Jicarilla Apaches from far to the northwest. Behind the store were barns crammed full with hay, and five apartments for overnight customers who had driven by wagon from outlying areas to buy six months of provisions. Bibo eventually opened stores in eight outlying locations, including Thornton (now Domingo), Bland, and San Ysidro (Liebert 2005).

In 1922 Joseph Bibo sold out his interest in the Merc to the Seligman brothers. Later a second generation of Seligmans took over operations of the store. The decades up to after WWII saw a decline in the fortunes in the general merchandise business. The railroads had first helped to establish the general store as a community's commercial center, the railroads now fundamentally changed the market, bringing about the emergence of the specialty store, which competed heavily with the rural general store. The last of the outlying stores, the one in San Ysidro, was closed down in the late 1940s (Chapter 6) and the Seligmans concentrated their efforts on the Bernalillo Merc. In 1965 the main street, Camino del Pueblo, was widened to four paved lanes (Figure 42B). In 1978 the Merc closed and was sold to Ken Craft, but the arrangement fell through in 1980 (Arango 1980a).

The Torres family now enters the story. In 1939 Benito Torres had opened up the Economy store at the corner of Camino del Pueblo and Avenida de Bernalillo (two blocks south of the Merc; Figure 35). When his son Lalo (one time Bernalillo mayor) took over the store it was renamed "Lalo's." Shortly after the Merc closed, Lalo was offered space in the old store. Lalo's two sons, Joe and Jack, both then recently graduated from college, had returned home for Christmas in 1980 and learned of the offer. They jumped at the opportunity and in 1981 opened the T&T Market in a section of the old building, allowing Lalo to retire (Arango 1980a). The building was renamed La Plaza and ultimately TA GR MO True Value Hardware (Figure 42C).

A. North view of Bernalillo's main street, the "Merc" in center distance, ca. late 1930s/early 1940s. (Photo courtesy Sandoval County Historical Society.)

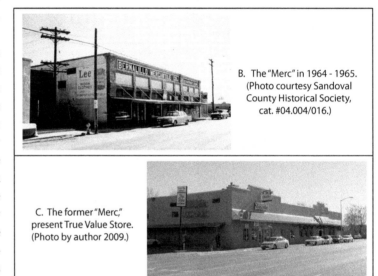

B. The "Merc" in 1964 - 1965. (Photo courtesy Sandoval County Historical Society, cat. #04.004/016.)

C. The former "Merc," present True Value Store. (Photo by author 2009.)

Figure 42. The "Merc."

No. 28. Las Cocinitas

According to Justin Renaldi, local Bernalillo historian, Las Cocinitas is a hodgepodge of small dwellings with rooms added through the centuries as needed to house growing families. (Sandoval County Historical Society, undated document). No one seems to have definite knowledge of the exact age of Las Cocinitas, nor of the origin of the name, loosely meaning, "Little Kitchens." Some believe that portions of buildings hidden within the rambling, tightly-packed adobes date back to the 17th century, but others argue that it mainly dates from the early 19th century (see Chapter 4).

No. 29. Lumbermill (site)

The mill operated from 1924 to 1973 (Figure 43A). After *el molino* was closed and dismantled the only traces of this bustling operation remaining were the old water tower, the rusting sawdust furnace, and the tumbleweed-filled "old mill pond." Part of the area has been almost now completely covered over by new homes. The mill pond is occupied by a park and recreation area. The water tower can be seen from all approaches to the town, a lonely and silent reminder that lumber was once a very big deal in Bernalillo (Figure 43B).

A. North oblique aerial view, 1968. (Photo courtesy of Sandoval County Historical Society and U.S. Dept. of Agriculture, Soil Conservation Service, cat. #NM-P618-6, 6.)

B. Northwest view of sole remains of lumber mill. (Photo by author 2004.)

Figure 43. Bernalillo lumber mill.

Nos. 30A and 30B. New Mexico Rail Runner commuter train stations.

These important features are listed last because they are brand-new additions. Two stations are located in Bernalillo, the first (30A) across the tracks from the L.B. Putney mill (No. 25), and the second (30B) just south of mile 0.4 on US-550. The commuter line is the culmination of a long, drawn-out process. The track had originally been laid by the Atchinson Topeka and Santa Fe Railroad (AT&SF) way back in the 1880s. However, on the last day of 1996, after a drawn-out series of negotiations spanning seven years, the Santa Fe Railroad quietly merged with Burlington Northern to become the Burlington Northern and Santa Fe Railroad (BNSF; Glischinski 1997).

The idea of a commuter rail line, using the BNSF trackage, first came up in 2003 when then-governor Bill Richardson announced he would pursue development of the system. During 2004 and 2005 detailed plans were made and the name, "New Mexico Rail Runner," was chosen to piggyback on the widely-recognized New Mexico state symbol, the roadrunner bird. In 2005 the state negotiated with BNSF and purchased from them the corridor between Belen and the Colorado border. Service began on a limited scale in 2006 between Albuquerque and Bernalillo, and by 2007 service had reached south to Belen. The second phase of the project was construction of an extension north from the existing trackage to the City of Santa Fe. Existing track is used to the base of La Bajada Hill, from where new track was constructed up La Bajada and along the I-25 median to Santa Fe. Service began to Santa Fe in December 2008.

Rolling stock consists of five Motive Power diesel-electric locomotives that burn biodiesel fuel, and ten passenger cars, each with a capacity of 200 (including 60 standing). The train operates on a push-pull configuration, with the locomotive always pointing south.

6

Rio Jémez Corridor

The valley of the Rio Jémez is the natural corridor leading west from the valley of the Rio Grande. It hugs the south side of the imposing masses of the Jémez Mountains and the Sierra Nacimiento and provides access to a vast open territory beyond (Figure 44A). As population pressures developed in the Rio Grande "mother valley" during the 18th century, settlers were compelled to migrate outward to areas with an adequate supply of water and pasture land. One of these choice areas was the rich grazing lands of the Valles Grandes in the heart of the Jémez Mountains, accessed via the upper Rio Jémez and San Diego Canyon north of the Village of San Ysidro. Another was the lush grazing land along the west flank of the Sierra Nacimiento. When copper and coal deposits became known in the 19th century, commercial outlets for these materials became essential if the materials were to become commodities. The railhead at Bernalillo became the economic anchor for these activities.

The earliest travelers naturally followed the course of the Rio Jémez itself. Before motorized traffic, all transport was driven by animal power, and that required a source of water and—hopefully—of forage. But the Rio Jémez had its problems too. The river did not always have the dependable flow needed to sustain the pack animals. Some travelers therefore chose to make a short-cut from the Rio Grande Valley by first traversing the top of the Santa Ana Mesa and then dropping down to the valley of the Rio Jémez when they felt it necessary. With the arrival of the Americans in the middle 19th century the military required communication and supply routes between Santa Fe and the western forts. The army therefore built wagon roads along the old trails. The town of Cabezón, on the Rio Puerco crossing, was established in 1872 to serve the increased traffic (more about this place in Chapter 10).

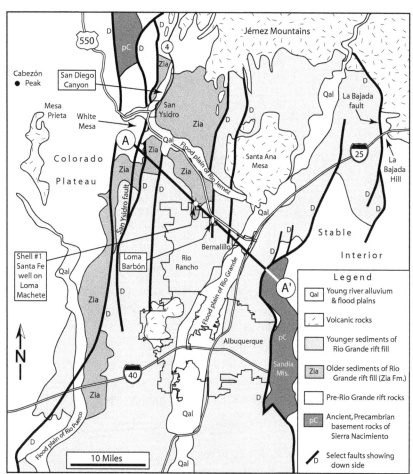

A. Geologic map. (Modified from NMBG&MR 2003.)

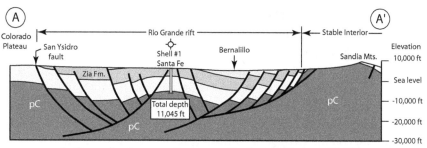

B. Generalized cross section across Rio Grande rift along US-550 (see location and legend in "A" above). Modified from Russell and Snelson 1994.)

Figure 44. General geology of Albuquerque-Bernalillo portion of Rio Grande rift.

When motorized vehicles finally arrived on the scene they simply followed the course of the earlier folks, but stuck to the south side of the river's flood plain. The track was sandy, unsteady, and subject to sudden obliteration by flash floods (Figure 45). But this was the way to the west. When the Santa Fe Northwestern Railroad was constructed in 1924 it too naturally followed the course of earlier travelers.

Figure 45. East view of primitive road toward old Santa Ana Pueblo (*Tamaya*), Río Jémez Valley, ca. 1920s. (Photo Palace of the Governors Photo Archives (NMHM/DCA), negative #158158; photographer J.L. Nusbaum.)

Geologic setting

For the first 21 miles west from Bernalillo the landscape is carved into unconsolidated beds of sand and gravel. These sediments are the basin-fill materials of what is called the "Rio Grande rift." The Rio Grande rift is a huge scar in the earth's crust caused where two large segments of the Earth's crust were pulled apart. Beginning about 26 million years ago (Ma) and continuing to a million or so years ago the Colorado Plateau—a huge, semi-circular block of relatively stable rock—was pulled and rotated clockwise away from the more vast terrain called the Stable Interior (Figure 44). The stretched zone between the two—the Rio Grande rift—gradually foundered and collapsed downward to form a topographic depression, into which thick deposits sand and gravel were washed by rivers eroding the flanking highlands.

The burgeoning City of Rio Rancho

The northern rim of the City of Rio Rancho fronts US-550 for the first few miles west of Bernalillo. Rio Rancho, which calls itself the "City of Vision," is today experiencing extremely rapid growth, particularly along this northern line (Figure 46). During my frequent drives past here on research outings for this book I felt sure that there were a few more houses in the afternoon than had existed in the morning! The south side of the highway is destined to be completely filled with new homes, access roads, traffic lights, and stores, in stark contrast to the undeveloped Santa Ana Pueblo land on the opposite, north side of the road.

Rio Rancho is one of the youngest cities in the state of New Mexico. The no-nonsense clarity and efficiency of its official logo (Figure 46A) nicely characterizes the city's can-do spirit. The city has no colonial roots. It was never a natural center for commerce or agriculture. It was simply invented. The text that follows is mainly from Ryan and Ryan (1996) and the City of Rio Rancho's internet web site.

The story begins with the Alameda Land Grant. The grant was issued by the Spanish Crown in 1710. The huge parcel stretched about 15 miles to the west from the Rio Grande. When the U.S. acquired New Mexico the land grants became subject to taxation. Apparently the heirs to the grant did not understand this new rule and in 1919 much of the grant had been sold to the San Mateo Land Company as an investment. In 1948 the company sold the land to Brownfield and Koontz, and the parcel became known as the Koontz Ranch.

In the 1950s two young entrepreneurs, Chester Garrity and Henry Hoffman, were operating a mail-order business in New York selling roses and other products. They eventually founded a corporation called American Real Estate and

Petroleum, or AMREP. In 1961 AMREP purchased the 55,000-acre Koontz Ranch for about $10 million. AMREP platted the ranch and sold lots under the name Rio Rancho Estates, mainly to eastern retirees via a clever and intense mass-marketing and mail-order campaign. The first home was completed in 1963, and by 1966 the 100th family had moved into the community. Rio Rancho Estates grew to 91,000 acres with the purchase of the King Ranch. In 1974 AMREP quit its mail order business to concentrate on home-construction and commercial development. By 1977 the marketing focus shifted from retirees to young families.

Because it was not yet incorporated, Rio Rancho received its services from Sandoval County. The county was accustomed to dealing with a mainly Hispanic and Native-American, rural population. Rio Rancho, in contrast, was dominated by urban Anglos, many of whom were not native New Mexicans. This produced some interesting friction. One little squabble involved a 500-acre parcel of land at the vital intersection of NM-528 and US-550. Both Bernalillo and Rio Rancho vied in the courts for this strategic parcel, but Rio Rancho won out narrowly (Jones 1988).

In 1981 about 18,000 acres of the total 91,000 acres were incorporated as the City of Rio Rancho. Population soon boomed, driven by business expansion, especially the addition of the huge Intel Corporation facility. By 1990 the city's population more than tripled to about 35,000. In 2002 and 2003 the city annexed a number of large ranches and increased the city's footprint to about 100 square miles. In 2003 the city had about 63,000 residents, and it is on its way to pass Santa Fe to become New Mexico's third largest city (Akers 2004). In 2003 the city's mayor predicted that Rio Rancho would eventually expand to about 300 square miles—three times the present size—with a population of 180,000. His scenario envisions Rio Rancho becoming New Mexico's Dallas, while Albuquerque would become its Ft. Worth (Velasco 2003).

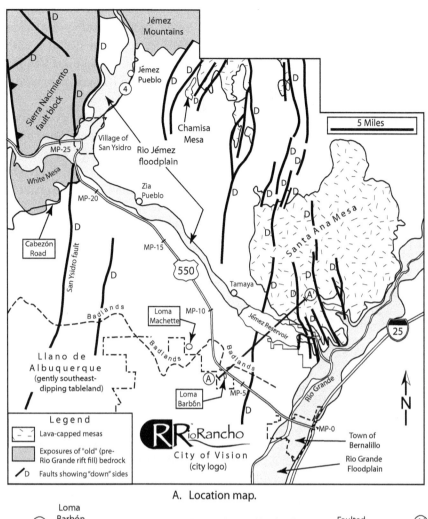

A. Location map.

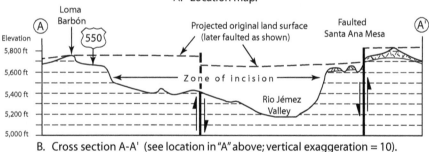

B. Cross section A-A' (see location in "A" above; vertical exaggeration = 10).

Figure 46. General geography of Rio Jémez Valley.

Loma Barbón with its mystery shaft, and Loma Machete

During a number of cold nights in early 2003 a homeowner in a remote part of far-northern Rio Rancho observed strange activity a few miles from his house, located just south of the high point on the West Mesa called *Loma Barbón*

Figure 47. "Mystery shaft." (Copyright *The Albuquerque Journal* 2003a, reprinted with permission.)

(Spanish for "Billie Goat Hill"), south of MP-6 on US-550 (Figure 46A). The gentle knoll, crowning a 600-acre parcel of State Land, stands at an elevation of 5,752 feet and provides an excellent view in all directions. For hours at a time a group of idling pickup trucks would focus their headlights on a general work area, but the homeowner couldn't see what the people were doing. What on earth was going on?

By April the bizarre activity caught the attention of the Sandoval County Sheriff's Department. When officials arrived at the site they came onto a vertical shaft, measuring 4 x 4 feet square and 70 feet deep (Figure 47). The shaft had been carefully lined with plywood and was clearly intended to last a while. One side of the shaft had planks jutting out to form steps, but otherwise it was a shear drop of seven stories straight down. At the bottom of the shaft was an eight-foot-long unlined corridor that dead-ended into dirt. It was obviously a work in progress. But for what? At least 50 cubic yards of earth (about ten single-axle dump-truck loads) had been laboriously removed from the shaft and carefully spread out around the site so that a pile would not be visible. The sheriff contacted the State Land Office (SLO), which owns the parcel, but the SLO knew nothing about the hole. An informant stated that in January there had been three or four men working out there, but he could not identify them because when the men saw they were spotted they hid in the brush.

The shaft is about a mile east of the New Mexico National Guard's Hawk Missile Training Facility site, but the Guard knew nothing about it (Vogel 2003). Was the hole intended to be a tunnel to the military site—a full mile away? How would the hole be ventilated during the work? How would the dirt be lifted out? Why was it so carefully lined? Was this a drug-smuggling/storage project? Why was the shaft 70 feet deep? The sheriff put out a request to the general public for information, but not surprisingly none was forthcoming. Whoever constructed the shaft was almost certainly operating outside the law and of course was not anxious to rush forward with an explanation. Ceding defeat, in December 2003 the SLO filled in the hole (*Albuquerque Journal* 2003b).

About two miles west-southwest of this spot, on another high point called *Loma Machete* (Spanish for "Machete" or "Cane Knife Hill"), is the site where a generation ago Shell Oil Company kicked off an ambitious oil- and natural-gas exploration program in the Albuquerque basin (Figures 44A and 44B). Shell believed that the ingredients for a successful oil and gas province, similar to those of the prolific San Juan basin to the northwest, might exist below the sands and gravels of the Albuquerque basin. The City of Albuquerque would provide a ready market. In 1972 the first well, the Santa Fe Pacific No. 1, was drilled on Loma Machete to a depth of 11,045 feet. At about 10,950 feet the drillers topped the ancient "basement" of igneous rocks—more than 15,000 feet lower in elevation than the outcrops of these same old rocks just below the very crest of the Sandia Mountains. By the end of 1980 Shell had drilled eight exploratory holes. All were "dry," and although a few flared small amounts of natural gas, the volumes were non-commercial. A great deal of money was spent, to no avail.

The big incision

As mentioned in the previous chapter, during about the past 800 ka the Rio Grande and its tributaries (including the Rio Jémez) have incised themselves hundreds of feet into the ancient landscape. West from the crest of Loma Barbón, from about mile 6.5 where the highway department has made a big cut through the hill to lower the grade, the incised nature of the landscape is particularly interesting (Figure 46A and 46B). The road cut provides an excellent view northward across rugged badlands down into the Rio Jémez Valley. A thickness of more than 400 feet of sediment has been excavated to create this valley and all the material hauled away by the downcutting Rio Jémez.

Volcanic ash beds

At mile 8.8, off to the west of the highway, are two thin but very conspicuous white beds incased within the usual buff-colored sands. They can be seen again at mile 9.1, down the hill to the east of the highway in the north wall of an arroyo (Figure 48), and at a recently-excavated spot on the east side of the highway at mile 9.25. These are volcanic ashes that had spewed out about 11 Ma from eruptive centers more than 1,000 miles to the northwest under southern Idaho (Pazzaglia et al. 1999). That date has been determined by analysis of certain radioactive minerals contained in the ash. Such dates are extremely useful because they allow geologists to assign absolute (numerical) ages to geologic events.

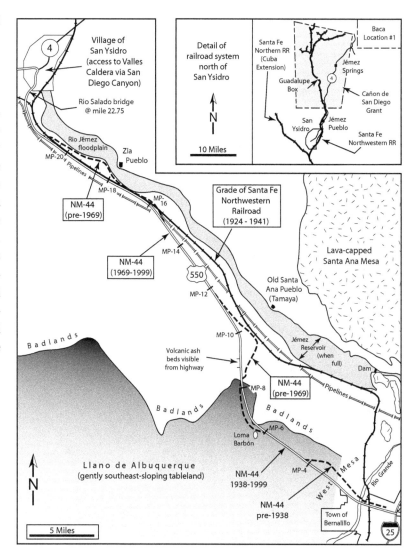

Figure 48. Busy Rio Jémez Valley transportation and utility corridor.

Old Highway NM-44

As with most highways, successive realignments over the years tend to both straighten the route and lower the grade. Old highway NM-44 was no exception. The road along this stretch was originally constructed in the early 1920s as a dirt track (University of New Mexico 1985). As mentioned earlier, the present bridge over the Rio Grande and a more direct road across the West Mesa were constructed in 1938 (Figure 48). Today, at MP-6, a short segment of this old alignment still exists just to the south of the present highway, but unfortunately the area is also used as an illegal dump site.

In late 1968 the two-lane, paved, rural highway NM-44 was constructed on its present, straighter alignment west from Loma Barbón (Figure 48). For the next 14 miles the more crooked, abandoned course of old NM-44 dances back and forth across the modern highway. At MP-8 the pre-1969 road can be seen to shoot off to the east in order to outflank the series of resistive sandstone ridges up ahead that would have required several expensive road cuts (see "Curbs and Gutters" below). The old road then rejoins the modern highway at MP-12. Just short of MP-15 the old alignment once more peals off to the northeast in order to skirt the edge of Rio Jémez flood plain, and finally rejoins US-550 at mile 20.7.

In 1999 the New Mexico Highway Department four-laned the entire 20.5-mile stretch of NM-44 from west of Bernalillo to the south edge of San Ysidro. Finally, as mentioned earlier, NM-44 was decommissioned in January 2000 and the entire reconstructed alignment redesignated as US-550.

At about MP-16 the old bed of the Santa Fe Northwestern Railroad merges with the northeast edge of US-550 and runs alongside it until mile 20.3, where the railroad shoots off again to the east (Figure 48). This is a relict of the rail line that once led from the timber country in the Cañon de San Diego Land Grant in the Jémez Mountains down to the lumber mill in Bernalillo (inset of Figure 48). This was a full-bore, standard-gauge railroad and not a little narrow-gauge affair.

Have you ever wondered why the standard gauge for American railroads is 4 feet 8.5 inches? That is a very strange number indeed. In short, it is based on the indivisible unit of width of the average horse's butt, times two. But then why two horses' butts? Well, the English first built their railroads that way and the American railroads were designed and built by English and Irish expatriates. The first railroads in England had been modeled after the early pre-railroad tramroads to service the mines, and the tramroads used that gauge. The people who built the tramroads used the same jigs and tools they had used to build wagons, built with that wheel spacing. The wagon wheels were spaced to match the "gauge" of the deeply rutted roads of England, and the ruts were spaced this amount. The long-distance English roads had been built by the Romans for their legions, who used war chariots drawn by two horses. And so, *voilà*, turning the entire sequence around we have the standard gauge of American railroads based on the original specifications of the Imperial Roman war chariot, i.e., two horses' butts.

The early American railroads were conceived as connectors to water transport. The completion of the Erie Canal in 1825 through the one relatively level crossing of the Appalachians had caused the port of New York to grow relative to the other major eastern ports. If the mountains were to be crossed at other points, only railroads would be practical. In the 1840s two railroad companies had demonstrated that it was possible to cross mountain barriers with railroads, and each had used the gauge 4 feet 8.5 inches. The predominance of this gauge on the major northeastern railroads made it the standard gauge for America. But the 4-foot 8.5-inch gauge didn't become "standard" overnight. In the Northeast the most common alternative gauge was 6.0 feet, most of the South adopted a 5.0-foot gauge, and west of the Mississippi 5.5 feet was common (Hilton 1990).

The 4-foot 8.5-inch gauge triumphed over broader gauges because of the need for compatibility and because it had the authority of precedence. Conversion from broad to standard gauge soon became common. By the 1870s the American railroad system was maturing and the 4-foot 8.5-inch gauge was finally completely accepted as the standard.

"Curbs and gutters"

Between miles 8.9 and 10.1 are four road-cuts through ridges of sand. On the east side of the highway, exposed in the cuts, are a number of bizarre structures sometimes called "curbs and gutters" because they somewhat resemble the curbing seen along modern streets (Figure 49). These strange, concrete-like shapes are developed within a sequence of clean, generally unconsolidated, wind-blown sand called the Zia Formation. Local portions

A. East side of US-550 at mile 9.25. (Photo by author 2005.)

B. Detail of photo in "A" above. (Photo by author 2005.)

Figure 49. "Curbs and gutters."

of the Zia—such as those here—have been cemented into rounded nodules and thick tabular bodies. They were formed by the movement of groundwater rich in dissolved calcium carbonate (or calcite, $CaCO_3$) that preferentially moved through zones within the sand that had slightly higher permeability. The fluids precipitated their loads of dissolved calcite cement in the spaces between the sand grains because of slight differences in fluid chemistry. Yes, groundwater once moved through this sand that now is high and dry, more than 150 feet higher than the present water table at the Rio Jémez just to the north.

The Zia Formation is the product of an early filling stage of the Rio Grande rift (18 - 15 Ma). At that time the Albuquerque basin portion of the rift was a closed depression, much like today's Death Valley in southeastern California. Westerly winds swept across the exposed old disintegrating sandstones on the Colorado Plateau, and deposited the loose sand as an extensive sand-dune field—called an *erg*—in the western part of the Albuquerque basin (Figure 50). The Zia Sandstone piled up to form a land surface hundreds of feet higher than it is today and the groundwater level was higher then as well. Only after the Great Incision, beginning about 800 ka, did the formations we see here become exposed.

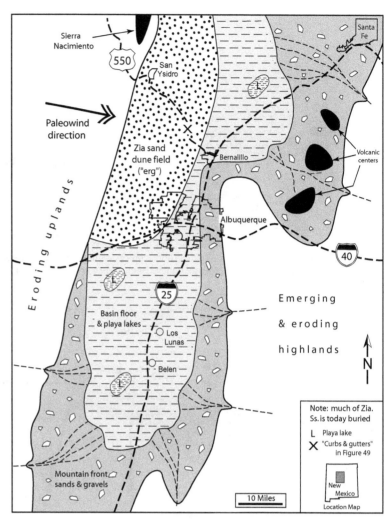

Figure 50. Paleogeography of Zia Sandstone, ca. 18 Ma.
(Modern geographic features shown for reference; modified from Connell 2004.)

Southern rim of the Jémez Mountains

Continuing on our journey to the north and northwest beyond MP-9 we see the southern rim of the Jémez Mountains on the horizon and its distinctive profile (Figure 51A). The two most prominent features in the profile are Chamisa Mesa to the west and Cerro Redondo to the east. The two, together with the high country between them, are the southern part of the Jémez Mountains, or what is sometimes referred to as the Jémez volcanic field. The flat-topped landform of Chamisa Mesa looms more than 1,200 feet above the surrounding broken country and 1,800 above the bed of the Jémez River. The mesa is capped by a thin layer of durable lava that protects the soft sandy material below. The lava bed is about ten million years old (10 Ma) and marks the volcanic field's initial outpouring of lavas onto an old land surface, which now is marked by the base of the lava flow (Goff et al. 1989; Figure 51B). Therefore the 10 Ma landscape was more than 1,200 feet higher than our landscape is today. Notice that another, slightly higher and more continuous mesa is located just to the east of Chamisa Mesa. This is Borrego Mesa. The two once formed a continuous surface but they have since been separated by an intervening, down-to-the-northwest fault.

The great rounded mass of Cerro Redondo, at 11,254 feet, is the second highest point in the Jémez Mountains and it surges out of the very center of the gigantic "Valles Caldera." (The caldera is a huge basin-shaped depression caused by a catastrophic volcanic explosion about 1.2 Ma). Cerro Redondo rising from its center is not a volcano, but rather it is the uplifted floor of the caldera itself (Figure 51B). The Jémez Mountains comprise a major subject all their own, but because our highway only skirts their southern rim I will merely touch on them with this short comment.

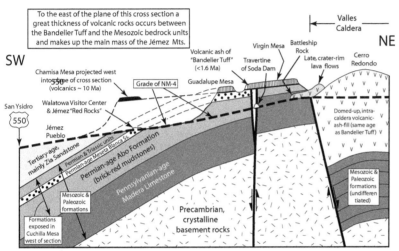

A. North view of Chamisa Mesa (far left) and partly-snow-covered Cerro Redondo (on right horizon) in Jémez Mountains from about MP-9 (see relationship between the two features in schematic cross section "B" below. Photo by author 2004.)

To the east of the plane of this cross section a great thickness of volcanic rocks occurs between the Bandelier Tuff and the Mesozoic bedrock units and makes up the main mass of the Jémez Mts.

SW

NE

Valles Caldera

Virgin Mesa | Battleship Rock
Volcanic ash of "Bandelier Tuff" (<1.6 Ma)
Travertine of Soda Dam
Late, crater-rim lava flows
Cerro Redondo

Chamisa Mesa projected west into plane of cross section (volcanics ~ 10 Ma)

Grade of NM-4 | Guadalupe Mesa

Walatowa Visitor Center & Jémez "Red Rocks"

Domed-up, intra-caldera volcanic-ash-fill (same age as Bandelier Tuff)

San Ysidro
550

Jémez Pueblo

Tertiary-age, mainly Zia Sandstone

Permian & Triassic units
Permian-age Meseta Blanca Ss.
Permian-age Abo Formation (brick-red mudstones)

Pennsylvanian-age Madera Limestone

Mesozoic & Paleozoic formations

Mesozoic & Paleozoic formations (undifferentiated)

Formations exposed in Cuchilla Mesa west of section

Precambrian, crystalline basement rocks

B. Schematic cross section north of US-550 along highway NM-4 (view to northwest, no scale).

Figure 51. US-550 and southern part of Jémez Mountains (location in Figure 47).

Rio Jémez utility corridor

In addition to the Rio Jémez Valley's historical role as a funnel for the movement of people, animals, and finally vehicles, the valley is also the conduit for five, virtually unnoticed highways of another sort. These are steel pipelines that move various types of extremely valuable energy commodities from the Four Corners area to the Rio Grande Valley (Figure 48). Large volumes of liquids and gases can only be transported to market economically via pipeline. The selected diameter of the lines varies with the anticipated fluid volumes to be transported and the cost of the pipe. The lines have an inherent flexibility in that the direction of flow can be reversed as needs change, and the type of fluid can be changed as conditions require. Once buried in their excavated trenches these pipelines become extremely valuable real pieces of property. The pipelines cross the highway from north to south at MP-16, and generally run just out of sight along a right-of-way on the south side of the road for the next five miles.

The first line was the 16-inch Texas-New Mexico pipeline built in 1957–1958 from the San Juan basin southeastward to refineries in West Texas, and then via connecting lines to other refineries on the Gulf Coast. In 2007 it was owned by Equilon (based in Houston). In the early 2000s Equilon announced plans to refurbish the line, reverse the flow from southward to northward, and pump refined products (jet fuel, diesel, and gasoline) from the Gulf Coast to the Four Corners and beyond. This is the same line that had caused a publicity ruckus in the Village of Placitas in mid-2000. The line runs through a part of the village and residents were concerned about safety, not only because of the line's advanced age but because of an unrelated but tragic natural-gas pipeline explosion near Carlsbad in August 2000 that killed ten people (*Oil and Gas Journal* 2000). The Placitas residents were demanding a full environmental impact statement before the flow-reversal would be authorized. The line had been sitting idle since 1998.

The second pipeline is a 36-inch carbon dioxide (CO_2) line operated by the Cortez Pipeline Co., a subsidiary of Shell. The gas is transported from Shell's McElmo Dome, one of North America's largest CO_2 fields located west of Cortez in southwestern Colorado. The inert gas is piped to the Permian Basin of southeastern New Mexico and west Texas where it is pumped into underground oil reservoirs to increase the subsurface pressure and hence recovery from tired old oil fields.

The last three lines are operated by Williams Co. (based in Tulsa), formerly MidAmerica Pipeline Co. (Mapco). An 8-inch line carries jet fuel, a 10-inch line carries refined products (alternatively gasoline, jet fuel, fuel oil and diesel fuel), and a 12-inch line carries a blend of natural-gas liquids, mainly ethane, propane, and butane.

Rio Jémez Indian Pueblos

This stretch of US-550 crosses the land of the pueblos of Santa Ana and Zia (Figure 52). We've already said something about Santa Ana in the context of the Town of Bernalillo in Chapter 4. These are two of New Mexico's seven Keres-speaking pueblos (the others are the eastern Keres pueblos of San Felipe, Santo Domingo, and Cochití in the middle Rio Grande Valley, and the western Keres pueblos of Laguna and Ácoma near Grants). People have lived in this area since at least 8,500 years ago, and by 500 CE they had established themselves as farmers along the Rio Jémez. The first Spanish visit to the valley was in 1541 by Captain Hernando de Alvarado, one of Coronado's officers. Later, in 1583, a colonial merchant named Antonio de Espejo visited the Rio Jémez and noted that the land of the [eastern] Keres, in the present vicinity of Zia Pueblo, had five pueblos housing as many as 15,000 people (Hammond and Rey 1966).

In 1689 the Spanish governor (in exile) Domingo Jironza Petriz de Cruzate supposedly granted blocks of land to a number of the pueblos. Each block consisted of four square leagues and measured one league (5,000 varas or about 2.3 miles) in each cardinal direction from the pueblo

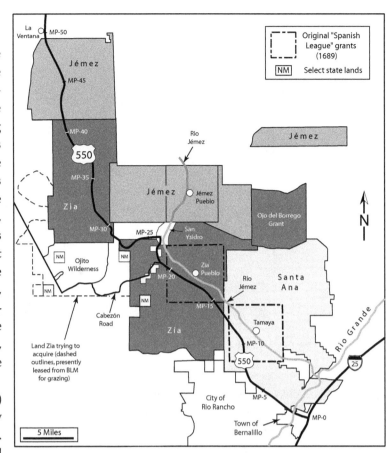

Figure 52. Rio Jémez Valley Pueblo Indian lands (shades of grey).

church (Figure 18). Each block contained just under 18,000 acres. Spanish law dictated that enough space should exist between each of the grants to avoid conflict. Where the grants occupied fertile land these gaps became chinks between the pueblos, which non-Indians were often quick to exploit. By the mid-18th century it was generally accepted that the pueblos each owned their Spanish-league grants and by the beginning of the Mexican period in 1821 the idea was accepted by all. In 1856 the U.S. government accepted the grants as valid and Congress approved them in 1858. In the 1890s, however, it was learned that the 1689 Cruzate grants were forgeries, but—too late—the deal had already been cut (Hall 1987).

Santa Ana Pueblo

The Santa Anas never found their Spanish document conveying to them their 1689 Spanish league grant, but they always believed that one once existed. The U.S. government agreed and issued a patent in 1883. Unfortunately the land was of poor quality and for many years was used only for grazing. As noted earlier (Chapter 4), this situation forced the Santa Anas to search for more arable land to the east along the Rio Grande. The old Spanish league pueblo, *Tamaya* (turnoff at mile 10.1), today is sparsely inhabited and is used mainly for ceremonial purposes. (We discussed the expansion of the pueblo in Chapter 4).

Zia Pueblo

At mile 15.5, at the point where US-550 tops a low crest and bends slightly to the left, our panorama changes significantly as we approach Zia Pueblo. A fortunate coincidence has aligned the pueblo, perched atop its river terrace overlooking the Rio Jémez, directly in line with the imposing geologic backdrop of the appropriately-named *Cuchilla Mesa* (Spanish for "knife mesa"). With a telephoto lens this is a real postcard shot (Figure 53).

Figure 53. West view of Zia Pueblo with its backdrop of Cuchilla Mesa. (Photo by author 2004.)

Zia Pueblo, like Santa Ana Pueblo, is Keres-speaking. It has been in its present location, atop an old ancestral Rio Jémez terrace since about 1300 CE (Figure 54). In the 1540s Coronado's men counted about 5,000 Zians living here, but in 1890 there were fewer than 100 (Chilton 1984). Today about 645 people live at the pueblo (2000 Census). The mission church of *Nuestra Señora de la Asunción* was built in 1692 after the re-conquest by De Vargas. It was built atop the ruins of the first church that had been built after the Oñate *entrada* in 1598 and destroyed during the Pueblo Revolt of 1680 (Sinclair 1980). In the mid-19th century the Zians did in fact produce a copy of their 1689 Cruzate grant document. In 1864 the U.S. government accordingly issued a patent for the Spanish-League grant of 16,282 acres.

Figure 54. North oblique aerial view of Zia Pueblo. (Photo by author 2001.)

The land is poor and requires extreme effort and diligence to eke a living out of it. Over the years the pueblo has judiciously been adding to its land base (Figure 52). First it added the eastern part of the San Ysidro Grant, and then the Ojo del Borrego Grant. In 1956 Zia Pueblo and Jémez Pueblo divided up the eastern part of the huge Ojo del Espiritu Santo Land Grant (more about this in Chapter 10). Also, in recent years Zia Pueblo has been buying parcels of private land west of the Village of San Ysidro. By about 2000 the pueblo had a land base of about 104,000 acres, most of which is used for grazing. In 2003 the pueblo stepped up its efforts to acquire an additional 14,000 acres of land from the Bureau of Land Management (BLM) surrounding the Ojito Wilderness, claiming ancestral ties. The BLM is resisting due to the area's rich dinosaur-fossil resources and to fears that the pueblo would close the area off to the public (Davis 2003).

Zia sun symbol.

Zia Pueblo is best known for its ubiquitous sun symbol, displayed most prominently on the New Mexico state flag. Herein lies an interesting tale. Antonio de Espejo had visited the Zia pueblo in 1583. In his report he referred to figures he saw painted on a wall of the pueblo that seemed to him to represent either the sun or maybe a sunflower. Much later, in the late 19th century, a Zia woman placed this symbol on a pottery vessel she had just made (Figure 55A). Somehow the water jar became part of the collection of the Museum of New Mexico in Santa Fe and later of the Laboratory of Anthropology. To the Zians the four rays symbolize the number four, which is sacred. It represents the four cardinal directions, the four seasons, and the four stages of life (childhood, youth, adulthood, and old age). The circle in the center links the four rays.

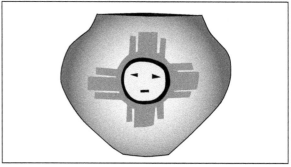

A. Zia Sun Symbol on 19th century pot. (Drawing by author.)

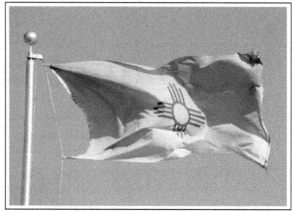

B. New Mexico state flag aflutter with its Zia Sun Symbol. (Photo by author 2006.)

Figure 55. Zia Sun Symbol.

In 1920 the New Mexico Chapter of the Daughters of the American Revolution proposed a state flag that would incorporate symbols that represented the state's unique character, and in 1923 the DAR accepted competitive designs for the flag. The winning submission was won by a Santa Fe physician and archeologist, Dr. Harry Mera. The design (actually made by the doctor's wife!) contained the red Zia sun symbol on a field of yellow — the colors of the Spanish queen that the conquistadors brought with them to the New World in the 16th century. In 1925, New Mexico governor Arthur T. Hannett officially adopted the Mera design for the new state flag with its now instantly recognized red Zia sun symbol against a brilliant yellow background (Figure 55B). Later, in the mid-1950s, amid considerable pomp and circumstance, New Mexico governor John F. Simms, Jr. presented the state flag to the Zia Pueblo governor (Sinclair 1980; Herrera 2008).

The Zians are justifiably proud that their sacred symbol has been chosen to represent the state of New Mexico to the world. They are less enthused about the symbol being used for miscellaneous logos on business letterheads or on New Mexico license plates. Since 1994 the pueblo has been pushing the state to compensate Zia for use of its symbol, but to avail. In 1999 hearings were held to determine if tribal insignias, including the Zia sun symbol, should be given copyright protection to prevent their commercial exploitation. Recognizing the sensitivity of the issue, in 2000 Southwest Airlines approached the pueblo and revealed its plans to decorate one of its new jets with the Zia sun symbol. The pueblo granted permission, for which the airline contributed a sum of money for the tribe's scholarship fund. Later that year Southwest unveiled its "New Mexico No. 1," a Boeing 737 aircraft, proudly sporting the Zia sun symbol painted on its side (Papich 2000).

Zian roads.

In the course of researching this book I consulted with a number of aerial photographs taken by the Soil Conservation Service (SCS) in the mid-1930s. My main objective was to identify previous alignments of the highway. However, while checking out the reach between Bernalillo and San Ysidro I stumbled into something very interesting. Five nearly straight dirt roads can be seen to radiate out from the center of Zia Pueblo, spanning the complete southwest quadrant like spokes of a wheel (Figure 56).

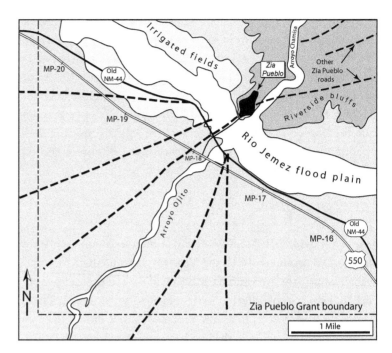

Figure 56. Zia Pueblo ritual roads (heavy dashed lines). (Based on 1935 Soil Conservation Service aerial photograph #306.)

These are eerily like the so-called Chacoan roads built nine centuries ago in the San Juan basin (dealt with again in Chapter 30). The width of the Zian roads is enough to accommodate a wagon, but no more. They seemingly go nowhere and end abruptly. They evidently serve no utilitarian purpose. However, it took planning and effort to put them on the ground so they clearly have important symbolic meaning to the Zians. In contrast, the Chacoan roads radiate outwards for many miles from a center in Chaco Canyon, but, strangely, only to the south, southwest, and north. It has been speculated by some that these roads commemorate the Chacoan people's migratory history to Chaco Canyon (Frazier 2005). Could this be a clue to the purpose of the Zian roads? The southwest-quadrant Zian roads converge on the pueblo itself—the Zians' central place. Perhaps the Zian roads are instead an attempt—like the Chacoans so long ago—to achieve cosmological harmony with the surrounding landscape. Their handiwork would remain hidden were it not for the intrusive medium of aerial photographs.

Jémez Pueblo

This pueblo sits off to the north of US-550 on state highway 4. The pueblo and the upper reach of the Rio Jémez that wends its way via San Diego Canyon up to the Valles Caldera are a story of their own. However, because they are peripheral to this book I'll barely touch upon them. Jémez Pueblo is one of the more conservative of the pueblos and the last Towa-speaking one. In 1838 the Pueblo of Pecos (near modern-day Pecos, east of Santa Fe), another Towa pueblo, was abandoned after a century of deprivation and population loss, and the few survivors relocated to Jémez. Today about 1,960 people live here (2000 census).

When *Don* Juan de Oñate visited this area shortly after the Spanish conquest of 1598 he found eight Jémez pueblos between here and *Giusewa* Pueblo at today's Jémez State Monument north of Jémez Springs. The present Jémez Pueblo is known as *Walatowa*, meaning either "The Place," "This is the Place," or "Home" in the Towa language. A Spanish mission, *San Diego de Congregación*, was established here in 1621. The Spanish were driven out during the Pueblo Revolt of 1680 but they returned in 1692 with an attitude, led by Governor *Don* Diego de Vargas. Fearing reprisal, the Jémez people abandoned Walatowa, and some fled to the Canyon Largo area to the northwest and mingled with the Navajo (see Chapter 21).

Eastern rim of the Colorado Plateau

As we pass Zia Pueblo and continue on our way northwestward we can make out the details of the distant "White Mesa" and the southern nose of the Sierra Nacimiento (Figure 57). White Mesa is our first glimpse of the old, solid rocks that form the floor the Rio Grande rift. As we approach San Ysidro, a series of deeply-buried giant stairsteps, composed of these older rocks, rise step by inexorable step until just south of San Ysidro they lunge to the surface at White Mesa. This point marks the boundary between the Rio Grande rift on the east and the Colorado Plateau on the west. Here the rather drab, sand-and-gravel-filled Rio Grande rift gives way to the utterly different Colorado Plateau with its colorful bedded rocks and rugged landforms.

Note: Horizontal dimension compressed to 50% of normal

Symbols: P = Permian; Tʀ = Triassic; J = Jurassic; K = Cretaceous

Figure 57. Northwest to north panoramic view of White Mesa and southern nose of Sierra Nacimiento. (Photo by author 2000.)

White Mesa and the buried stairsteps leading up to it are separated by a series of down-to-the-east faults. (A fault is a fracture in the rocks along which measurable movement has taken place.) The most prominent of these faults is the so-called San Ysidro fault, well-displayed at MP-22 on the east side of White Mesa. The downward movement along this fault is about 900 feet (Woodward et al. 1989; Figures 44 and 58).

The top of the appropriately-named White Mesa is a brilliant white and can be seen for miles around. New Mexico's largest gypsum mine, operated by the Centex American Gypsum Company, is located on the very top on land leased from Zia Pueblo. The gypsum weathers to a dull gray but is white when fresh. The mineral gypsum is a hydrated calcium sulfate (chemical formula $CaSO_4 \cdot 2H_2O$). The material is used in the manufacture of wallboard and plaster. A steady stream of huge dump-trucks rumbles down the slope from the mine and hauls the ore to Centex's relatively new wallboard plant in Bernalillo and to an older one in Albuquerque. At the present rate of extraction

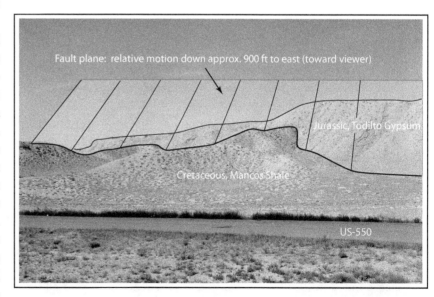

Figure 58. Southwest view of San Ysidro fault. (Photo by author 2000.)

93

the White Mesa deposit has reserves for several more decades (Pazzaglia et al. 1999). The mine provides a nice source of royalty income and employment for the pueblo.

Off to the south of US-550, just past MP-21, is the unpaved Cabezón Road (Figure 46A). About 0.1 mile along it is a fork. The right branch goes up the hill to the gated gypsum mine. The left one, the main Cabezón Road, is marked by a big sign that says "No Trespassing, Zia Indian Reservation." In actual fact, access is permitted to the public lands beyond. Four miles west down Cabezón Road is the border of Bureau of Land Management (BLM) land and the fascinating Tierra Amarilla anticline (more about this in Chapter 7). An additional six miles west is the Ojito Wilderness and the trailhead to the dinosaur quarry where fragments of the 110-foot-long sauropod dinosaur *Seismosaurus* were recovered in the early 1990s.

View from the Rio Salado

At mile 22.75 the highway crosses the bridge over a sluggish Rio Salado (Figures 48 and 59). This is an excellent vantage point at which to introduce the interesting geology yet to be seen to the north (Chapters 7 and 8). Most obvious is the stunning Cuchilla Mesa, the upturned eastern rim of the Sierra Nacimiento that commands the view on the right. In the left distance, in the center of the bulging mass, is Pajarito Peak, at 9,042 feet the highest point in the southern Sierra Nacimiento. Less obvious but more interesting are the "underfit" canyon seen end-on in the left middle distance and the flat-topped terrace in the foreground north of the river. Remember the last two items because I will return back to them in the next two chapters.

Figure 59. North view from the Rio Salado bridge south of San Ysidro. (Photo by author 2006.)

Village of San Ysidro

This mostly Hispanic village was established in 1699. It was named after Saint Isidore, the patron saint of farmers and a Spanish churchman who died in 636 CE (Julyan 1996). The village was originally located in the 2 x 12-mile San Ysidro de los Dolores Land Grant issued in 1786, a skinny east-west parcel sandwiched between the square Zia and Jémez Spanish-league grants (Figure 60A).

San Ysidro is one of those communities that the opportunistic Spanish established in the gaps between the pueblo grants. As mentioned previously, the Spanish government purposely left these gaps to minimize the potential for conflict between the pueblos. Because the pueblos were generally located along choice, irrigatable land, the gaps were usually choice spots as well, and Spanish settlers predictably scarfed them up. The grant was surveyed in 1877 as a prelude to receiving a patent. A resurvey was requested due to conflicting claims by Zia and Jémez Pueblos, but it was never carried out and a patent was never issued (University of New Mexico 1985).

The land ownership issue raised its head again in the 1920s. The people farmed long narrow strips of land extending out on either side of the Rio Jémez and paid taxes on that land. The eastern and western parts of the "grant" however were held in common, but for some reason the taxes due on this land had not been paid. The San Ysidro Land Company, incorporated in 1927 to buy, sell, and convey real estate, paid all the taxes on the common lands for the following ten years. In 1936 the company claimed ownership of the common lands by the principle of adverse possession, then sued to quiet claim in State District Court, and finally received a patent to the common lands. Later that year the company sold the common land to the U.S. government. In 1938 the government placed the eastern part of the grant in trust for Zia Pueblo (Figure 60B). The western part is now under the administration of the Bureau of Land Management (BLM). In 1967 San Ysidro incorporated as a village. Its initial shape consisted of a corridor of land on either side of highways NM-4 and NM-44 (now US-550). Finally, in 1974, the village acquired the adjacent irrigated lands and assumed its present, fuller shape (University of New Mexico 1985).

Throughout the 19th century San Ysidro had remained an agricultural and sheep-raising community. The town's men developed a reputation for their expert knowledge of the Navajo country to the northwest. In the late 19th century the place was a transportation hub astride the access to the Jémez Mountains to the north, and the east-west military supply route linking Ft. Union (near Las Vegas in northeastern New Mexico) with Ft. Wingate in western New Mexico and Ft. Defiance in eastern Arizona. Later the town was crossed by the main mule-drawn, buckboard stage line between Santa Fe and Prescott, Arizona, that was operated by the Star Line Mail and Transportation Company. The road from Bernalillo to Farmington was built in 1920 (University of New Mexico 1985). Normally one would expect such a "gateway" to the popular hot springs of Jémez Springs to the north and the San Juan basin to the northwest to sport at least a few restaurants and motels. Instead what we find here is a sleepy village of some 238 people (2000 Census, down about 100 from 1990) and the smallest incorporated village in Sandoval County.

Water

The principal reasons this village is, and has remained, so small are the unreliability and questionable quality of its water supply. The village gets its drinking water from an infiltration gallery located near the city hall on NM-4. From there the water is piped up to a tank perched on the low mesa north of US-550 on the west side of the village. The volume thus produced is at times less than what the community needs and there have been times when the water table has dropped to such levels that the gallery has not been effective. For example, in the late 1970s the National Guard had to truck water to the village from Bernalillo (*Albuquerque Journal* 1979).

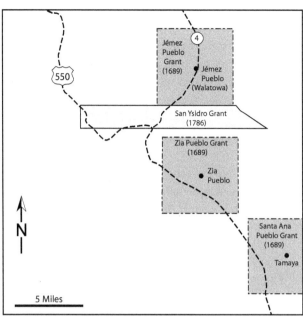

A. San Ysidro Land Grant and adjacent Indian land, 1786.

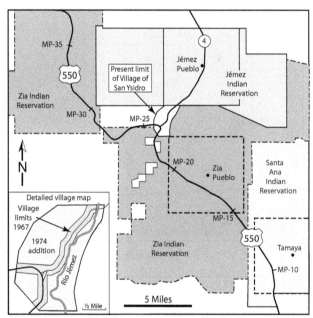

B. Village of San Ysidro and adjacent Indian land, present.

Figure 60. Evolution of Village of San Ysidro.

Water quality is a more serious problem: it contains significant amounts of arsenic! Since 1974 the Safe Drinking Water Act set the acceptable level of arsenic in drinking water at 50 parts per billion (ppb). In 1996 the Environmental Protection Agency (EPA) began a review of the standards, amidst much controversy and political posturing. In January 2001 the EPA lowered the acceptable level to ten ppb (Reid et al. 2003), and the ruling took effect in January 2006. The village's ultimate source of water is the Rio Jémez and the springs bleeding into it, especially the thermal springs in the Jémez Springs area, 17 miles upstream. These springs, issuing from the geothermal groundwater system buried beneath the volcanic field of the Jémez Mountains, are rich in naturally-occurring arsenic. At the Soda Dam area (1.5 miles above Jémez Springs) about 400 gallons per minute of water from hot springs enters the Rio Jémez. Recent sampling of the waters entering the river near the springs indicated arsenic concentration of 1,770 ppb just south of Soda Dam, and 830 ppb from Travertine Mound within the Town of Jémez Springs. At the Rio Jémez bridge, just north of the Village of San Ysidro, recent arsenic levels from the river ranged from 40 to 300 ppb (Reid et al. 2003). Presently the residents of the village rely on special filtering devices in their homes to remove the arsenic. Such a situation is hardly conducive to economic development.

Site of old Seligman store

On the west side of the main drag, between a boarded-up former Circle K store and a new Family Dollar store is a brushy, somewhat derelict patch of ground. This is the site of the old Seligman store (Figure 61A). As mentioned in Chapter 4, a short time after 1903 the Bibo family moved to a new location in an existing dry goods store in Bernalillo and named it the Bernalillo Mercantile, or simply "the Merc." Bibo then opened outlying stores in six locations, including here in San Ysidro (Figure 61B). In 1922 Joseph Bibo sold out his business interests to the Seligman brothers. The decades up to after WWII saw a decline in the fortunes of the general merchandise business. The Seligmans closed down the San Ysidro store—the last of the outliers—in the late 1940s and concentrated their efforts on the Bernalillo Merc (Arango 1980a).

A. Southwest view of derelict site (center left) of store. (Photo by author 2009.)

In about the early 1980s the junction of old NM-44 and NM-4 was rebuilt. Sections of the old abandoned alignment can still be seen. At the intersection is the CWW Feed Store and More, built in the 1950s, and since the early 1990s owned and run by entrepreneur Connie Collis.

B. Interior of store, 1930s(?). (Photo courtesy of Sandoval County HIstorical Society.)

Figure 61. Seligman store in San Ysidro.

I informally call the short stretch of US-550 that extends from the Village of San Ysidro at mile 23.5 west to about MP-28 where it turns the corner around the southern nose of the Sierra Nacimiento the "Rio Salado Pass" (Figure 62).

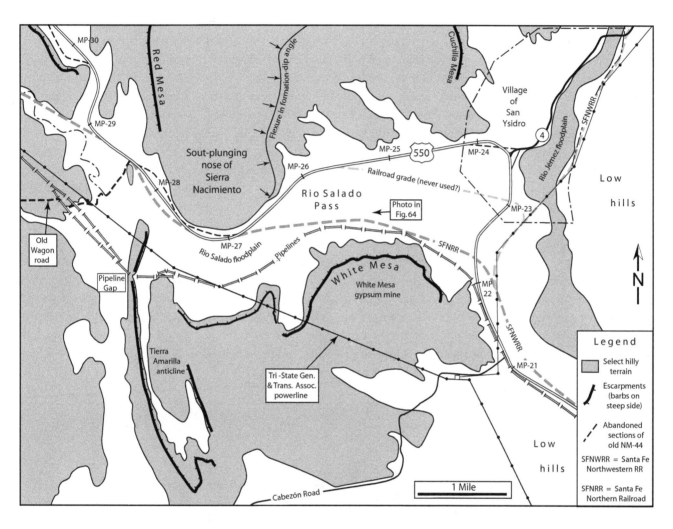

Figure 62. Geography of Rio Salado Pass area.

The sluggish *Rio Salado*, or "salty river," flows through the pass from west to east. The river is appropriately named. Its banks are covered by a thin crust consisting of whitish sodium chloride (NaCl, or halite, common table salt) and lesser amounts of sodium carbonate (Na_2CO_3, or soda) that issues from a series of nearby springs draining into the river. Casual observers often have simply called the white crust "alkali." Up to at least the 1930s the Jémez Pueblo Indians would sack up the material and haul it up into the mountains to salt their sheep (Counselor and Counselor 1954).

The coincidence of a somewhat dependable water supply for animals (although barely, if at all, potable for humans) and an easy grade guaranteed that Rio Salado Pass would become an important gateway. This natural corridor has been used in turn by Navajo raiders pillaging Spanish livestock, the Spanish military on punitive expeditions against the Navajo, American military wagon traffic west to Ft. Wingate, American railroad surveyors seeking an ideal transcontinental route to the Pacific, settlers of the Rio Puerco hauling their grain to mills at Jémez Pueblo and purchasing supplies at San Ysidro, cattle drives from the upper Rio Puerco to the railhead in Bernalillo, and later, railroaders seeking access to the coal deposits near La Ventana.

One noted user of the pass was James H. Simpson, who in 1849 camped with a large party in the pass. Simpson was a lieutenant in the Army's Corps of Topographical Engineers and a graduate of West Point. In the first half of the 19th century survey parties were under the Army's jurisdiction because there was no civilian service organized for exploration. However, most military parties at the time did have technical people in tow (Fenton and Fenton 1952). After ten years of distinguished service in the East, Simpson joined Captain Randolf Marcy's expedition near Ft. Smith, Arkansas, which at the time was escorting a wagon train of immigrants and gold seekers to Santa Fe. In August 1849 Simpson received orders to accompany Lt. Colonel John Washington on a punitive expedition against the Navajo Indians. The expedition consisted of 55 artillery troops, 120 infantry troops, and a gaggle of officers, interpreters, and assistants. The party left Ft. Marcy at Santa Fe and journeyed to Jémez Pueblo. They camped at the pueblo for three days while they transferred their supplies from wagons to mules. They also gathered up some 60 Pueblo Indians and 50 Mexican volunteers. By this time the expedition had grown to almost 300. From there the party descended the Rio Jémez to the tiny settlement of San Ysidro, turned west along the north bank of the Rio Salado and camped about three miles west of the junction with the Rio Jémez, probably at about MP-27. Simpson was a perceptive traveler and left lucid written accounts of his observations, but, strangely enough, he wrote nothing about this interesting place. From here he continued on to Chaco Canyon and points beyond, naming Mt. Taylor after then-President Zachary Taylor on the return home (Kues 1992).

In the 1870s the army constructed a wagon road through the pass (Figure 5). Until the early 1900s the road served as the main route connecting Santa Fe and Ft. Wingate via the town of Cabezón on the Rio Puerco. The wagon route generally followed the course of the modern highway west from San Ysidro on the north side of the river (Figure 62). It was used for troops, military supplies, heavy freight, a Star Route mail line, and passenger stagecoaches. The old wagon road hooks off to the west from US-550 at mile 28.4 (just past the ridge of grayish gypsum on the left, at the gate leading to BLM land), and once led to Cabezón and points beyond. That road is not passable today.

As mentioned earlier, in 1924 the Santa Fe Northwestern Railroad (SFNWRR) was completed from Bernalillo to San Ysidro and north into the Jémez Mountains (Figure 62). The line's purpose was to supply the lumber mill in Bernalillo with logs harvested from upland forests. From 1926 to 1927 a spur line, the Cuba Extension Railroad, was constructed from a junction at the main SFNWRR line in San Ysidro to the coal deposits at La Ventana, 27 miles to the northwest. To reach La Ventana the line had to pass through the Rio Salado Pass and to bridge the Rio Salado. During the construction the railroad was renamed the Santa Fe Northern (SFNRR). Today little vestige remains of the SFNRR in the pass. Part of the line's trace has been obliterated by the modern highway and part washed away by floods. The best place to see the old railroad grade is on the south side of the river near the base of White Mesa (Figure 63) and on aerial photographs. As stated before, state highway NM-44 was constructed in the 1920s and finally paved in the late 1930s. In the late 1960s the highway was realigned and straightened around the southern nose of the Sierra Nacimiento.

Figure 63. West view of SFNRR grade. (Photo by author 2001.)

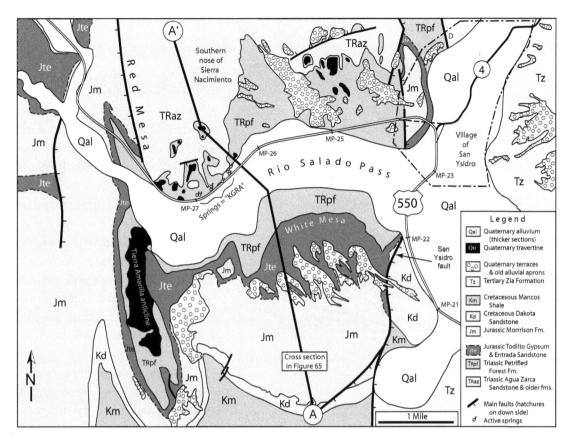

Figure 64. General geology of Rio Salado Pass area. (Modified from Woodward and Ruetschilling 1976.)

Just past MP-24 on the north side of the highway (Figure 62) is a mobile home, above which is the Village of San Ysidro's water tank perched on a low, flat-topped platform. On the south-facing wall of the platform a sharp eye will spot a color change on opposite sides of an imaginary vertical line—from reddish, Triassic-age mudstones of the Petrified Forest Formation to the west to buff to tan sandstone of the Cretaceous-age Dakota Formation to the east (Figure 64). This is the northern trace of the same San Ysidro fault we saw earlier near Cabezón Road on the approach to San Ysidro (Figure 58). Recall that the rocks on the east, Rio Grande rift side have been downdropped about 900 feet along the plane of the fault.

The platform (seen earlier from the Rio Salado bridge, Figure 59), is capped by Pleistocene-age gravel that is the remnants of an older, and therefore higher, river level. Note that the gravel cap is not cut by the San Ysidro fault, indicating that the gravel was deposited after the faulting had ceased. At MP-25 is an official highway department sign supposedly marking the eastern boundary of the Colorado Plateau. One could persuasively argue that the boundary is really located back at the San Ysidro fault.

On either side of the highway spectacular exposures of Mesozoic-age rocks form the southern terminus of the north-south-trending Sierra Nacimiento (Figure 64). The part of the Sierra located on the north side of the highway is configured into a broad, flat-bottomed, south-plunging syncline, and is upfaulted and curled up on the east and west sides (the east side was the Cuchilla Mesa we saw on the approach to San Ysidro; Figures 57 and 59). The Mesozoic strata exposed at White Mesa to the south have been stripped off the uplift on the north by progressive southward cliff-retreat—a process that continues today (Figure 65).

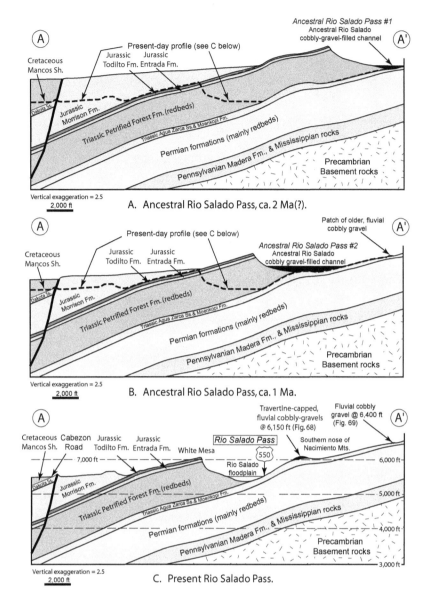

A. Ancestral Rio Salado Pass, ca. 2 Ma(?).

B. Ancestral Rio Salado Pass, ca. 1 Ma.

C. Present Rio Salado Pass.

Figure 65. South-to-north cross section across Rio Salado Pass (see Figure 64 for location).

Origin of Mesozoic formations

The essential architecture of this area is produced by five geologic formations of Mesozoic age (Figure 4) and some very recent accumulations of fresh-water spring deposits. The standard method to understand an area's geology is of course to show the distribution of the formations on a geologic map (Figure 64) and on a geologic cross section (Figure 65C). An additional, very instructive method is to portray the ancient geography that existed at the time the formation was being deposited on a "paleogeographic map." Such a map provides a snapshot of the area's geologic history during a specific time period. Figure 66 shows paleogeographic maps of four of the five Mesozoic geologic formations responsible for the scenery present in the pass.

For the moment let's skip past the oldest of the five units (the Triassic-age, Agua Zarca Sandstone) and go to the second oldest. The Petrified Forest Formation, also of Triassic age but slightly younger, is the brick-red to maroon, soft, and easily-eroded mudstone and volcanic ash that forms the soft slopes on either side of the highway. The name comes from its type section in the Petrified Forest National Park in Eastern Arizona, where the equivalent unit contains the famous petrified wood. The formation was deposited about 220 million years ago (Ma) by rivers flowing northwest across a broad, low-lying alluvial plain. The area of the future New Mexico was then located just a few degrees north of the equator (Figure 66A). The Triassic rivers filled in their basin with sediment containing volcanic ash and the remains of tree trunks. The ash reacted with the buried stumps, and *voilà*—petrified wood.

By 165 Ma, in the Middle Jurassic, the world had changed and the future New Mexico had moved northward from the equator to about 15 degrees North Latitude. Building atop the now-buried Triassic rocks, the area became the site of an extensive sand-dune field. Such an expanse of wind-blown sand is called an *erg* (*a la* the Zia Formation mentioned earlier) from the Hamitic word describing large parts of the modern Sahara in north Africa. The depositional product is today's Entrada Formation—a hard, buff-colored, resistant cliff-former that occurs just below the top of White Mesa (Figure 66B). The name for this very extensive unit is taken from its type section at Entrada Point in southeastern Utah.

By 160 Ma, later in the Middle Jurassic, a portion of the Entrada erg subsided and allowed a body of brackish water from the open sea (which at the time was located to the north) to flood across the erg in southwestern Colorado and northern New Mexico. This shallow body of brackish water soon became effectively cut off from the open sea and became what is referred to as a *salina* (Figure 66C). Evaporation of the brackish water from the salina produced a brine, from which a sequence of the calcium sulfate ($CaSO_4$) mineral anhydrite precipitated. Many millions of years later, as the anhydrite was uplifted and put in contact with surface waters, the material absorbed water and converted to the less-dense mineral gypsum ($CaSO_4 \cdot 2H_2O$). (It should be noted that some geologists believe that the original salt was gypsum rather than anhydrite.) This altered stuff, coupled with the underlying Entrada sandstone, forms the imposing cliffs on the north-facing White Mesa on the south side of the pass.

As already mentioned, the gypsum is presently being mined from the White Mesa Mine. The miners strip the gypsum back a layer at a time with their heavy equipment. Early on they discovered a series of open cracks in the gypsum that became narrower as they stripped deeper and deeper. In 2005 they made a fascinating discovery in one fissure at a depth of 40 feet below the original top of the gypsum: the fossilized remains of extinct Pleistocene-age mammals, including two camels, a bison, a horse, and a deer (Rinehart et al. 2006). Camels? When notified, paleontologists from the New Mexico Museum of Natural History rushed to the site. They cut around and under the fossils, and transported the exca-

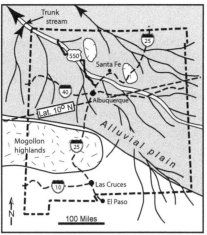

A. Upper Triassic, Petrified Forest Fm., ca. 220 Ma.

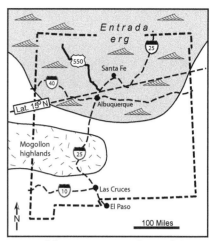

B. Middle Jurassic, Entrada Sandstone, ca. 165 Ma.

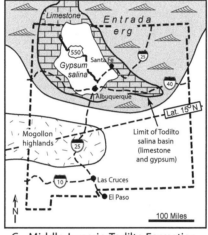

C. Middle Jurassic, Todilto Formation, ca. 160 Ma.

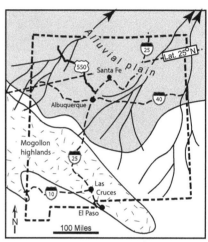

D. Late Jurassic, early part of Morrison Fm. ca. 150 Ma

Figure 66. Paleogeography of Triassic and Jurassic formations in New Mexico. (Modern geographic features shown for reference; modified from Lucas and Anderson 1997; Lucas 2004.)

vated blocks containing the fossils to the museum for treatment and eventual display. Evidently these large grazing critters had stumbled into the fissures (they might have been covered over by brush at the time) and became entrapped. After their lonely deaths their bones settled deeply into the wedge-shaped fissure. The very existence of the remains of large Pleistocene mammals clearly indicates that this part of New Mexico was a very different place 100,000 to 200,000 years ago (more about Pleistocene "megafauna" in Chapter 31).

By about 150 Ma, in the Upper Jurassic, this part of New Mexico had moved north to about 25 degrees North Latitude and the geography had again dramatically changed (Figure 66D). Major northeasterly-flowing rivers drained a vast flood plain. This was the time of the giant dinosaurs, and the flood-plain environment provided them an ideal habitat. The deposits were to become the Morrison Formation, named for exposures at Morrison, Colorado, near Denver. Some of the most spectacular dinosaur fossils in North America have been recovered from the Morrison Formation. In fact the awesome dinosaur fossil, *Seismosaurus*, (part of which is on display at the New Mexico Museum of Natural History) was quarried during the 1980s from the Morrison in the Ojito Wilderness, about six miles west of the pass (Gillette 1994). The Morrison Formation forms the broken country on the south side of White Mesa and west of the pass, but it is not as visible to the motorist as are the older formations.

Sierra Nacimiento and Laramide compression

A useful piece of vocabulary when considering the history of New Mexico and the Rocky Mountains (and a zinger to drop at a cocktail party) is the term "Laramide." Its name comes from the Laramie Formation exposed near Laramie, Wyoming. The term refers to a compressional episode of mountain building that affected the western half of North America between about 75 and 40 Ma, and produced the Rocky Mountains. The Sierra Nacimiento was also produced during this important event and is therefore technically part of the Rocky Mountains. As we've seen, the east face of the Sierra Nacimiento near the Rio Salado Pass consists of the curled-up Mesa Cuchilla. The not-yet-seen western face — Red Mesa — is similarly curled up (Figure 67).

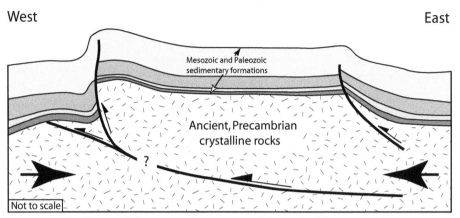

A. "Laramide" compression event and thrust-faulting, ca. 55 Ma.

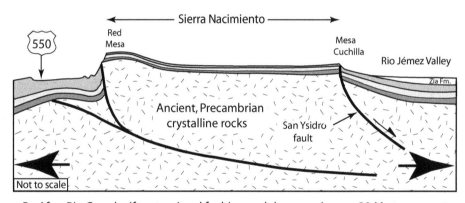

B. After Rio Grande rift extensional faulting and deep erosion, ca. 20 Ma to present.

Figure 67. Schematic cross section showing possible development of Sierra Nacimiento.

The entire block has clearly been folded, at least on its edges, and certainly uplifted along faults with respect to adjacent rocks to the east and west. As we'll see on our journey north along US-550 (Chapter 8), the western side of the Sierra Nacimiento follows an almost straight north-south line. This remarkable boundary consists of a steep, up-to-the-east, east-dipping fault. The eastern boundary of the Sierra Nacimiento is somewhat obscured by the onlapping volcanic of the Jémez Mountains.)

In a sense the entire block has literally been "popped up." The question that naturally comes up next is: How can a block of the Earth's crust be pushed upwards without leaving a hole down below? Well, it can't. Years of careful study in the Rocky Mountains have revealed that the steep faults that are exposed at the surface often flatten out with depth, and that slabs of deeply-buried "basement" rocks (crystalline rocks of Precambrian age) have been thrust under other slabs — a reverse shingling effect. This caused the crust to thicken and parts of it to pop up like a watermelon pit between wet fingers. I have therefore used such a conceptual model to force a guess about the deep structure and origin of the Sierra Nacimiento (Figure 67A). Much later, after 20 Ma, the entire area was subject to regional tension — the exact opposite force — causing the eastern part of the Sierra Nacimiento to founder and become part of the downdropped Jémez Valley portion of the Rio Grande rift (Figure 67B).

Red Mesa

The southwestern, curled-up rim of the Sierra Nacimiento is sometimes called Red Mesa (Figure 67B). The western face of the Sierra (a few miles up ahead) spectacularly exposes Permian-age, reddish sedimentary rocks called "redbeds," hence the name Red Mesa. From the east we can see that the Sierra, with its curled up western rim, plunges down to the south and disappears beneath the alluvium of the Rio Salado. However, an interesting interruption of the southward plunge can be seen looking west from the highway (Figures 65C and 68). What is that strange bump on the ridge?

Figure 68. West view of southern end of Sierra Nacimiento with its anomalous bump in center. (Photo by author 2005.)

The "normal" top surface of the Sierra consists of the massive sandstone of the Agua Zarca Formation, with a few scattered preserved remnants of the overlying Petrified Forest Formation (Figure 64). The little bump is in fact a small cap of travertine, about ten feet thick. (Travertine is calcium carbonate, $CaCO_3$, deposited by fresh-water springs.) The original extent of the travertine has been much reduced by the breaking off of its edges and slumping of the material down from the bump.

Most interesting though is that just below the little travertine cap is an outcrop of cobbly to bouldery sediment, perhaps ten to 30 feet thick (its top and bottom are obscured), and below it is a remnant of the Triassic, Petrified Forest Formation bedrock Figure 64). The cobbles and boulders are made of "exotic" Precambrian rocks, including much granite and much less abundant quartzite. I call the chunks exotic because they are out of place — the nearest source of granite bedrock is located to the north, on the other side of the deep canyon of the Arroyo Rio Peñasco (more about this in Chapter 8). The cobbles and boulders are well-rounded and clearly have been worked and abraded by vigorous stream action. What are stream-deposited exotic cobbles and boulders doing way up there on an isolated bump on the Sierra Nacimiento? Our landscape has obviously changed a great deal since this material was first deposited.

But there's more. Even higher up on the Sierra Nacimiento, at an elevation of 6,400 feet (about 1,000 feet above the level of the present Rio Salado) is another deposit of rounded cobbles and small boulders, made up mainly of quartzite and only secondarily of granite (Figures 65 and 69). Where did the quartzite chunks come from and how did they ever get way up there? The nearest source of quartzite bedrock is about 100 miles away, in the high country north of the Jémez Mountains.

Note boot for scale

Figure 69. Deposit of quartzite cobbles high up at 6,400 feet on Sierra Nacimiento. (Photo by author 2005.)

This is a point where we need to stir up our creative juices, and where geology becomes fun. In our mind's eye we must restore the landscape to an earlier form. We must envision a scenario in which a large, high-energy river can somehow flow west completely around the Jémez Mountains and transport its load of exotic cobbles via a precursor, ancestral Rio Salado across the southern nose of the Sierra Nacimiento. We can do this with a sequence of geologic cross sections showing the progressive shift of the Rio Salado Valley downward and southward (Figures 65A and 65B). As mentioned earlier, the formations have been stripped off the Sierra Nacimiento from north to south, and the north-facing White Mesa on the south side of the pass is merely the latest, transitory position of the receding cliffy landform (Figure 65C).

Travertine springs

Between mile 26.5 and MP-27 on the north side of the highway is a series of warm springs that are part of a so-called "Known Geothermal Resource Area" (KGRA in Figure 64; Rogers et al. 1996). However, water temperatures are generally less than 75° F and are not sufficiently high for the thermal generation of electricity. Most of these springs have constructed small mounds of travertine (more about this below). The constant oozing has created marshy terrain around the springs. The highway engineers were forced to built culverts to channel the spring waters under the road and to flow downslope to the Rio Salado. The springs have been known for a very long time and once provided water for desperate livestock, but the mineralized water is generally not potable for humans (Renick 1931).

The spring waters contain a high concentration of natural arsenic. In fact Craigg (1992) reported an arsenic concentration of 520 parts per billion (ppb) from one of these springs, 50 times the mandated maximum level of ten ppb that took effect in January 2006. The springs were also valued for their medicinal qualities by the local people. An early report noted that, "the waters of the springs are cold [i.e., not hot]. They have medical properties; and throughout the summer months, the Mexicans bathe in them" (Reagan 1903).

Travertine, mentioned above, is a relatively dense to vuggy, finely crystalline deposit of calcium carbonate ($CaCO_3$) deposited at the earth's surface or in the shallow subsurface by freshwater springs. It forms in caves as flowstones and dripstones as stalactites and stalagmites, and also at or near the mouths of springs such as here. The term comes from the Latin name *lapis tiburtinus*, the "stone of Tibur." The Italian corruption of tiburtinus is *travertino*. The city of Tibur, today's Tivoli, is located east of Rome. The extensive deposit of travertine there was formed in extensive, shallow Pleistocene lakes fed by acidic hot springs. Travertine has been quarried there since Roman times. In Rome, travertine from Tivoli can be found in the Colosseum and the colonnade of St. Peter's basilica as well as in many other buildings (Chafetz and Folk 1984).

Two commercial-grade occurrences of travertine are known to exist in New Mexico. Commercial travertine — i.e., that which is extracted for dimension stone — is dense, hard, massive to banded, and can take a polish. The New Mexico Travertine Company has an extensive quarry operation about 20 miles west of Belen in Valencia County, and the Apache Springs Company has a small quarry about eight miles west of Radium Springs in Doña Ana County (Austin and Barker 1990). An excellent example of dimension-stone travertine can be seen at the Albuquerque Sunport on the veneer covering the base of the Indian statue at the entrance to the boarding-gate wings, and magnificently displayed inside the rotunda of the state capitol building in Santa Fe.

Tierra Amarilla anticline

Beginning at about MP-26, a white wall abruptly jumps into view to the southwest. This is the western rim of the so-called "Tierra Amarilla anticline" (Figure 64). The strange notch sliced through the wall is the portal for five pipelines (mentioned earlier in Chapter 6). This part of the Rio Salado pass was far too narrow to accommodate the highway, the river, and the pipelines, so the latter had to be routed right through the wall.

The Tierra Amarilla anticline, sometimes called the "T-A anticline," is one of the very most interesting geologic features and hiking destinations in the state. It is so spectacular and so accessible that the University of New Mexico conducts geologic field-mapping classes there. The term "anticline" describes a sequence of folded rock strata, in which the flanks of the structure are inclined downward away from a central axis. The structure therefore is convex upwards. It is not possible to fully appreciate the shape of the structure from the level of US-550, but from the air it is quite spectacular (Figure 70).

Figure 70. North low-oblique aerial view of Tierra Amarilla anticline. (Photo by author 2001.)

The northern half of the feature lies on State land (Figure 71). State land is not public land. When I have taken people to these springs from the north (the easiest access) in the late 1990s I had to acquire a permit from the State Land Office ($60/day) to legally set foot on the land. The most hassle-free access is from the south, but it involves more hiking. The southern half of the anticline is located on Bureau of Land Management (BLM) land and is therefore open to the public.

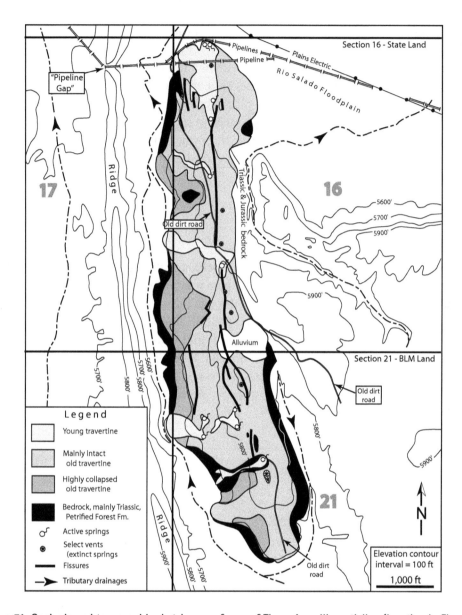

Figure 71. Geologic and topographic sketch map of core of Tierra Amarilla anticline (location in Figure 64.) (Bedrock geology and travertine outline modified from Woodward and Ruetschilling 1976.)

The quality of the geologic features found on the T-A anticline is on a par with that of many a National Monument. However, upgrade to a National Monument would require access roads, trails, and administration (translation=lots of money), and would guarantee that the area would be overrun with people. As it stands now, it is often possible to explore the place in utter solitude. The T-A anticline is a New Mexico treasure!

The surface and near-surface part of the structure is composed of folded Mesozoic-age rock formations (Figure 72). Like the structurally-related Sierra Nacimiento, the anticline was formed by the Laramide compression that peaked in intensity at about 55 Ma. During the same episode a fault ripped north-south through the center of the structure and crumpled and crushed the core (Figure 72A). Movement along this fault was probably lateral rather than up-down. The present overall shape of the structure is caused by deep erosion of the interbedded soft and resistant rock layers. Sharp ridges formed by the hard, Jurassic-age Todilto Gypsum and Entrada Sandstone surround a deep valley cut into soft Triassic-age mudstones of the Petrified Forest Formation that flank the inner core of the structure (Figure 72B and Figure 72C).

At deeper, unexposed levels of the structure is the massive Agua Zarca sandstone—the same unit that forms the backbone of the south-plunging Sierra Nacimiento seen on the north side of US-550. Sometime after erosion exposed the soft, ruptured core of the structure, groundwater that was saturated with dissolved calcium carbonate ($CaCO_3$) made its way up through the crushed core, probably from the porous underlying Agua Zarca sandstone. As the water reached the surface the dissolved carbon dioxide bubbled away, making the dissolved $CaCO_3$ unstable and causing it to precipitate out as layered travertine. Sustained outflow and precipitation built up an outward-sloping carapace of travertine above the soft and wet Triassic mudstones below (Figure 72D). The carapace attains a maximum thickness of 30 to 33 feet at the center and thins downslope. When that full thickness was reached the carapace became physically unstable because it was perched atop the soft, wet mass of Petrified Forest mudstones. Gravity acted to pull the carapace outward and downward from the center over the lubricated substrate and created a series of widening fissures along the crest. The sliding mass on the western side foundered and broke up into huge collapse zones (Figure 72E).

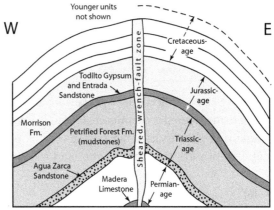

A. Initial folding and wrench-faulting along longitudinal shear zone, ca. 55 Ma.

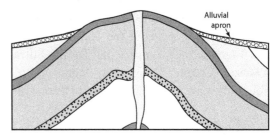

B. Deep incision down to top of Todilto Gypsum and formation of alluvial apron

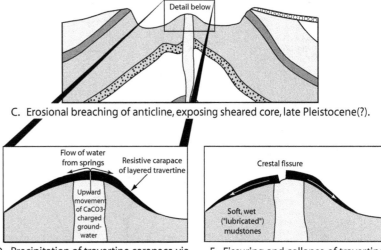

C. Erosional breaching of anticline, exposing sheared core, late Pleistocene(?).

D. Precipitation of travertine carapace via springs on core of anticline, late Pleistocene.

E. Fissuring and collapse of travertine carapace, late Pleistocene to present.

Figure 72. Schematic cross sections showing formation of Tierra Amarilla anticline travertine.

Into portions of the widening crestal fissures, a new generation of travertine springs developed, some of which are active now, especially on the northern end of the anticline.

The age of the travertine carapace is not known for certain. A tentative correlation can be made to another large occurrence of travertine a few miles to the north, on Zia Pueblo land between US-550 and the west face of Red Mesa (we will not see these from the highway). These travertines have been dated radiometrically to be between about 270 and 60 ka (Formento-Trigilio 1997). It is reasonable to suspect that the travertine carapace of the T-A anticline is of a similar age. Compared to the ages of the underlying Mesozoic rocks (about 150 Ma) the travertine was formed yesterday.

8

Rio Salado Anticline

The Rio Salado anticline is a pocket of wild and colorful geology. Remember that an anticline is a geologic structure where the layered rock units are folded in a convex-up pattern. The outcrop pattern of the structure's upturned resistant outer ridges has focused the erosion of an elongated topographic bowl inside the anticline. The bowl has a front door or entrance of sorts, the "South Gate," and an exit, the "North Gate" (my informal terms).

"South Gate"

At mile 29.5 we enter the Rio Salado anticline via the "South Gate" (Figure 73). The gate consists of a gap through the sharp cliffs formed by two of our old Mesozoic-age friends (Figure 74). The upper formation is the whitish to light gray gypsum of the Todilto Formation, and below it is the buff-colored sandstone of the Entrada Formation. Together they comprise the Todilto-Entrada cliff-forming couplet—the same one we saw back at the top of White Mesa in Chapter 7. The valley floor is underlain by the soft red mudstones of the Petrified Forest Formation. Once through the gate we find ourselves in a zone of geologic chaos.

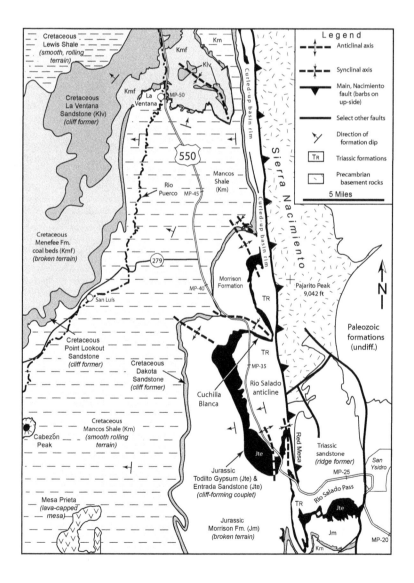

To our west, and north into the distance, the Todilto-Entrada cliff couplet sternly disciplines the course of the highway. To our east the couplet becomes highly contorted and terminates abruptly at the foot of Red Mesa—the southern part of Sierra Nacimiento—which towers overhead. Between the crumpled couplet and the mass of the Sierra Nacimiento is an up-to-the-east fault, along which the rocks of the Sierra have been pushed up nearly 1,500 feet. That event clearly took considerable geologic effort. In a very real sense we are amidst a collision zone that is in a state of suspended animation.

Figure 73. Generalized bedrock geology, San Ysidro to La Ventana. (Modified from Woodward and Schumacher1973; Woodward and Martínez 1974; Woodward and Ruetschilling 1976.)

Figure 74. East, high–oblique aerial view of the "South Gate" (center right) to the Rio Salado anticline. (Photo by author 2001.)

Collision zone

Now let's take a closer look at this jumbled mess. The rocks seen here deform in two distinctly different ways: the layered formations west of the Sierra Nacimiento block are highly foldable, while the rocks making up the core of the Sierra Nacimiento are not. To appreciate the difference, consider a telephone directory vs. a block of wood of equal size. The phone book (roughly analogous to the layered formations of the Rio Salado anticline) is easy to bend (or fold) because the pages can slide across each other, i.e., they have parallel planes of weakness. The block of wood (analogous to the Sierra Nacimiento) lacks parallel planes of weakness and therefore cannot be bent to any significant degree. Now, if we were to shove the phone book edge-on against the wooden block, the book would be deformed (folded) and the block would just sit there undisturbed. In simplistic terms, such is the scenario here.

Let's now tweak this general model a bit. First, the sequence of Mesozoic geologic formations are not exactly like a phone book, but rather they consist of the durable Todilto-Entrada cliff-forming couplet and the underlying, weak Petrified Forest Formation. If this sequence were to be compressed the couplet would break loose from and slide over the top of the Petrified Forest, like a phone book modified by inserting a thick piece of cardboard within the pages. The sequence of Mesozoic formations is indeed profoundly folded here, *a la* our modified phone book. The net result has been a wavy pattern of ridges surrounding a central valley, across which highway US-550 wends its way.

A map of the Rio Salado anticline and its vicinity, showing the axes of the folds (lines drawn along the crests of the anticlines and the low points of the synclines), reveals a northwest-orientation to these axes (Figure 73). How and why this trend? We can learn the answer by returning to our phone book analogy. For simplicity take a single page, place it on a desktop, hold the right side fixed, and push the left side away from you a little. The page will ripple and the crests and saddles of the ripples will trend to the upper left—i.e., the northwest. Multiple lines of evidence suggest that the layered formations of the east rim of the Colorado Plateau (which includes the area of the Rio Salado anticline) collided with the rigid Sierra Nacimiento block and wrenched northward along the bordering Nacimiento fault. This event was the Laramide compression, mentioned earlier. Fully a ten-mile-wide zone on the west side of the Sierra was thus deformed to produce its belt of northwest-trending folds. A one-mile-wide collision zone right against the Sierra was most intensely deformed into a series of sharply curled-up formations, which in turn forms a narrow zone of ridges and valleys.

Since the Laramide event, mainly during the last few million years, erosion has been gouging its way downward into the deformed pile of sedimentary rocks, stripping the uplifted formations off the Sierra Nacimiento and creating the present landscape. As the flowing streams progressively eroded downward they preferentially dug into the soft stuff and flowed around the hard stuff and modified their courses accordingly. Some of that modification was due to stream piracy.

Stream piracy

The shifting of a stream course as it entrenches its bed, seeking the easiest path downhill, sometimes leads to an intriguing phenomenon called "stream piracy." A colorful term, "piracy" invokes images of lawless brigands taking what they want, when they want. The term is quite appropriate in that it refers to a stream that erodes its way headward (upstream) into the drainage basin of another stream and steals its water. The aggressive stream is often unable to fully erase its earlier nefarious activity and leaves vestiges of the crime (abandoned or nearly abandoned channels) behind to be plotted by the U.S. Geological Survey on its 7.5-minute topographic quadrangle maps.

The vicinity of the Rio Salado anticline provides some excellent examples of stream piracy. First, at mile 29.5 we cross the bridge over the Arroyo Peñasco (Figure 75). This pathetic little rill doesn't look like much, but it's a pirate. But let's continue on our way a little farther up the road. At mile 31.7 the highway crosses a bridge over the Arroyo Cuchilla (not to be confused with the Cuchilla Mesa near San Ysidro). To our west this ephemeral stream exits the anticline via a 200-foot-deep gash sliced right through the western cliff. How could this pitiful little streamlet have cut its way through this rampart? Such a stream is called "underfit," in that it does not seem capable of having cut the valley it inhabits. The nearly empty valley is a hallmark of stream piracy.

To unravel the geologic story we must recall our earlier stop at the Rio Salado south of San Ysidro (Figure 59). There we saw a 150-foot-deep canyon coming south straight at us down the Sierra Nacimiento. It had been cut into the massive bedrock of the Agua Zarca Sandstone. Today the canyon contains a miserable little stream that is ephemeral at best. Like the canyon at the Arroyo Cuchillo this grossly underfit stream is a mere shadow of the one that actually did the work of cutting the canyon.

To answer what's going on here we must go back some 300,000 to 200,000 years (ka) ago into the past with our as-ever fertile geological imagination to retrace the tortured history of the area's drainage. At one time powerful streams, supplied by copious precipitation in the higher elevations (above 8,000 feet) to the north, flowed directly south down to an earlier Rio Salado at the Rio Salado Pass (Figure 76A). How do we know this?

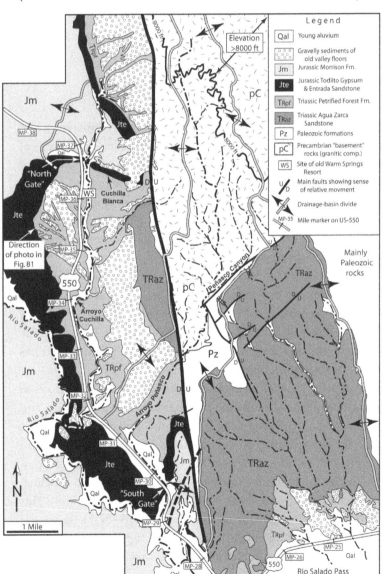

Figure 75. General geology and drainage of Rio Salado anticline area. (Modified from Woodward and Martínez 1974; Woodward and Ruetschilling 1976.)

Recall again that back at the Rio Salado bridge south of San Ysidro we saw a low terrace on the north side of the Rio Salado (Figure 59). This terrace (on which the village water tank is perched) is capped in part by a once-continuous sheet of naturally-cemented coarse sediment containing boulders and cobbles of granite. The nearest source of granitic bedrock is located in the headwaters of these ancestral streams on the Sierra Nacimiento (Figure 75). The streams that delivered chunks of granite to this terrace therefore once had to flow directly from this granitic source area. Today this source area is cut off and separated by the upstream canyon of the Arroyo Peñasco ("Peñasco Canyon," Figure 76C) which today drains the higher elevations. The west-flowing Arroyo Peñasco drainage has captured, i.e., pirated, the headwaters of the streams that once flowed south to the Rio Salado with their granitic debris.

But there's more. The underfit canyon of the Arroyo Cuchilla seen west of mile 31.7 has yet to be explained. An earlier, more energetic stream must have cut this canyon, but when and from where? Based on the published literature (especially Formento-Trigilio and Pazzaglia 1998) and the USGS topographic maps, I conclude that soon after the capture of the headwaters of the south-flowing streams on the Sierra Nacimiento by the Arroyo Cuchilla, the newly-reinforced pirate stream used its new-found energy to cut down through the western wall of the anticline at mile 31.7 (Figure 76C). Then, sometime after that, the Arroyo Peñasco drainage cut headward to the north and pirated the pirate. Now the drainage scooped up at Peñasco Canyon by the Arroyo Cuchilla flows south along the Arroyo Peñasco (Figure 76D).

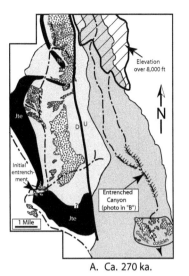

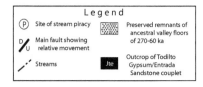

A. Ca. 270 ka.

B. North view of entrenched canyon shown in "A."

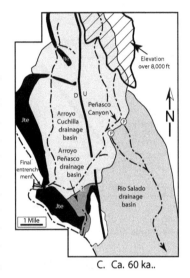

C. Ca. 60 ka..

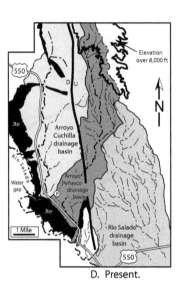

D. Present.

Figure 76. Generalized evolution of drainage basins (shades of gray) in Rio Salado anticline area (see Figure 75 for geology).

So a mystery has presented itself and a possible solution has appeared. It might be said that before I brought up the subject of stream piracy the motorist was untroubled and blissfully ignorant of such complications. Sorry, that's the price of curiosity!

West side of the Sierra Nacimiento and Red Mesa

At about mile 33.5, where US-550 is confined by the cliffs to the west and the Arroyo Cuchilla to the east, the character of the high Sierra Nacimiento off to our east changes abruptly. The layered mass forming Red Mesa is replaced northward by what seems to be an ugly, structureless mass (Figure 77A). The latter rocks are of Precambrian-age and are mainly of granite-like composition (actually granites that have been metamorphosed to form a modified type called "gneiss," pronounced NICE). They provided the granitic cobbles and boulders mentioned earlier.

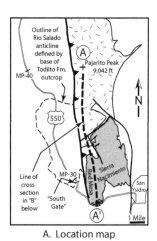

A. Location map

Nacimiento fault | Red Mesa portion of Sierra Nacimiento on upthrown side of Nacimiento fault | Down-thrown fault block at foot of Red Mesa (west and out of plane of cross section below)

Photo panorama looking east along part of cross section below

A
Elevation
9,000 ft — Pajarito Peak 9,042 ft
8,000 ft
7,000 ft — Precambrian basement rocks
6,000 ft
5,000 ft — Vertical exaggeration = 3

Triassic Agua Zarca sandstone cap (dark gray)

Remnant of Triassic Petrified Forest Fm.

Permian Yeso Fm.
Permian Abo Fm.

550 Rio Salado Pass

A'

1 Mile

B. Schematic cross section and photo panorama along crest line of Sierra Nacimiento (location in "A" above). (Photos by author 2001.)

Figure 77. North-to-south profile of western rim of Sierra Nacimiento, east side of Rio Salado anticline.

The Sierra's irregular north-to-south profile is due to the fact that the entire mountain mass is plunging down to the south, and that the 1,200-foot-thick sedimentary section seen on the face of Red Mesa has been progressively stripped off the uplift to the north, exposing the Precambrian core (Figure 77B). What is not readily apparent is about 3,000 feet of younger, Mesozoic-age sedimentary rocks, some of which form the cliffs to the west, have been stripped off as well. Pajarito Peak (9,042 feet) is the prominent high point on the craggy hill of Precambrian rock that sports a number of communication towers. A road leads to the summit of the peak, but don't take it—it's on Zia Pueblo land and is off limits.

Cuchilla Blanca

This great white slash, the *Cuchilla Blanca*, (Spanish for "white knife") is the northern limb of the Rio Salado anticline, just east of the exit point of the anticline, the "North Gate" (Figures 75 and 78). It is formed by the resistant Todilto-Entrada couplet, dipping away from us to the north. It appears to be a solid white mass because the rubble from the Todilto Gypsum at the top tumbles down and completely carpets the underlying Entrada sandstone and Petrified Forest Formation.

Figure 78. North view of Pajarito Peak and Cuchilla Blanca, "White Knife."

The *cuchilla* long served as an attention-getter and landmark along the road to Cuba. It also heralded the approach to a well-known spring (see Warm Springs Resort below). In 1930, sheep rancher Jim Counselor was bringing the object of his affection, the girl who would soon become his wife, from Albuquerque to see his ranch in Canyon Largo northwest of Cuba (more about them in Chapter 20). Years later, in her smooth style Ann wrote:

> "Our car climbed to the summit of the ridge that he had pointed out to me, and there, spilling away into the distance, lay a long, narrow, twisting valley of such wild beauty and unbelievable colors that the sight took my breath away. In the distance, at the upper end, a high fortress-like wall of pure white rock [Cuchilla Blanca] damned [sic] off the valley in that direction. A dark-green carpet of sagebrush floored the valley, through which a small, meandering creek had cut its way. Sloping upward from the fringed edge of the green carpet to the foot of the rocky sides, the bare earth was a rusty red color. The shoulder of the Jémez Range [Red Mesa] that formed the east side of the valley was a sheer cliff of red rock rising a thousand feet, on top of which a few leaning pines grew, looking as though they were peeking over the rim for a view of the beautiful valley below. On the lower side, the rocky terraced and domed sides were of different colors, ranging from grays and whites to pale yellow. Along this side, too, large pieces of rock as big as boxcars had fallen away from the cliffs and lay scattered along the slopes as though some giant hand had once started to build a grand amphitheater and then had neglected to finish it" (Counselor and Counselor 1954).

Warm Springs Resort (site)

At mile 35.7, just off the east side of the highway ("WS" in Figure 75), is a old access road leading down to a graded pad. This is the site of the one-time Warm Springs Resort, and it is now on Zia Pueblo land. During the four-laning of US-550 in 2000 this place was used by the contractors for storage, but after the work was done the site was "restored" to its original condition. Not a trace of the old resort remains. But the site has a rather convoluted and quite interesting history.

In 1926 a well was drilled here in the search for oil. The 1920s was a time of furious exploration activity in the San Juan basin to the northwest near Farmington, and most successful wells at the time were drilled on "closed anticlines" that effectively trap the oil and gas. As mentioned earlier, in an anticline the formations dip downward in at least two opposite directions from a central axis. In a closed anticline the formations dip downward in all directions. However, the Rio Salado anticline has formations dipping down in only three directions and is therefore unclosed. Despite this inconvenient fact the structure attracted interest early on as a exploration target. At the time the land was owned by the Ojo del Espiritu Santo Company (OESC), which was in turn owned by the powerful Catron family of Santa Fe, and the tract was sometimes called the Catron Ranch (more about the OESC in Chapter 10).

In 1925 the Catrons had leased the area to an Albuquerque business man/politician/entrepreneur by the name of George Ambrose Kaseman. He was born in Shamokin, Pennsylvania in 1868. He completed a course in business college but apparently did not graduate. Poor health compelled him to move to New Mexico in 1887 with a new wife in tow. Shortly afterwards he attended Stanford University where he shared a room with future president Herbert Hoover. Continued bad health forced him to return to New Mexico. He became involved with merchandising, the telephone business, coal mining, and banking. He was very good at these activities and became quite wealthy (Gill 1994). In1924 he founded the Albuquerque National Bank. Then he got into the oil business.

In early 1926 Kaseman started drilling a well in a valley bottom near this site, but lost the hole while drilling. He moved the rig over and started a second well but lost that one too. He then moved the rig to this site, a little more than a mile to the north, and began a third well, the Kaseman No. 2. It was drilled to a depth of 2,008 feet. It failed to find oil or gas but it did find warm, "artesian" water (artesian water has the energy to flow to the surface on its own without pumping). In the strange lingo of the oil business this was a "dry hole," i.e., "dry" of oil or natural gas. After casing was cemented in the hole, warm water (>120 degrees F) flowed at about an impressive rate of 2,000 gallons/minute.

In 1927 plans were made to plug the well, but by then the casing had corroded and water was coming out of the ground around the casing. That summer an unsuccessful attempt was made to plug the well. By then thoughts turned to using the well as a source of stock water because there was no other supply in the area. Cattle would not drink the water near the well due to the temperature, but some distance away, where the water had cooled down, they eagerly gulped it down. That year the well had in effect become a spring, flowing about 2,500 gallons/minute (Summers 1976).

Kaseman moved on and continued his endeavors elsewhere in the oil and natural gas business. In 1938, as he was witnessing a procedure on a well near Hobbs, in southeastern New Mexico, he was killed by an explosion (Reeve 1961). His wealth, through his wife, was instrumental in founding the Anna Kaseman Hospital in Albuquerque (Gill 1994).

The soon-to-be Mrs. Ann Counselor, on the same drive in 1930 that was mentioned above wrote:

> "Away up in the valley, toward the white dam [Cuchilla Blanca], I traced the outline of the creek bed. Then I rubbed my eyes for better vision, for a light, smoky cloud was rising from the water all along the creek! . . . As we drove on, the mystery of the smoking water was simply explained: near the head of the valley, beneath the white-rock dam, a big spring gushed forth, emptying its boiling hot water into the creek. When this hot water came in contact with the cool atmosphere, it produced a cloud of steam" (Counselor and Counselor 1954).

In the mid-1930s boys from the Civilian Conservation Corps (more about the CCC in Chapter 11), whose camp was located five miles to the north, dammed the flow from the spring for an *ad hoc* swimming pool (Civilian Conservation Corps 1936). At least by the late 1930s the locals regularly availed themselves of the thermal waters. Johnny Hernández, present owner of the Young's Hotel in Cuba, tells this story: "When I was a kid, 12 or 13 years old [1939–1940] I had a Model A, and my granddad [John Young] had this sciatic rheumatism so bad. We'd load up the old Model A and we'd go down there and dig a hole and we'd take a hot bath in that mineral water" (Hernández 2000b).

At least by the early 1950s a large swimming pool had been constructed and the warm water directed to the pool via a channel for the Warm Spring Resort, managed by Mr. and Mrs. W.W. "Pete" Chapman (Bryan 1953). There was a two-story hotel with a nice lobby, and a tiled swimming pool (Hernández 2000b). In the early 60s they there was a clinic and a doctor on hand, and occasionally a local family went there to have a baby delivered (Johnson 2003). They also had a little restaurant that featured home-made pie. It was a big deal to get a piece of it after a day's swimming (Wiese 2001).

Around 1970 the place was abandoned and it fell into disrepair. Then hippies moved in and turned it into a nude bathing spot. This caused the predictable traffic snarls and travel hazards as people would stop and gawk. Finally, Zia Pueblo, via the Bureau of Indian Affairs, fenced the site off and bulldozed everything flat. For years nothing remained except cement slabs and a water tank (Figure 79). Now even they are gone.

Figure 79. East view of remains of Warm Springs swimming pool before final obliteration. (Photo by author 2002.)

Figure 80. North view of ancestral valley floor, from mile 33.5. (Photo by author 2005.)

Ancestral valley floor

Heading north at about mile 33.5 the observant motorist will see something amiss. Whereas the cliffs on the west side of the highway slope down to the west, a prominent flat surface directly ahead is seen to defy the rules and slope in the opposite direction (Figure 80). This surface is the remnant of the old valley floor as it existed sometime between 270 ka and 60 ka. Recent erosion has removed the upper reaches of the surface where it once seamlessly merged with the rim of the Todilto-Entrada cliff (see "Photo" indicator in Figure 75, and Figure 81).

Figure 81. Northeast low–oblique aerial view of eroded remnants of ancestral valley floor. (Photo by author 2001.)

The nature of this surface can best be seen on the west side of the highway between mile 34.6 and a little past Warm Springs. A few feet below the top of the surface is a whitish band (seen nicely where the highway department has cut through it). This is a layer of calcium carbonate-rich material called "caliche." It is part of the natural soil profile and is characteristic of soils formed in the arid and semi-arid Southwest. Rainwater collects very fine calcium carbonate dust from the atmosphere, transports it down to the land surface, and flushes it via infiltration a few feet into the subsurface, where it precipitates as caliche. If the surface is allowed to remain "stable" (i.e., if it is not buried or eroded away and therefore maintains a constant distance below the land surface), over many thousands of years the concentration will build up to a rock-like consistency. The texture and calcium carbonate concentration of such a zone provide the geologist one measure of the zone's geologic age.

"North Gate"

Upon exiting the Rio Salado anticline at the "North Gate" the modern alignment of US-550 follows that of its immediate predecessor NM-44. In 1935, however, the "old" graveled NM-44 veered well off to the right and struggled through the broken topography of the Morrison Formation (Figure 82).

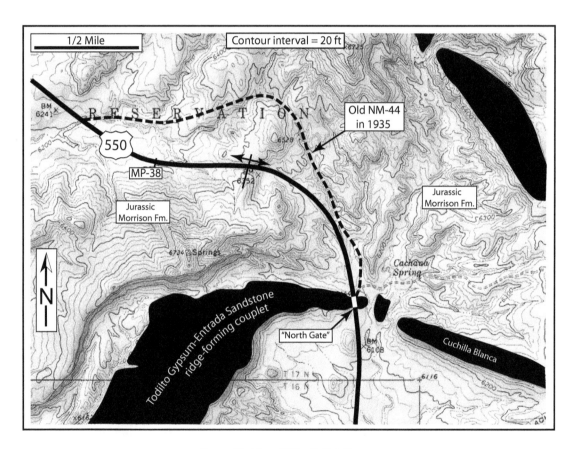

Figure 82. Map of "North Gate" area.

At mile 37.6 the highway crosses a divide (that the road engineers lowered during the four-laning of US-550 in 2000–2001) on the rugged Morrison topography (Figure 82). The Morrison Formation, dated at about 155 Ma, is a series of sandstones and mudstones deposited by rivers. It is famous for its huge dinosaur fossils in the Ojito Wilderness and for its deposits of uranium in the Grants Uranium Region. Non-commercial amounts of uranium were mined about 0.75 miles north of this spot during the Cold War uranium frenzy of the late 1950s. From the divide we begin a long decline into the very different world of the middle Rio Puerco Valley.

9

Middle Rio Puerco Valley

Beyond about mile 38.5 we leave the cliffy, broken topography of the Morrison Formation behind and enter a very different landscape formed on sedimentary rocks of Cretaceous age (Figure 83). This landscape is characterized by broad open flats underlain by soft "marine" (deposited in an open sea) mudstones, interrupted by thick, cliff-forming sandstones. The mudstones give the land a light brownish cast. Their original color was a dark gray rendered by traces of iron that existed in a reduced chemical state. A falling water table has allowed the atmosphere to interact with the iron in the mudstones and to oxidize (rust) it to its present brownish color. And then, suddenly, off to the west at mile 38.7, surges the impertinent form of Cabezón Peak.

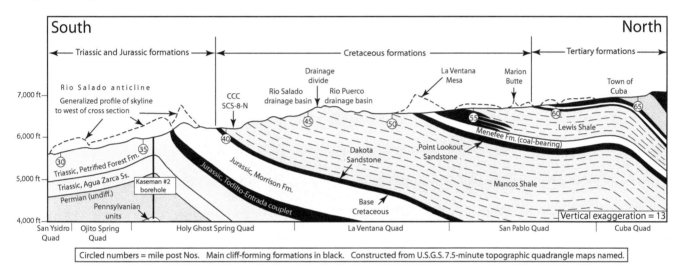

Figure 83. Longitudinal south-to-north geologic cross section along graded bed of US-550 between MP-30 and MP-65.

Cabezón Peak viewshed

We've entered the extensive viewshed of Cabezón Peak (Figure 84). I've coined the word "viewshed" for this book. It analogous to "watershed"—a river's drainage basin, and "airshed"—a collection area for an area's air flow, and a term used in environmental studies to denote source areas for air pollution. A viewshed then is an area "bathed" in the view from a certain prominent feature.

Cabezón (Spanish for "Big Head") Peak, an extremely photogenic feature, is the most spectacular of a group of volcanic features collectively called the "Rio Puerco Volcanic Necks." At an elevation of 7,785 feet it soars some 1,300 feet above its elevated cliffy platform and about 2,000 feet above the valley floor. It is very real competitor to the much more famous Devil's Tower in northeastern Wyoming, which is a full-blown National Monument. In fact, because Devil's Tower is such a recognizable image, it starred in Stephen Spielberg's very successful 1977 science fiction movie thriller, *First Encounters of the Third Kind*. Cabezón Peak would have served just as well (Figure 85).

Rio Puerco volcanic necks

The volcanic necks of the middle Rio Puerco Valley are the resistant cores of volcanic vents, most of which were constructed between 3.0 and 2.7 Ma, atop a landscape of Cretaceous-age rocks (Figure 86; Hallett et al. 1997). The absolute, radiometric age of the Cretaceous rocks here is about 80 to 90 million years—very much older than the igneous necks that penetrate them. The hard, resistant volcanic cores have since been exhumed by stripping away of their once-enclosing, poorly-consolidated cindery covers (Figure 87).

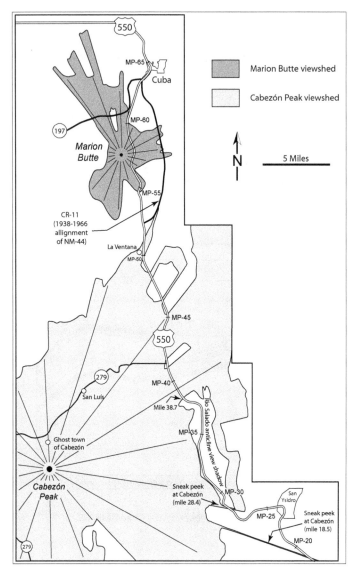

Figure 84. Select viewsheds.

A. West, low-oblique aerial view showing path to the summit.
(Photo courtesy of freelance photographer James Creager 2002.)

B. Northeast view of rather chaotically-developed columnar joints.
(Photo by author 1997.)

Figure 85. Cabezón Peak.

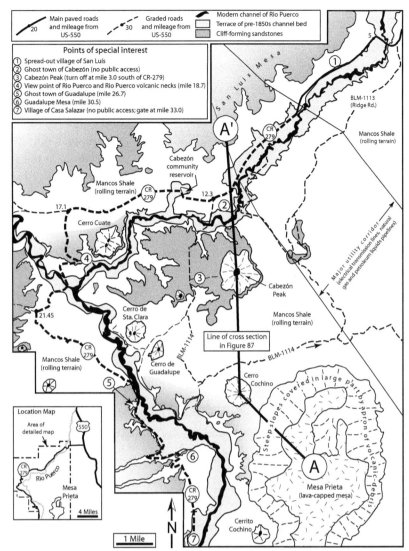

Figure 86. Middle Rio Puerco Valley and Rio Puerco volcanic necks.

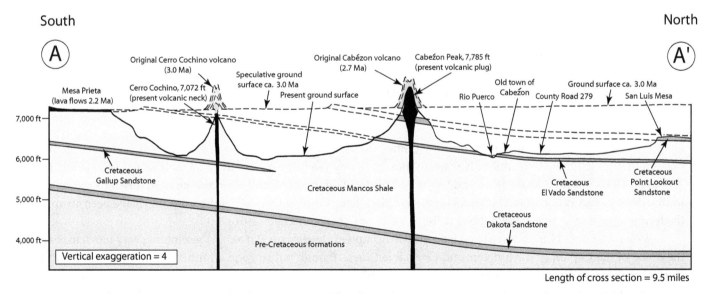

Figure 87. South-to-north cross section across portion of middle Rio Puerco Valley (location in Figure 86).

119

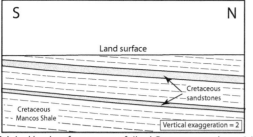

A. Original land surface on top of tilted Cretaceous rocks, ca. 3 Ma

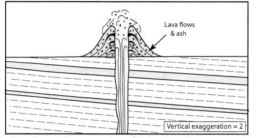

B. Eruption and construction of Cabezón volcano, ca. 2.7 Ma

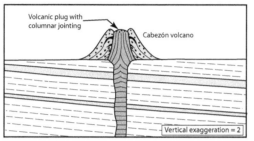

C. Injection of volcanic plug into throat of volcano, ca. 2.7 Ma.

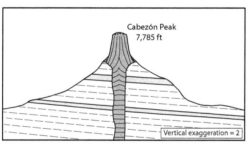

D. Erosional shaping of modern Cabezón Peak

Figure 88. Origin of Cabezón Peak. (Modified from Hallett 1992.)

Six out of a total of more than 15 of these volcanic edifices dominate the valley. Of these six, Cabezón Peak is the most accessible from US-550 (Figure 85A). Its geomorphic evolution is generally representative of the other Rio Puerco necks (Hallett 1992). About three million years prior to the formation of the necks, the landscape consisted of terrain sculpted onto Cretaceous-age sedimentary rocks that dipped gently to the northwest (Figure 88A). At about 2.7 Ma molten material surged upwards through the Cretaceous rocks, spewed out at the surface, and constructed a volcano (Figure 88B). As the short-lived volcanic activity abated, a plug of lava filled the throat of the volcanic vent and chilled in place (Figure 88C). Like most solids, the just-consolidated and hot neck began to shrink as it lost heat sideways to the cindery encasing rock and upwards toward the atmosphere. Tension developed through the entire body of the rock, but rather than the whole mass contracting as a unit, shrinkage occurred around many equally-spaced centers. Cracks between the centers, typically arranged in a hexagonal pattern, resulted in what is termed "columnar jointing." As the rock body continued to cool from outwards to inwards and downwards, the joints developed inwards and downwards parallel to the direction of cooling. Unlike Devil's Tower, which lost its heat mainly upwards, Cabezón Peak's columnar joints formed irregularly due to heat being lost in variable directions (Figure 85B). Finally, deep erosion during the past 2.7 million years has carved up the necks and produced the interesting landscape we see today (Figure 88D).

The Navajo inhabited this region for a long time prior to the middle 19th century, and the peak had great ceremonial importance to them. According to Navajo mythology, many years ago a giant lived on their sacred Mt. Taylor. The Twin War Gods pounced on the giant and lopped off his head. They rolled and bounced the head down the mountain until it came to rest as Cabezón. The Navajo called it "Black Rock Coming Down." The Spanish may have been aware of the former appellation when they dubbed it "Big Head," or Cabezón (Jackson 1996).

When Lt. James H. Simpson passed here in 1849 (Chapter 7) he described it as " . . . resembling very much in shape the dome of the Capitol at Washington; and *Cerro de la Cabeza*, though not so good a representation, yet cannot fail to suggest to the traveller [sic] a like resemblance" (Kues 1992).

In addition to Cabezón, three of the six main volcanic necks have colorful Spanish names: the low-lying *Chafo* (roughly translated as "squashed"), *Cuate* ("cousin" because there are two summits), and *Cochino* ("boar" or "hog"). The other three are more conventional: *Guadalupe* (from the nearby village named after Nuestra Señora de Guadalupe, Jesus' mother who reportedly appeared as a vision to a Mexican Indian man in 1531), and *Santa Clara* (from St. Clare of Assisi who assisted St. Francis in founding an order of cloistered nuns) (Julyan 1996).

The huge, battleship-shaped (in map view) mesa southeast of Cabezón Peak (Figure 86) is *Mesa Prieta* (Spanish for "Dark Mesa"). The mesa is capped by flow of basaltic lava that spewed out onto the ground at about 2.2 Ma, i.e., only about 0.5 million years after the intrusion of Cabezón Peak. The surface onto which the lava flowed, now some 1,500 feet above the present bed of the Rio Puerco, was the level of the land surface at that time. The mesa forms an impressive barrier that makes travel in and out of the middle Rio Puerco Valley difficult.

The modern valley

There is some confusion about the name "Rio Puerco" in New Mexico because there are three such-named rivers: 1) the Rio Puerco of the West, in the western part of the state, runs southwest through Gallup into Arizona, 2) the Rio Puerco of the North, a tributary of the Rio Chama on the north side of the Jémez Mountains, and 3) our Rio Puerco of the East, which from here on I'll simply refer to as the "Rio Puerco." The valley of the Rio Puerco is informally subdivided into a lower valley, generally south of I-40, a middle valley between I-40 and the village of San Luís (No. 1 in Figure 86), and an upper valley between San Luís and Cuba. However, this terminology is not always consistent.

The Spanish word *puerco* means "dirty," and the river is appropriately named. The Rio Puerco begins as a spring-fed stream in the San Pedro Parks sector of the Sierra Nacimiento east of Cuba (more in Chapter 15), but for most of its length it is a ephemeral stream. It is the largest tributary to the Rio Grande and provides about 4% of the main river's water flow near the confluence near San Marciel. However, the Rio Puerco supplies an amazing 70% of the suspended sediment load to the Rio Grande. In fact, the Rio Puerco has the fourth highest suspended-sediment concentration in the world, and only two stations on China's huge Yellow River and one on a tributary to the Yellow River have recorded higher concentrations (Gellis et al. 2000). This enormous sediment load potentially presents a significant eventual threat to the storage capacity of Elephant Butte Reservoir, constructed in 1916 on the Rio Grande in southern New Mexico. The sediment load is testimony to the erosion wreaked by the river in the past and continuing in the present. For these reasons the Rio Puerco has attracted a great deal of attention and concern from government agencies.

Where is all this sediment coming from and why is this valley such a prolific producer of it? The answer lies in the valley's unusual geology. Very little sediment comes from the volcanic necks because they are hard and resistive to erosion. They do fall apart though and the debris forms broad sloping aprons surrounding the necks. The Cretaceous-age bedrock is more easily eroded. The valley consists of areas of low-lying topography underlain by soft mudstones

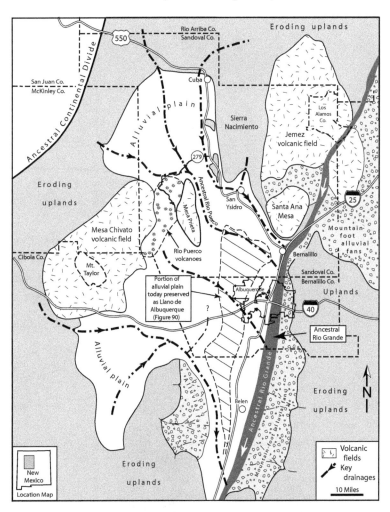

Figure 89. Generalized paleogeography and paleodrainage before deep incision, ca. 2 Ma. (Modern geographic features shown for reference.)

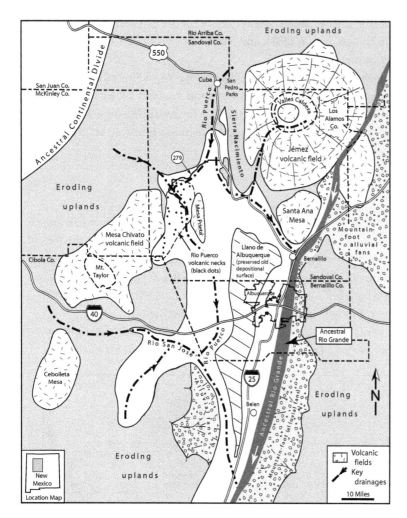

of the so-called Mancos Shale, and it is partitioned into three belts by the cliffy sandstones (Figure 86). The valley occupies a position on the eastern rim of the much larger San Juan basin, and these stratified bedrock units dip gently (< 5°) to the northwest toward the basin's center.

After the emplacement of the volcanic necks and the extrusion of the lavas of Mesa Prieta, and before major entrenchment of the present drainage, the ancient rivers flowed southeastward to the Rio Grande basin (Figure 89). A remnant of this old river-produced an alluvial surface that has been preserved as what is called the "Llano de Albuquerque," on which much of the City of Rio Rancho and the west side of Albuquerque are built. Since about 2.5 Ma until sometime around 180 ka the ancestral rivers eroded down about 600 feet to excavate a valley into the Cretaceous-age bedrock. At the same time a tributary of the Rio San José cut headward (northward) into the soft mudstone bedrock and pirated the east-flowing streams one by one. Thus the modern Rio Puerco was formed by stream piracy committed by the ancestral Rio San José (Figure 90).

Figure 90. Generalized paleogeography and paleodrainage during deep incision, ca. 1 Ma. (Modern geographic features shown for reference.)

The subject of piracy came up earlier in this book and will come up again. It is important to understand that as a landscape underlain by layers of inclined or "dipping" sedimentary rock is eroded downward, any ridges of resistant beds, such as sandstone, will shift their positions on the surface through time (Figure 91). As the ridges shift, the rivers carving the landscape will flow toward those more erodible areas of softer rock between the ridges and often will shift course to do so. Such constant shifting and exploitation of soft rocks lead to stream piracy.

By about 1850 the channels of the Rio Puerco and its tributaries had back-filled their bedrock valleys with about 80 feet of alluvium (Love and Connell 2005). This alluvium in effect was placed into temporary storage, awaiting only a change in climate. That change came at the end of the Little Ice Age (ca. 1300–1850 CE). During the cooler years of the Little Ice Age precipitation

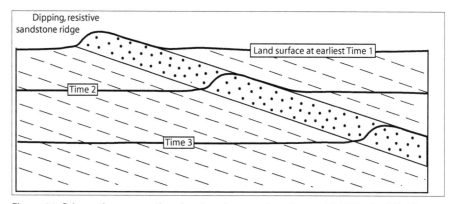

Figure 91. Schematic cross section showing downward erosion and "shifting" of dipping sandstone ridge through time.

occurred mainly in the winter and was of relatively low intensity. Rainfall then was able to soak into the ground and maintain a high water table. During the 17th and 18th centuries, for example, the river meandered about on the top of its flood plain through a lush, fertile valley. Spring runoff caused the Rio Puerco to overflow its banks and to water the entire valley to the depth of several inches. The muddy waters spread silt and nutrients over the flood plain. Small lakes, or *lagunitas*, and springs were abundant. Grass was in places belly high and vast *bosques* of cottonwood trees abounded (López 1980). In fact, in 1692 De Vargas referred to the Rio Puerco as "La Torriente de los Alamos" (Spanish for "Torrent of the Cottonwoods") (Aby et al. 1997).

At the onset of the warmer and more arid conditions characteristic of our present era since 1850, the rainfall came mainly as intense summer monsoonal downpours (Figure 15B; Pazzaglia 2005). During such deluges less water was able to soak into the ground and more flowed laterally across the land surface. The energized flows coalesced to form a vigorous and highly destructive river system that scoured down into and ate up the alluvial valley-fills. The fragile fills are composed of very fine sandy to silty material that is rich in gypsum, alkali, salts and lime—all soluble (López 1982). Repeated wetting and drying of this stuff causes expansion and contraction and the development of cracks. The cracks allow penetration of water and further development of "piping holes." The pipes may extend underground for significant distances before discharging their water content and sediment from the pipe walls into lower arroyo bottoms (Widdison 1959).

Overgrazing, beginning in the 1870s, in morbid concert with the changing climate with its destructive deluges and expanding pipes, created the perfect storm. In short, it is the unstable alluvium of the valley fills that is supplying the huge sediment load in the main river channel. During the past century the Rio Puerco has cut a new channel some 30 to 50 feet down into this alluvium. The mid-19th century river level is now an abandoned surface, a "terrace," perched above the present, active river channel below. The old cottonwoods that once studded the valley floor are history, and only the occasional "fossil" remains (Figures 92A and 92B).

Initial settlement

There are indications that people lived in the middle Rio Puerco Valley sometime before 600 CE or perhaps even as early as 500 CE (Hurst 2003). Significant settlement was first accomplished by the Anasazi people, who lived in the valley periodically for more than five centuries from about 785 until about 1300 CE (Baker 2003). The term *Anasazi* is a Navajoan word best translated as "ancestors of our enemies" (Stuart 2000). Given that today's

A. Remains of enormous old cottonwood tree at foot of Cerro Santa Clara, evidence of a once-high water table. (Photo by author 2003.)

B. Northwest view of Rio Puerco showing 30 feet of entrenchment. (Photo by author 2003.)

Figure 92. Entrenchment of Rio Puerco and coeval lowering of water table.

puebloan people have developed a proprietary feeling about the ancient builders of such places as Chaco Canyon, the word Anasazi has fallen into disfavor. Although a multitude of book titles would have to be changed, the preferred term for these ancient people is now "Ancestral Puebloan."

The valley contained the necessary ingredients for a successful agricultural economy: water, arable land, construction materials, "lithic resources," firewood, gatherable natural vegetation, and game, all within reasonable travel distance from habitation sites. Water and arable land were the number one concerns. The valley's high temperatures and low humidity insured that the natural vegetation adapted to the arid conditions. Dry farming was not a viable option. The early inhabitants established small farms watered by "flood irrigation." This type of agriculture focuses on localized areas within the valley that are subject to sporadic flooding, such as the mouths of arroyos and side canyons (Bryan 1929).

Other necessary ingredients for settlement by a stone-age people were an adequate supply of construction material, i.e., rock—there was plenty of that, and "lithic"material from which to fashion stone tools. Numerous flattish areas flanking the valley and suspended above it are carpeted with course cobbly gravels. This material was deposited at these elevations by energetic Pleistocene streams that flowed at this high level before the modern valley was incised. A minor but very important component of the gravels is a distinctive rock called the "Pedernal Chert." (The term is redundant, akin to the term "pizza pie," because *pedernal* is the Spanish word for "chert.") Chert is a microcrystalline variety of quartz (SiO_2). It is extremely hard, breaks with a curved "concoidal" surface, and thereby makes exquisite cutting tools. One such quarry of this chert was worked on the cliff top overlooking the river about a mile north of the townsite of Guadalupe (No. 5 in Figure 86; Brett 2003).

The Ancestral Puebloans (a.k.a. Anasazi)

The habitation of the Ancestral Puebloan people in the middle Rio Puerco Valley was concentrated in two areas on the west side of the Rio Puerco (Figure 93). The northern of the two was inhabited from about 785 to 1275+ CE. It extended generally from north of the Hispanic townsite of Guadalupe to that of Casa Salazar on the south. The main Ancestral Puebloan site in the northern area (see Guadalupe Ruin below) is located at the mouth of two side canyons cut into the sandstone cliffs where flood irrigation could be effectively practiced. The southern of the two areas was inhabited from about 950 to 1275+ CE (Baker 2003). The maximum population in the two areas taken together was reached about 1200 CE, and totaled something on the order of 1,500 (Durand and Baker 2003). Today this number of people living in this desolate place is almost incomprehensible.

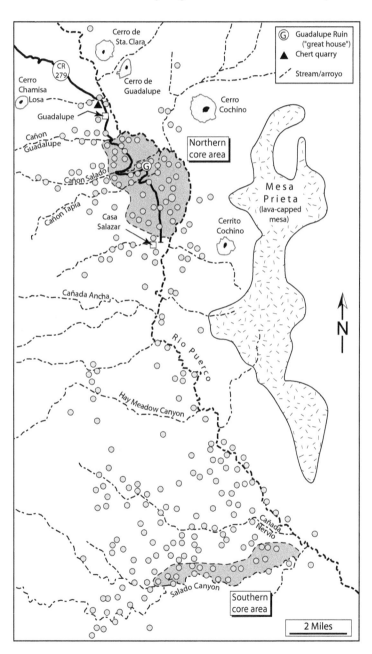

Figure 93. Ancestral Puebloan sites (gray circles) in middle Rio Puerco Valley. (Modified from Baker 2003.)

124

Prior to the middle of the 12th century the middle Rio Puerco Valley was an integral part of what is termed the Chacoan heartland (more in Chapters 25 and 26). Shortly after the Ancestral Puebloan culture centered at Chaco Canyon began to lose its vigor, ca. 1130 CE, the culture in the middle Rio Puerco Valley began to follow an independent course (Hurst 2003). As at Chaco Canyon, shortly after 1275 CE the Ancestral Puebloan culture in the valley came to a rather abrupt end. The late 13th century was characterized by regional drought. The agricultural carrying capacity of the valley was soon exceeded. The more arid climate, coupled with the precipitation occurring as violent summer storms, initiated the incision of arroyos and the drop of the water table, destroying the agricultural economy (Durand and Baker 2003). The valley was doomed. By the end of the 13th century both Chaco Canyon and the middle Rio Puerco Valley were abandoned. The valley would remain empty for the next 350 years. During this time the arroyos that had been cut in the late 13th century were backfilled to produce a restored level flood plain for the next wave of settlers, the Navajo.

The Navajo in the middle Rio Puerco Valley

In the mid-16th century Athapaskan-speaking people trickled into what would become New Mexico (Jackson 1996), and as early as 1639 people later known as Navajo appeared in the middle Rio Puerco Valley (Jersig 2003). The valley was a special place to the Navajo and it figures prominently in their creation myth. They established farms watered by flood irrigation on the flood plain, and raised corn, beans, and squash. They would plant their crops, then leave, and return in the fall to reap whatever had grown. By the mid-18th century the Navajo were raiding the Spanish settlements on the Rio Grande for livestock, and the Spanish returned the favor, with interest. Despite the mutual animosity, overcrowding in the Rio Grande Valley during the late 18th century caused the Spanish settlers to set their sights outward to the main river's tributary valleys, such as the middle Rio Puerco. Prominent Spaniards petitioned the governor for land grants, and a number were duly issued. The new settlers moved in and established small ranches.

Early Spanish villages and Miera y Pacheco's wonderful map

Four small villages were established in the valley in the 1770s as land-grant seats. They appear on the 1779 map of Miera y Pacheco (Figure 94, and section below). From north to south they were: *Ranchos de los Mestas* ("Ranches of the Mestas," an early settler family), *Porteria* ("Gatekeeper's House") at the foot of the volcanic neck named *Cabezón de los Montoias* (name taken from the Montoya family), *Guadalupe de los Garcías*, (name from the García family), and *Lagunitas* ("Little Lakes").

Miera y Pacheco's map is a story of its own. *Don* Bernardo de Miera y Pacheco was a Spanish army engineer, merchant, Indian fighter, government agent, rancher, artist and map maker. Originally from northern Spain, he had settled in El Paso by 1743. In 1756 the provincial governor, Juan Bautista de Anza (1778–1788), was concerned about the haphazard layout of the towns and settlements that

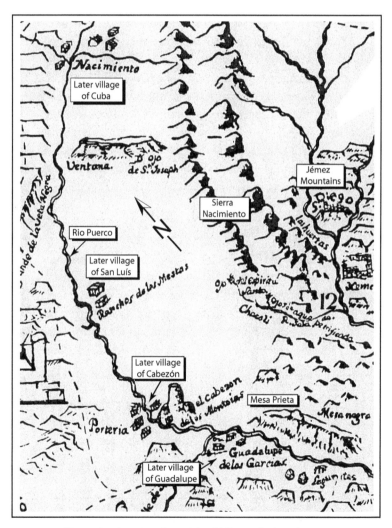

Figure 94. Annotated portion of Miera y Pacheco's 1769 map. (Modified from Kessell 1975.)

made it next to impossible to defend against raids from the nomadic tribes. However, no map of the area existed at the time on which to base policy. Recognizing *Don* Bernardo's talents, de Anza appointed him to map the province of New Mexico. Bernardo made maps of the province until he died in 1785. His most famous map, that of the *Interior Province of New Mexico*, was completed in 1779 (Figure 94; Kessell 1975). Sometime after its issue the map was lost. In the 1930s a Hispanic shepherd found it carefully sealed in a tin can in a cave beside the Rio Grande near the Civil War battlefield of Valverde, about 30 miles south of Socorro. The shepherd suspected that the document had value so he rushed it to the local Works Progress Administration (WPA) office, from where it made its way to Santa Fe (Forrest 1989). The map is a priceless find.

Each of the four valley communities cultivated a small patch of barely-marginal bottomland. At this time the river had not yet incised the modern channel, but rather flowed near or at the level of the flood plain, now a terrace. Diversion for irrigation could have been a relatively simple task, but due to undependable sporadic flows the area was used mainly for raising cattle (Rittenhouse 1965). The settlers immediately went head to head with the Navajos, who considered the Spanish as intruding on their turf. The Indians repeatedly raided the farms for livestock and the settlers appealed to the governor for help. A pattern of abandonment and reoccupation of farms ensued (Jackson 1996). By the late 1770s the settlers had had enough, and they left.

Later Hispanic villages

In 1846 the U.S. acquired New Mexico from Mexico. The U.S. military established a network of forts to contain the hostile tribes, but in defiance the Navajo increased their raids and terrorized the Territory. By 1864, after a scorched-earth campaign by the army under Colonel Kit Carson, the Navajo were defeated and marched off into four years of miserable exile. After that date the threats in the Rio Puerco became a thing of the past. In 1868 the Navajo were released from their internment. They moved to their new reservation in northwestern New Mexico (Chapter 24), and Hispanic settlers, many of them descendants of the original families, returned to the valley. They came from such places as Albuquerque, Atrisco, Los Ranchos de Albuquerque, Alameda, and Bernalillo, and they brought with them a 150-year-old farming/pastoral way of life (García 1994). The four Rio Puerco villages seen today all date from about 1872 and were established near their old 18th century sites on the river flood plain: *San Luís* at the former site of Ranchos de los Mestas, and sometimes called "Domínguez" after a local family; *Cabezón* at *Posteria*; *Guadalupe* (or *Ojo del Padre*, "Spring of the Father") at Guadalupe de los Garcías; and *Casa Salazar* (from the Salazar family) at Lagunitas.

Successful agriculture in the valley during this period required irrigation and irrigation required construction of diversion dams. The success of the settlements absolutely depended on the ease at which the flood plain could be irrigated. The elevation of the river bed relative to the flood plain therefore became a factor of paramount importance. In the early days one man could divert the river into the community ditch by placing a single cottonwood log across it (Jackson 1996). But times had changed. One dam at the ghost town of Cabezón was 34 feet high and 155 feet across and was built of logs, rock, and brush (Rittenhouse 1965).

The settlers raised beans, chili, corn, wheat, alfalfa, oats, hay, and vegetables. They grazed herds of sheep and cattle on huge parcels of open hinterland above the flood plain. Sheep thrived. For example, in 1891 the wool clip from Cabezón alone was 40,000 pounds. The huge load took 17 wagons drawn by 67 horses to get it to the wool warehouse in Albuquerque (Rittenhouse 1965). The farmers converted the valley to a lush garden. The vegetation in places was so dense and tall that farmers had difficulty finding their milk cows and work horses. In the late 19th century the valley was known as the "breadbasket of New Mexico" (Strong 1957).

For several years the communities did quite well, especially Cabezón. Two roads crossed at that townsite. One very old road—an Indian road—led from the Rio Grande pueblos, to Jémez Pueblo, and west to Zuni Pueblo. Later, in 1868, when the U.S. Army built Ft. Wingate near Gallup, the same route was used as a military route, as a stage route connecting the territorial capitals of Santa Fe and Prescott in Arizona, and, by 1875, a Star Line mail route (Figure 5). Cabezón had a post off ice and was sometimes called "La Posta." The town almost became a stop on a transcontinental rail line. In early 1880 the Atcheson, Topeka, and Santa Fe (AT&SF) Railway Company was advancing toward Albuquerque. The company planned a western branch from central New Mexico to California, but the exact take-off point and route were uncertain. A survey party was dispatched to determine a route from near Bernalillo on the Rio Grande to San Ysidro, Cabezón, and hence to San Mateo, northwest of Mt. Taylor. However, before they could finish their task they were called back because the railroad company was unable to secure the necessary land in Bernalillo for a diversion point (see Chapter 4). Instead the California branch was eventually built west from Belen, and the completion of the line in 1881 made the route through Cabezón obsolete. A second, much younger road was a mail route from Albuquerque to Nacimiento (now Cuba). It led northwest from Albuquerque to Casa Salazar. From there it followed the Rio Puerco to Guadalupe, crossed the river north of Guadalupe, proceeded to Cabezón, San Luís, and on to Cuba.(Rittenhouse 1965).

Loss of the range

But trouble loomed over the horizon. The Rio Puerco was changing. By the late 1880s the results of overgrazing and intensifying monsoonal rains led to severe arroyo entrenchment and lowering of the water table, accelerating a process that had probably begun about 1850. Irrigation ditches silted up and farmland on the river terrace was eroded away as arroyos widened their channels. Flash floods coupled with drought worsened the problem. The Rio Puerco channel at La Ventana was eight feet deep in the 1870s, but by the 1950s it was more than 50 feet deep.

The ruinous winter of 1928 destroyed about half of the local Hispanics' sheep. In 1905 the sheep rancher Frank Bond had begun operations in the Cuba area and by the mid-1920s had become extremely powerful. After 1928 he gained control of the range and rented sheep out on shares to the destitute sheepmen (Calkins 1937). Grazing intensified, erosion increased, the water table continued to drop and irrigation became even more difficult.

By the late 1920s grazing and wage labor had displaced agriculture as the primary component of the economy. By the Great Depression most wage labor had been wiped out and many of the inhabitants became dependent on government work programs (Forrest 1989). The town of Cabezón received a brief shot in the arm in 1930, when the first natural gas pipeline from the San Juan basin to the Rio Grande Valley was built by Southern Union a few miles to the north (see Utility Corridor below, and Figure 86). Many of the workers stayed at the town during the project (Looney 1968).

By the 1930s the land was in such bad shape that the federal government stepped in. The Taylor Grazing Act of 1934 (see below) restricted grazing on public lands. One of the large land grants used by the settlers of the valley for grazing was the huge Ojo del Espiritu Santo Grant. It soon became off-limits. Life in the valley was becoming tougher.

The Infamous Taylor Grazing Act (TGA)

For many years western stockmen used the unappropriated lands of the public domain as grazing pasture for their livestock. Although this land was the property of the U.S., the government did little to control its use. In time the stockmen came to believe that the public domain was theirs. By 1932 rumblings began coming out of Washington that the government was about to flex its considerable muscle and restrict access to the public domain. This thoroughly alarmed stockmen all over the West, including the settlers of the middle Rio Puerco Valley. Echoing a suggestion made earlier in 1929 by then-president Herbert Hoover, they countered that the public domain should be ceded to the states who could better care for the land. They suggested that the states then sell the land to the stockmen, with preference for those already using it. Some vocal, well-connected interests had been clearly eying the area. The New Mexico Cattle Growers' Association took a different approach by supporting the TGA, but wording it such that preference be given to those who had invested so much in the range and who had based their livelihoods on it—again those already using it. As the Depression worsened the stockmen had their financial backs to the wall and worried that a fee to use the public domain would ruin them. The tide was turning in favor of the growing conservationist movement in the country (especially with the inauguration of the conservationist from New York— Franklin D. Roosevelt). The concept of government regulation of the public domain had taken hold (Stout 1970).

The purpose of the Taylor Grazing Act of 1934, amended in 1936, was 1) to stop destruction of the public grazing lands via overgrazing and soil erosion, 2) to regulate the orderly use, improvement and development of the lands, and 3) to stabilize the livestock industry dependent upon the public range. The act authorized the Secretary of the Interior to establish grazing districts on the unappropriated and unreserved public domain and to issue grazing permits for those districts upon the annual payment of a reasonable fee. Recognizing that demand for the permits could exceed the carrying capacity of the range, Congress ordered the Secretary to give preference to those settlers, residents, and other stock owners who are within or near a grazing district. The permits were to be for a period of not more than ten years and renewable such that stock growers could plan their operations over a period of years and even secure loans using the permits as collateral.

Destruction of grazing land

Until 1934 the Ojo del Espiritu Santo Land Grant (Chapter 10) had been the property of the Catron family. For years the settlers of the middle Rio Puerco had grazed their animals on the grant, to the great irritation of the Catrons. In 1934 the federal government acquired the grant. After passage of the TGA the government repaired the fences around the grant and dispatched rangers to the valley to remove all their livestock from the grant.

The valley's residents found themselves owning more livestock than they could graze and had to drastically reduce their herds. Inability to pay grazing fees effectively excluded them from the grant in favor of those ranchers—often outsiders—having the necessary financial resources. Government "rangers" would come in and shoot cattle, with little explanation, when the residents were slow in adhering to the order (García 1992). With so few animals and the worsening condition of the range the settlers' livelihoods were doomed. The "great humiliation" foisted on the residents of the valley left a legacy of deep and enduring bitterness (García 1994).

The final straw

A major dam, constructed of brush and timber and serving the communities of Guadalupe and Casa Salazar, was washed out in 1936 (Widdison 1959). One story has it that a campfire got out of control and set fire to the flammable structure (García 1992). Irrigation thus was eliminated and from that point on the local residents had to depend on the capricious rainfall for their crops. The last dam was built later in 1936 at San Luís by the Civilian Conservation Corps from their nearby camp, SCS-8-N, under the supervision of the Soil Conservation Service (see Chapter 11). This was the only dam constructed in the valley with government assistance. It served only the San Luís area, and it was washed out for good by a large flood in 1951 (Widdison 1959).

During the late 1930s, men from the valley found employment with the Works Progress Administration (WPA), building roads around the town of Cabezón, but this only postponed the inevitable. By the end of 1957 all the settlements but San Luís had been abandoned. The arroyos are today rapidly widening and laterally eating up the old Rio Puerco terrace that was originally the rich bottom land that had sustained the settlers (such as a site one mile north of Guadalupe, Figure 95). In a century or possibly less the terrace may be completely stripped away.

A. Step by step.

B. The fence builder didn't envision this.

Figure 95. Rapid lateral destruction of valley-fill sediment of middle Rio Puerco flood plain. (Photos by author 2003.)

128

Side trip down into the valley via CR-279

At mile 41.6 on US-550, Sandoval County Road (CR) 279 takes off to the west and accesses the middle Rio Puerco Valley (Figure 86). This turnoff is one of those obligatory detours from our main route due to the unique historical and geologic nature of this area. The valley contains some of the state's finest (and in a way, saddest) scenery, four abandoned or nearly abandoned towns, and an Ancestral Puebloan ruin. The fence line along the north side of CR-279 just west of US-550 is the boundary between the Zia Indian Reservation (south) and the Jémez Indian Reservation (north). At mile 1.5 (there are no mile posts along CR-279), crossing the road from southeast to northwest is the bed of the abandoned Cuba Extension (Santa Fe Northwestern) Railroad from San Ysidro to the coal fields at La Ventana. At mile 4.5 the road crosses the entrenched Rio Puerco and at mile 4.9 it swings sharply to the south at the Red Mountain Ranch and begins its southerly trajectory down the valley.

Village of San Luís

For more than two miles south from mile 5.3 is the dispersed settlement of San Luís, strung out along the community's main, abandoned irrigation ditch (Figure 86). The flood plain here was once rather continuous and the town's layout was therefore little restricted. Today some of the buildings are derelict but others appear to be weekend hideouts. During the last few years the population of San Luis has increased a bit from only two families in the late 1950s. Much of the new "development" is in the Neo-American style, i.e., mobile homes. At mile 7.4 is the community church, *Inglesia de San Luís Gonzaga de Amarante*, built in 1917. The simple structure across the road on the right is the village's *Penitente morada*.

Utility corridor

From mile 8.1 to 11.2 a major utility corridor crosses the road (Figure 86). This three-mile-wide swath of power lines and pipelines is a vital artery of energy and energy resources coming from the coal-fired electric-generating plants and oil and natural-gas fields of the San Juan basin to the northwest to the Albuquerque metropolitan area and beyond. The lines consist of three high-voltage power lines, a natural-gas pipeline, and a liquid-petroleum gas (LPG) pipeline. The pavement of CR-279 ends at mile 8.5, where a side road takes off west to Torreón. At mile 11.7 the road swings sharply to the right (southwest) at the watering hose. Most of the residents of this area have to haul water to their homes from hoses such as this. At mile 12.3 there is a fork in the road. The road south, BLM-1114, goes past the ghost town of Cabezón and south to Cabezón Peak, the huge volcanic neck which dominates everything around here.

Ghost town of Cabezón

Figure 96. South view of ghost town of Cabezón. (Photo by author 2003.)

The ghost town of Cabezón is on private land (Figures 86 and 96). The Cabezón church, *La Inglesia de San José*, was built in 1894. The church's bell was hauled in by railroad and wagon from St. Louis, Missouri. By the 1950s the town was almost abandoned and the church became unprotected. For some time a number of property owners had maintained their homes for weekend or summer use, but while they were gone acquisitive visitors descended on the town. Around 1960 a resident surprised a group of armed young men while they were in the process of lowering the bell from the roof. They were driven off, and the bell moved to the church at San Luís for safe keeping. Also around 1960 a writer for the *Albuquer-*

que Tribune received a package in the mail. It contained a small metal crucifix. An enclosed note was written by a woman who was a former New Mexico resident. She said that she had read about the ghost town in the writer's column (probably in 1955) and later visited the place. She had taken the metal crucifix from the cemetery as a good-luck souvenir. However nothing but bad luck hounded her steps and she blamed the crucifix. She begged the writer to return the item to Cabezón and the writer did so (Bryan 1964). In the early 1960s the town was locked up to protect the town from the collectors, and it has remained so ever since.

In 1970 several scenes were shot here, with the unmistakable profile of Cabezón Peak in the background, for the western movie *The Hired Hand*, directed by and starring Peter Fonda. The movie, released in 1971, followed on the heels of Fonda's blockbuster flick *Easy Rider*. Evidently the latter film was too psychological for the times and it quickly bombed. After languishing for three decades in the photo archives the work was "rediscovered." It was carefully restored and re-released in 2003.

Cabezón Peak and Cabezón Wilderness Study Area

Cabezón Peak lies at the center of the newly-created, 210,000-acre Cabezón Wilderness Study Area (WSA). The WSA was established amidst a great deal of controversy. It is the first wilderness area established in New Mexico since 1984 when the Bisti and De-Na-Zin areas were set up in San Juan County. The very first place set aside to protect wilderness values had been in the Gila National Forest in southwest New Mexico, but the movement really gained traction after the Wilderness Act of 1964. The new WSA is generally bounded by the dirt BLM roads that encircle the mountain (Figure 86).

In 1992 the first Bush administration proposed the area as a WSA, but the idea laid inactive for almost a decade. The proposal was championed by such groups as the Forest Guardians and the New Mexico Wilderness Alliance. The proponents argued that because the land is part of the public domain administered by the Bureau of Land Management (BLM), policies should reflect the needs of all Americans and that it should be preserved untouched for future generations.

Opponents included the Sandoval County Commission and some 40 ranching families who live around the area and rely on grazing leases on the public land. Members of the Sandoval family have been particularly eloquent in their resistance. Helen and her brother Restie

Sandoval trace their roots in the old village of Cabezón back for generations. Helen, a former Sandoval County commissioner, argued that the proponents " . . . who are pushing it don't want it for the people of Sandoval County. They want it for the people of Bernalillo County and Santa Fe County—a place where they can go hiking with no one around." Restie Sandoval headed the Cabezón Water Pipeline Association, which maintains 29 miles of water pipelines that supply a series of stock tanks. His worries were that if the government eventually restricts access to the pipelines the ranchers will be driven out (Armijo 2001b).

Access to Cabezón Peak is via BLM-1114, exactly three miles to the south of CR-279 (Figure 86). At that point a dirt track leads east up the hill to a small parking area at the west foot of the peak. The route to the top begins from the parking area, swings around to the south side to the top of the intermediate platform, and then around to the southeast side. From there the summit is reached via a slot in the volcanic plug (Figure 85A). The ascent is certainly not for the faint-of-heart. I've been to the top several times, but would never attempt it alone because there are just too many opportunities for serious mishap. A broken ankle could mean your end, and only your bleached bones would be there to greet the next group of climbers! But with care, ample time, water, and reliable companions, the ascent is a real "high."

A world-class scenic overlook

At mile 12.3, where BLM-1114 forks to the south, the main road CR-279 continues west (Figure 86). At mile 17.1 there is another fork at which CR-279 veers to the south (left). At mile 18.7, just before CR-279 road begins its descent into the valley, is located on the left a blocked-off primitive road. A 250-yard walk on this road takes you to a wonderful view point overlooking the valley of the Rio Puerco and the main volcanic necks. This is one of the most impressive views in the state of New Mexico and superb picnic spot! However, the beauty of the moment fades somewhat when one ponders the fact that this geologic wonderland represents the remains of a destroyed agricultural province and a crushed people.

Ghost town of Guadalupe

This little ghost town, located at mile 26.7 on CR-279, was founded at the source of a dependable spring, the *Ojo del Padre*, and the buildings were all clustered near the spring (Figures 86 and 97A). The large, two-story structure on the east side of the road is the Juan Córdova house and store, built in 1900 (Figure 97B and 97C). The building once had a dance hall attached to the back, east side (García 1994). The pond on the north side of this building is fed by a pipe leading from the spring. Other nearby structures include the church, the dance hall, and the school, as well as a number of residences. All the adobe is returning to the soil. A few more modern structures seem to be occasionally occupied. According to the census of 1880 the towns of Guadalupe and Casa Salazar (down the road a bit) had 161 and 200 inhabitants, respectively. The 1910 census recorded 357 people for the two towns combined, so the population at that time was stable.

The community had a post office during two periods. From 1898 to 1905 the post office was named Miller, from John Miller, a cattle rancher from Casa Salazar (García 1994). Later, from 1940 to about the time of the town's final demise in 1958, the post office was called Ojo del Padre (Julyan 1996). However, the name Guadalupe persisted above the others.

Most of the people were dispersed up and down the valley, and would come to the plaza only for special occasions. The economy was based on farming and livestock. A strategically-placed dam across the Rio Puerco north of here (until its demise in 1936) permitted irrigation of such crops as corn, wheat, pinto beans, chile, melons, and pumpkins. The inhabitants of the immediate community got their drinking water from the Ojo del Padre. Those at a distance either gathered rainwater from their corrugated metal roofs or hauled water in barrels from the Ojo del Padre or from the Ojo del Espiritu Santo (Widdison 1959). The people dried food for the winter, and later, by the 1940s, canned their food. Wheat was taken to Jémez Pueblo for milling into flour. Surplus crops were sold to the merchant Richard Heller in Cabezón. And of course there was meat. Even during the Great Depression the people were never hungry. Guadalupe carried on a spirited rivalry with their nearby neighbor Casa Salazar on feast days, and there were dances in the dance hall (García 1992). Religious life was devout and rich. Life was good.

But then conditions deteriorated. The erosion of the land, the drought of the 1930s, the Taylor Grazing Act of 1934, the loss of the dam in 1936, the "great humiliation" caused by the enforced stock reduction, and the failure of many WWII vets to return to the land all took their slow-motion toll. In 1957, Sandoval County refused to provide a teacher and the school had to close (García 1992). In 1958 the post office and grocery store were also closed. The town was dead.

Dr. Nasario García, retired professor of languages at New Mexico Highlands University, was raised in this tiny village (but born in Bernalillo because his parents wanted him to be born in a "real" town!). He has written five delightful little books about Hispanic life in the Guadalupe area during the first half of the 20th century (Garcia 1987, 1992, 1994, 1997a, and 1997b).

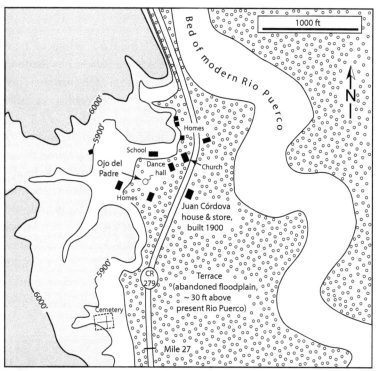

A. Town map. (Modified from García 1994.)

B. West view of Córdova House. (Photo by author 2000.)

C. North view of Córdova House and its backdrop of Sta. Clara Peak (left) and Cabezón Peak (center). (Photo by author 2000.)

Figure 97. Ghost town of Guadalupe.

South of Guadalupe, at mile 28.9 the road swings to the west to go up Cañon Salado (Figure 86). The relentless, post-1950s headward erosion of the Arroyo Salado has made this detour necessary. At mile 29.5 the road steeply descends and crosses the drainage at the bottom of the arroyo. This spot must be crossed with care. I personally would not try it without a four-wheel drive, high-clearance vehicle, especially when alone. It is the only place along CR-279 where a special vehicle is recommended.

At mile 30.5 there is a small parking area on the east side of the road at the foot of Guadalupe Mesa, just past the outcrop of a nicely-exposed, vertical igneous dike (Figures 86 and 98A). The dike was intruded into the Cretaceous-age sedimentary rocks, probably between 3.0 to 2.7 Ma, and exhumed via subsequent deep erosion. An Ancestral Puebloan ruin called the Guadalupe Ruin, not visible from the road below, sets on top of the mesa. This is BLM land and the little signs indicate that the ruins are "protected." The ruin is a one-storied and E-shaped, and consists of about 50 rooms and three kivas (Figure 98B). The site was first identified as "Anasazi" in the mid-1950s. Excavation was conducted from 1973 to 1975 by Eastern New Mexico University and stabilization was completed in 1982.

The "big house" was first built about 960 CE in the midst of an existing community and served a ceremonial purpose rather than habitation. The site is surrounded by many smaller structures that likely served as domiciles for the regular population (Stuart and Gauthier 1981). This "great house on the hill" was used until sometime after the Chacoan collapse ca. 1130 CE. From the late 12th century until abandonment of the valley after 1275 the site was occupied by a different people, probably from Mesa Verde (Baker 2003).

The little unexcavated kiva on the mesa's south, lower-down promontory was constructed of basaltic (dark, fine-grained volcanic rock) cobbles and boulders. The companion mesa on the west side of CR-279 is capped by a deposit of about ten feet of similar cobbles and boulders. The top of Guadalupe Mesa was also once capped by this material but most of it was removed by the Ancestral Puebloans before they erected their structures. The tops of the two mesas were once part of a continuous land surface on which these cobbles and boulders were deposited. The basaltic debris was transported and rounded by an ancient stream system that predated the Rio Puerco and flowed at this high level. These streams drained the volcanic terrains near Mt. Taylor.

It is quite possible that the Guadalupe-area Ancestral Puebloan community was a link between the turquoise mines near Cerrillos (northeast of Albuquerque) and the main power center at Chaco Canyon (Figure 99). By 1000 or perhaps as late as 1020 CE, Chaco Canyon had been firmly established as the center for finished turquoise, serving much if not most of the San Juan basin. The finished product thus became an integral part of the Chacoan economy. To secure its control of the industry, Chaco Canyon would have had to control the source of the material. The Guadalupe area may have served as the conduit for the transport of bulk turquoise from

A. East view of Guadalupe Mesa, with Cerro Cochino in left distance and Mesa Prieta in right distance. (Photo by author 2000.)

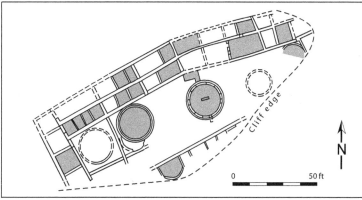

B. Plan of Guadalupe Ruin on top of Guadalupe Mesa. (Modified from Baker 1984; Stuart & Gauthier 1981.)

Figure 98. Guadalupe Mesa and Ruin.

Cerrillos to Chaco Canyon (Frazier 2005). Today it is simply impossible for us to imagine sandal-clad men hauling bulk rock on their backs all these many miles. There must have been an impressive organization in place to supervise the human caravans, maintain records of some kind, to supply the necessary water and food for the men, and to keep the men moving when they wanted to kick back.

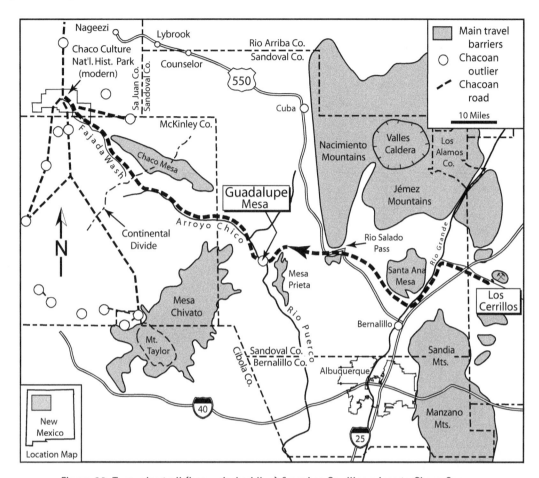

Figure 99. Turquoise trail (heavy dashed line) from Los Cerrillos mines to Chaco Canyon.

During the height of Chacoan culture, the "Chaco Phenonemon" (ca. 900–1125 CE), Chaco Canyon was the hub of a vast network of more than 80 "outlier" villages, including Guadalupe. In many cases the hub connected and communicated with its outlier settlements via an array of roads radiating from the canyon (Figure 99). However, there is no trace of a Chacoan road connecting Guadalupe Ruin with Chaco Canyon. It is quite possible that none was needed because of an existing, quite natural route. Travelers from the Guadalupe outlier could ascend the Rio Puerco to the junction with the Arroyo Chico, ascend the Chico and its North Fork nearly to the continental divide, cross the divide to the headwaters of the Fajada Wash, and descend the wash directly into Chaco Canyon.

Casa Salazar

At mile 33.0 is a locked gate (Figure 86). CR-279 ends here. In the near distance is the tiny village of Casa Salazar. Miera y Pacheco's 1789 map shows this community as *Lagunitas*, or "Little Lakes" (Kessell 1975). A century later it was renamed after the Salazar family. The town was located on both sides of the river and was rather spread out because the arable land was also spread out. The community was served by a post office from 1888 to 1919 (Julyan 1996). Today only a few people live here full time and tend to their livestock. On the hilltop overlooking the village on the west are the home of a Mr. Romero, since deceased, and his family's one-room chapel.

10

Twisted Saga of the Ojo del Espiritu Santo Land Grant

At mile 28.7 (south of the Rio Salado anticline), US-550 enters the southeast corner of the old Ojo del Espiritu Santo Land Grant. Today this is the eastern border of a segment of Zia Pueblo land, and it is appropriately marked by a highway sign saying "Entering Zia Reservation." However, there's also a little item of historical interest here. About 20 feet east of the highway sign, just inside the right-of-way fence, is a lonely white pillar (a "mini Washington Monument," lower inset in Figure 100). It was installed during the 1930s by the Highway Department to mark the end or beginning of major, usually federally-funded construction efforts. On its sides, in fading paint, the vertical inscription "NRS - 203D" probably indicated a New Deal, National Recovery (NR) project.

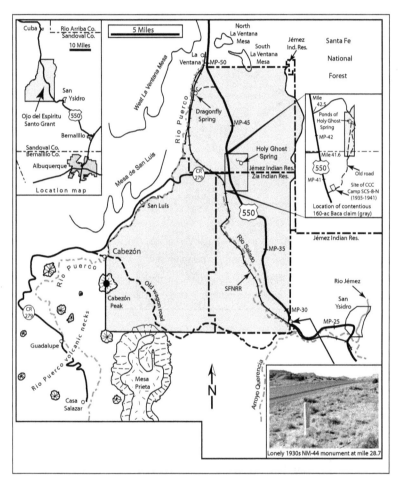

Figure 100. Ojo del Espiritu Santo Land Grant (gray). (Modern boundaries of Indian lands shown for reference.)

At mile 42.5 a road off to the east leads to Holy Ghost Spring on Jémez Pueblo land, at the foot of the Sierra Nacimiento (upper right inset in Figure 100). According to Julyan (1996), many years ago a member of a Spanish expedition saw columns of steam rising from a body of standing water and stumbled back to camp muttering, "El Espiritu Santo" (Spanish for "the Holy Spirit" or "Holy Ghost").

The *Ojo* (literally the Spanish word for "eye," but in New Mexico more commonly used for "spring") del Espiritu Santo became the focal point for a huge, undefined chunk of grazing land. In fact the value of the land rested fully on this dependable source of water. The spring eventually was to give its name to a huge and infamous land grant. The struggle for control of this grant raged for over a century and a half, beginning in the 18th century. In 1766 the Spanish governor granted the grazing rights to the Ojo del Espiritu Santo area to Jémez, Zia, and Santa Ana Pueblos (McKenna 2006). The name "Espiritu Santo" appears on *Don* Bernardo Miera y Pacheco's famous map of 1779 (Figure 94). Clearly, at that early date it had been a well-known source of sustenance.

Cabeza de Baca family

The family name of Cabeza de Vaca, goes back to 1212 CE. At that time the Spaniards were involved in their campaign to drive the Moors from Europe's Iberian Peninsula. To guide his men through a mountain pass a soldier posted a sign consisting of the head of a cow on a stake. For this deed the king of Navarre knighted the soldier and gave him the name, *Cabeza de Vaca* (Spanish for "Head of a Cow," Perrigo 1985). The family became a favorite of the royalty and many members became wealthy.

In 1528 a certain Álvar Núñez Cabeza de Vaca accompanied the Narváez expedition to Florida. After a shipwreck he and three others made an epic eight-year trek across the American Southwest and eventually got back to Mexico City. His tale attracted a great deal of attention. In 1539, the Negro slave Estevan, one of the members of the first expedition, led the Franciscan monk Marco de Niza north to Arizona. From afar the impressionable de Niza saw the sun-lit pueblo of Hawikuh, on the present Zuni Pueblo in west-central New Mexico, and rushed back to New Spain with wild tales of the fabulous Seven Golden Cities of Cíbola. This provided the irresistible bait that would lure Coronado north in 1540–1542 for the Spanish *Entrada* into New Mexico.

In the early 17th century Cristóbal Vaca arrived in New Mexico to join Oñate's colony. In Spanish the "v" and "b" are pronounced very much alike and soon "Baca" replaced "Vaca" in common usage. The original name, Cabeza de Vaca, was not used by the family for the next two centuries. Around 1670 a group of Bacas and others founded the town of Peña Blanca (Chávez 1992). Luís María Baca was born in 1754 in Santa Fe. In time he became a *rico* (a rich guy) involved in land and livestock activities in Peña Blanca. He eventually ran through three wives and in the process fathered some 23 children. By 1803 he was an influential man. Also in 1803, possibly to augment his prestige, Luís María evoked his lineage and from that year on he signed his name, "Luís María Cabeza de Baca" (Chávez 1992).

Ojo del Espiritu Santo Land Grant

Luís María must have been quite a character, and certainly had a colorful background. In 1815 he applied for the Ojo del Espiritu Santo Land Grant and he was granted it that same year (Figure 100). He moved to the site with his third wife and children but they were driven off by the Navajos. Instead he opted to look elsewhere for grazing pasturage. In 1821 he and multitude of sons and sons-in-law applied for a huge tract, some 500,000 acres, east of Santa Fe on the Pecos River, and received the grant in 1826. He built a house in Las Vegas and moved there with his family, but raids by Plains Indians forced them back to Peña Blanca, where in 1827 he was killed during a row with Mexican soldiers (Perrigo 1985).

The Ojo del Espiritu Santo Land Grant was approved by the New Mexico Surveyor General for the Baca heirs in 1860. The huge tract, with 113,141 acres, was defined by the following limits:

On the east by the Sierra de Jemez [Sierra Nacimiento];

On the west the Puerco river and the point of Prieta table land [=Meśa Prieta];

On the north the table land commonly called La Ventana [South La Ventana Mesa?];

On the south by the cañon of La Querencia and the boundary of the lands of *Don* Antonio Armenta (U.S. Court of Private Land Claims 1860).

Struggle of Cabeza de Baca vs. Catron

Beginning a few years before the approval of the grant, the plot thickens. I'm sure a lawyer would revel in this issue, but I will simply attempt to simplify here the twisted tale based on the correspondence of the Ojo del Espiritu Santo Company (OESC) housed at the Center of Southwest Research at the University of New Mexico (Ojo del Espiritu Santo Company, undated file box).

In 1857 several claimants made power of attorney to Tomás Cabeza de Baca. He was charged to employ an attorney named James S. Watts to argue the heirs' case for the grant, in exchange for a one-third interest in the grant in lieu of cash. In 1864 a second power of attorney was made by some heirs and grantees authorizing Tomás to "attend to and manage all matters or business relatory to the interests of said heirs of *Don* Luis M. Baca . . . " and to deed the one-third interest to Watts. The deeding was carried out later that year.

Meanwhile the grant was being used by others. By the mid-1850s *Don* José Leandro Perea, powerful *patrón* of Bernalillo (Chapter 4), ran 50,000 head of sheep on the land during the winter and by the mid-1860s he and others tripled that number (Bayer 1994). In 1872 the large interest in the grant controlled by Tomás was acquired by Thomas B. Catron (more about him below). The next year Watts, by now a judge on the New Mexico Supreme Court, sold his interest to a third party who, in 1881, sold it to Catron. From this point on the name Catron would be closely intertwined with the Ojo del Espiritu Santo Land Grant.

In 1874, Tomás Cabeza de Baca and his wife gave their interest in the grant to their three sons, and in 1877 Catron took a mortgage on the interest from the three children in lieu of a fee to represent them in a legal case. In 1881 the sons conveyed their interest back to their widowed mother, in 1885 the widow conveyed the interest to Catron's law partner, and that same year the partner conveyed it to Catron. This issue was now becoming very murky.

In 1877, José Leandro Perea and Thomas B. Catron had filed a lawsuit, requesting partition of the grant. (In an "unpartitioned" grant no one heir could claim title to a specific parcel or fixed percentage of the whole, but rather owned a percentage "in common" with the other heirs.) The case lingered until 1899. In June of that year a deed was recorded by Tomás Cabeza de Baca as attorney to Catron. Catron claimed that the deed had been executed by Mr. and Mrs. J.S. Watts and by Tomás Cabeza de Baca under power of attorney almost 27 years earlier, in October 1872.

Thomas B. Catron had dominated business and politics in New Mexico for 50 years. He was born in Missouri in 1840. He fought with the South during the Civil War and acquired his mental toughness from that experience. After the war he studied law. In 1866 he traveled the Santa Fe Trail to New Mexico and studied Spanish on the way. His father had given him two prairie schooners, loaded with flour, with two eight-mule teams. In Santa Fe he sold the wagons, teams, and flour for $10,000. In 1866 Catron moved to Mesilla in southern New Mexico, where he learned the legal business and began a career of land acquisition. In 1875 he met a school teacher in Mesilla, 17 years his junior. They married in 1877 and moved to Santa Fe. They had four sons and spared no expense in educating them with the aid of carefully selected tutors. His young wife loved to travel and in 1892 took her sons with her to Europe. For the next 16 years Catron's family spent long periods in Europe and he rarely saw them. In 1908 his wife returned home to Santa Fe, her health broken, and she died the following year (Westphall 1973).

By this time the older son, Charles C. Catron (1879–1951) had been admitted to the New Mexico Bar and by 1914 became treasurer of the Ojo del Espiritu Santo Company. A few years later Charles became vice president and his younger brother, Fletcher A. Catron (1880–1964), became secretary. The other two brothers, John S. Catron (1878–1944) and Thomas B. Catron, Jr. (1888–1973) were part owners of the company but did not assume active roles in its business.

Thomas B. Catron, Sr. could be magnanimous to his friends and ruthless to his enemies. He was a member of the so-called Santa Fe Ring, a loosely organized group of lawyers, politicians, and businessmen—both Anglos and Hispanos—with similar interests. Catron used his legal knowledge to leverage his way into the acquisition of some 3,000,000 acres of land in New Mexico, most of it in land grants. His aim was to create an empire for himself.

Throughout his life he loaned money to others. Frequently land was put up as collateral on these loans and he then had to accept the collateral in lieu of a cash fee. Whenever the opportunity offered itself he acquired interests in land grants. He was loath to sell property and, finding himself often land-poor, constantly had to sell off tracts for cash (Westphall 1973). Due to his strong-arm methods and questionable ethics, his name to this day is anathema to the people of small villages of northern New Mexico.

In 1854 Congress had created the office of Surveyor General, and later that year the first Surveyor General arrived in New Mexico. His first order of business was to sort out the territory's convoluted land-grant issue. Claimants to land grants were required to prove their ownership to the Surveyor General, who would then approve or disapprove them. If the Surveyor General approved the grant it would be recommended to Congress, and Congress would then make the final decision. During all this time the claimants were forced to hire attorneys to argue their case in court. Often they lacked the cash for legal fees and instead paid the lawyer with an interest in the grant. Via this method savvy lawyers, such as Catron, acquired huge land holdings in the territory.

About 1869 one of the land grant heirs, Diego Cabeza de Baca, probably a grandson of *Don* Luís María, had built five houses and a chapel for himself and his kids near the Ojo del Espiritu Santo spring. He believed that he owned a $1/128^{th}$ unpartitioned interest in the grant. Another player, Ciriaco Cabeza de Baca, joins the cast here and was probably a son of Diego. The OESC maintained that Diego, through his attorney, had deeded his interest in the land to Catron in 1872, but Ciriaco insisted that the deed was never recorded and that Diego probably never knew of it. Ciriaco and his father had paid all the taxes on the property. They made a lake out of the spring, constructed and maintained a road part of the way from San Ysidro to Cuba, and constructed an acequia extending out from the Sierra Nacimiento. Ciriaco and his father owned only a small part of the grant but they emphatically claimed a 160-ac area around the spring. The importance of that well-watered site cannot be overemphasized. It was the most important source of water on the entire grant and its control ensured control of the whole thing.

Ojo del Espiritu Santo (Catron) Ranch

In 1903 *Don* Pedro Perea (nephew of José Leandro) and his wife revived the partition suit, which had been pending since 1877. The grant was finally partitioned and sold to the highest bidder—Pedro Perea and his wife—and the funds distributed to the many owners. Also in 1903, the Ojo del Espiritu Santo Company was incorporated and established its main office in Bernalillo (the office was moved to Santa Fe in 1915). The Pereas promptly deeded the grant to the new company. Among the owners of the company, in addition to T.B. Catron, were the senior Catron's sons Charles C., Fletcher A., and Thomas B., Jr.

Diego died in 1910, leaving no will, and Ciriaco continued to occupy the land near the spring, claiming ownership of the area based on the principle of adverse possession, which required a ten-year occupancy. However, questions arose whether Ciriaco truly occupied the site. OESC insisted that, unlike Diego, Ciriaco did not, and in fact lived in Rito de los Pinos (La Jara), north of Cuba and moved onto the site only in about 1915. Ciriaco, for his part, insisted that he had occupied the property continuously.

Now the Sánchez family enters the picture. Manuel Sánchez y López had married one of Ciriaco's daughters in 1908. Manuel (according to a handwritten note by an unknown author in the OECS files) claimed that when he married, Ciriaco had been living in Rito de los Pinos and had been there for more than ten years. He added that Ciriaco came to the spring shortly after Diego died in 1910, and that Ciriaco had grazed sheep all over the grant between 1915 and 1917. The Cabeza de Baca and Sánchez families never gave up on their claim to the land immediately surrounding the all-important spring. The families paid taxes on the tract and persistently maintained that the 160 acres around the spring was theirs (upper right inset of Figure 100).

The pace of events quickened in 1917 when OESC filed an injunction to Ciriaco, in effect kicking him off the grant. The company charged Ciriaco of trespass and insisted that his inheritance had been sold. Ciriaco said that OESC did not physically possess the tract and that therefore he could not be trespassing. He also brought up the awkward question of why T.B. Catron had waited 27 years before recording the supposed deed from Tomás Cabeza de Baca. Catron summed up what he considered his history of acquisitions thusly: one third of the grant from

J.S. Watts, the inheritance of Tomás Cabeza de Baca, and about one-half of the grant from several of the grantees and heirs via power of attorney.

The case went to the District Court in Santa Fe in 1917, and a jury ruled in favor of OESC. Baca requested a new trial, citing numerous irregularities in the proceedings. In 1918 the case went to the New Mexico Supreme Court and in 1923 the high court upheld the earlier ruling. That should have been the end of the matter, but as we'll see, it festered.

Even before the trials, the OESC began upgrading the ranch. Apparently the property was never intended to be a working ranch. The Catron brothers were not ranchers. The two most active in the ranch matters were Charles and Fletcher, both of whom were lawyers and had attended prestigious universities. Thomas was a graduate of West Point and was a career military man. During this time the health of T.B. Catron, Sr. was failing, and his sons, especially older son Charles, struggled to save his crumbling land empire. T.B., Sr. died in 1921, but from some time prior to his death, son Charles had run the ranch. Charles saw the ranch in quite different terms than his father. T.B., Sr. envisioned the grant as part of an empire, but Charles saw it merely as a potential source of income via exploitation of the ranch's natural resources. These resources were grass for grazing, gypsum, and perhaps oil and natural gas.

Of immediate importance was to protect the grant's rich grass resource from others. Trespassing livestock were gobbling up the grass, but by 1916 about ten linear miles of the grant had been fenced in. In 1918 a five-room adobe ranch house was constructed at the Rito Semilla (the acequia that the Cabeza de Bacas had built) near the spring, and the following year the ranch manager, Ted Sparr, from San Ysidro, moved in. (He would remain the manager until 1934.) By 1920 the grant had been completely fenced and the company was advertising grazing land in the *El Paso Morning Times*.

Despite Charles' best efforts the company was always in dire monetary straits. The ranch was not earning a dime. Income-tax filings and payments were constantly late. In 1922 Sandoval County placed a lien on the property for unpaid taxes. OESC chafed at this because they believed that the tax rate was twice what grazing land was worth. The company was borrowing money to pay expenses. That year the OESC began shopping the ranch to oil concerns for leasing, and even offered it for sale, at $5/

acre. Several geologists had studied the area and detected three anticlines, or "prospects" (anticlines make excellent drilling targets for oil and natural gas). Two wells, the Conkling well on the "Ventana" structure on the northern edge of the grant, and the Kaseman No. 2 well on the "Jémez" structure, a.k.a. the Rio Salado anticline, were drilled during 1925–1926 (Chapter 8). The company believed in the prospects, and the Catron family and family friends invested heavily in them. However, both wells were unsuccessful. A description of the company in 1925 read: "OESC is an inactive company which leases grazing land. Such lands now have no value. OESC has never had a net income. Value of company stock is speculative. No stock has ever been sold or trafficked." The company continued its efforts to shop the ranch out for oil exploration but they never again received a serious nibble.

By 1926 the financial problems of the ranch, as well as other investment activities, were getting to Charles. He wrote to his brother Tom: "Sometimes when I think the whole matter over I get a sickening feeling in the stomach, not a belly ache, but just a sickening feeling. At other times prospects look good, but I assure you that working and living under the oppression of debt such as I have been doing for the last five years, using my own funds to carry matters on is making an old man of me fast" (letter, Charles C. Catron to Thomas B. Catron, Jr., September 22, 1926). In 1928 Charles mortgaged the ranch to a wealthy businessman named Bernard B. Jones of Washington City (as Washington D.C. was then called). From that day on the Catrons struggled mightily to keep Jones at bay. In 1929 Charles left the company to became a justice on the New Mexico Supreme Court where he served through 1930. Meanwhile Fletcher saw to the affairs of the company until Charles returned.

In late 1929 the nation entered the Great Depression. Everyone was desperate for cash. By 1930 even the Bernalillo Mercantile store was begging the ranch for payment on debts, and Sandoval County threatened another lien for payment of back taxes. By 1931 Charles began a major effort to unload the ranch. The company didn't even have money to pay poor Ted Sparr, but allowed him to live in the ranch house for no salary. Charles became desperate to sell the ranch to pay off the mortgage and interest to B.B. Jones. Jones was rapidly losing patience and strongly suggested that the OESC sell the ranch at a reduced price to pay off the $120,000 due by that time. Charles deftly held him off by claiming he was trying as hard as he could to sell the ranch. Charles dropped the asking price from $5/ acre to $4, then to $3 and then to $2. Price didn't seem to be the issue. The real issue was that no one had cash. Now thoroughly exasperated, he suggested that Jones foreclose on the mortgage and they could fight it out in the courts for a year or more, or that he could assent to sale of the grant for $50,000.

Brother Tom B. Catron, Jr., never really actively involved in the operation of the ranch and now a major in the U.S. Army stationed in Ft. Benning, Georgia, interjected in early 1932 with a bizarre plan to "save" the ranch. He suggested turning the place into a health and recreation resort. He claimed to have an army buddy with disposable cash who was ready to settle down. Proposed activities near the "main plant" (the area near the ranch house) would be riding, polo, golf, tennis, hot and cold swimming, flying, bowling, etc. (letter, TBC Jr. to CCC, January 29, 1932). Charles was not amused. We can imagine his curled upper lip when he responded dryly to brother Tom: "Personally, I am not particularly interested in devoting thought to your plan. I am willing to accept any plan that anyone else has worked out, provided the damn Jones mortgage can be paid off" (letter, Charles C. Catron to Thomas B. Catron, Jr., February 2, 1932).

Charles, now desperate, pulled out all the stops. In early 1932 he wrote a letter to the famous Will Rogers in Hollywood, whom he had met before. Rogers had acquired some ranch land in the Roswell area. Charles broke the ice thusly: "Hello Cowboy. Say man, I been reading some of your skin brands and you sure keep your iron hot. Hog-tie them and burn em deep and you wont lose no stock. Hair brands ain't good."

Then he cut to the chase: "Well, Pal, you ain't seen nothing yet. I am sittin on 113,000 acres of sweet grass and water, all stock wired with good ranch house, barns and corrals, and all ready for some good she stuff. Am holding this down for some city mavericks what dunno what they got or maybe they're plumb scared. . . . Now, Bill, you just come out here and let me show you this layout, and you kin buy her so cheap you'll think they're all crazy. That's just what I think, only I ain't got the money to buy her . . . Well, come on and look er over, and if you buy, I'll make me some money too and get me a small ranch of my own. Maybe we'll be neighbors . . . Let me know when you are comin and I'll show you plenty. An we can set on the fence and whittle" (letter from Charles C. Catron to Will Rogers, January 16, 1932). This folksy spiel from the pen of a worldly city lawyer and ex-justice of the New Mexico Supreme Court! Evidently Rogers didn't take the bait, but he must have grinned mightily, or probably even enjoyed a really good guffaw.

At about the same time Charles wrote to a Mrs A. Semple McPherson of the Angeles Temple in Los Angeles, offering the ranch for $2.50/acre to establish her gospel city, which apparently she was proposing at the time, on the grant (letter from Charles C. Catron to Mrs. A.S. McPherson, February 15, 1932). Nothing came of that bizarre idea either. An attempt to entice the medical community to establish a tuberculosis recuperation facility on the grant also came to naught (letter from Charles C. Catron to Dr. W. R. Lovelace, May 24, 1932). As Charles frantically attempted to sell the grant into 1934, the financial pressures increased, debts and taxes went unpaid, tempers flared, and poor Ted Sparr struggled to keep trespassers off the grant. Things were rapidly falling apart.

The federal government takes over

With Charles' anxiety increasing to fever pitch, the U.S. government entered the picture and expressed interest in acquiring the grant. In 1933 the newly-appointed Commissioner of Indian Affairs, John Collier, began a campaign to acquire eroded lands for Indian use. Relying on the Taylor Grazing Act of June 1934, which authorized the government to regulate the use of public rangeland in the West, Collier began to look at sub-marginal, degraded Spanish land grants to benefit various Indian pueblos. Collier contacted Charles and by February the BIA was a prospective buyer of the grant. In October the BIA approved purchase of the grant for $2.50/acre, and Charles used the cash to pay off his creditors. Out from under the onus of the ranch's debts, Charles C. Catron went on to enjoy a distinguished legal career. He died in Santa Fe in 1951. His brother and junior partner in the OESC, Fletcher, also a prominent attorney, died in Santa Fe in 1964. The military brother, Thomas, died in Baltimore in 1973. Ranch manager, Ted Sparr, died in San Ysidro in 1939. The Catron era was over.

Final return of grant to pueblos

In 1938 the government set aside a large portion of the Ojo del Espiritu Santo Land Grant for Indian use. In 1954 the *New Mexico Stockman* published an article that stated that a huge amount of land acquired by the government in the 1930s might be disposed of by the Department of Agriculture. This greatly alarmed the Jémez and Zia Pueblos (Santa Ana Pueblo had long ago directed their attention east to the Rio Grande Valley), because part of this land included the Ojo del Espiritu grant, which they had understood had been transferred by the BIA to the Department of Agriculture and had been set aside for them for grazing. The ruckus that developed caused the government in 1956 to transfer two segments of the grant in trust to each of the Pueblos of Jémez and Zia.

Thus ended almost two centuries of struggle for this tract by the Indians. They had recovered the Ojo del Espiritu Santo Land Grant (McCarty 1969). What goes around (sometimes) comes around! Jémez Pueblo received the spring, and Zia Pueblo the Rio Salado anticline. With today's windmills and stock tankage the grazing value of the land is no longer dependent on the spring itself. Jémez Pueblo has converted the spring into fishing ponds, known today as the Holy Ghost Spring Recreation Area, but it has been closed to the public for several years. For many years the Sánchez family tenaciously maintained that they still owned 160 acres around the spring. Until about 2000, ruins of the Baca/Sánchez houses could still be seen near the spring, but sometime after that they were torn down.

11

Forgotten Civilian Conservation Corps (CCC) camps

Two Civilian Conservation Corps (CCC) camps were once located along old NM-44 (Figure 101). Each was once a beehive of activity, with supply trucks and hundreds of men coming and going. Barely a trace remains today, and the camps are all but forgotten, except for a few old-timers who once worked there.

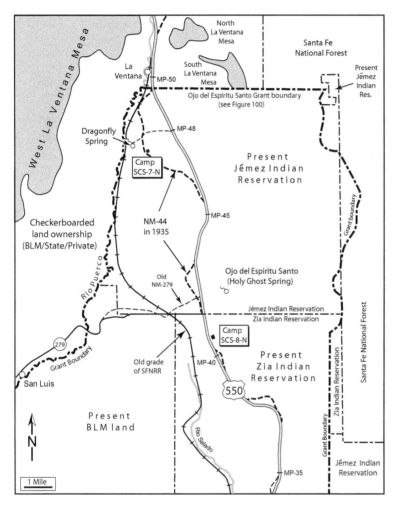

Figure 101. Location of 1930s Civilian Conservation Corps (CCC) camps.

What was the CCC?

The CCC was one of those very rarely successful results of federal government involvement in American life. It was one of many government programs enacted during the Great Depression of the 1930s that were known mainly by their initials, thus the term "alphabet soup" to characterize the lot. The CCC was in existence for 9.25 years, from early 1933 until mid-1942 when the U.S. geared up big time for its involvement in World War II.

The precursor of the CCC was established in 1933 by the newly-inaugurated president Franklin Delano Roosevelt (FDR) during the depths of the Depression. He realized that there was more at stake than the nation's fallen economy, as terribly serious as that was. He also realized that the nation's self-confidence was at risk. A generation of destitute young men was discouraged, undernourished, and undereducated, and in danger of being squandered. It must be pointed out that the CCC was a work program—not a welfare or make-work program. It was designed to marshal the latent energy of its young men, to give them a sense of purpose, to build them physically and morally, and at the same time to restore the public lands that were in dire shape due to years of erosion and neglect. It was the quintessential win-win situation.

Congress approved funding on March 31, 1933 for a period of two years. The new organization was called the "Emergency Conservation Work" (ECW), and only later, in 1935, would it become known as the "Civilian Conservation Corps" (CCC). The program was intended to be temporary and not to interfere with normal employment. Later, when the program proved to be so popular and successful, Congress approved additional two-year funding cycles.

To avoid creating a new bureaucracy FDR used existing federal entities. The U.S. Army was given the task of operating and supervising the camps, which eventually numbered some 4,500 throughout the nation. At the time the country was divided into nine U.S. Army "Corps" areas. New Mexico, Arizona, Colorado, Texas, Oklahoma and most of Wyoming made up the 8th Corps area. The enrollees were organized into "companies" of 200 men. The New Mexico companies were given sequential numbers, preceded by the corps number (e.g., 836). The number usually stayed with the company despite moves and rotation of enrollees through its ranks.

Enrollees came from the ranks of poor, unmarried, and for the most part white young men, age 18 to 25 (later changed to 17 to 28), whose families were on relief. They were called "Juniors" or "boys." They were paid $30/month, $25 of which was sent home. This caused some friction with the Regular Army because a soldier at that time made only $21/month. A second group of enrollees consisted of local experienced men, or LEMs, who served as project leaders in the Junior camps. A third group consisted of WWI veterans. In general, where black men were present the camps were racially segregated. This was the 1930s. The CCC was not an utopia.

Enlistment was for a six-month "period," with the option of reenlisting, for a total of two years. The initial Period 1 ran from April 31, 1933 to September 30, Period 2 ran from October 1, 1933 to March 31, 1934, and so on for a total of 19 periods.

After receiving initial training, vaccinations, and clothing, each 200-man company was sent by train and/or truck to an assigned camp location. Once on site the responsibility for planning and supervising the work programs was given to those existing national agencies within the Departments of Interior and Agriculture responsible for the public domain. These included the U.S. Forest Service, the U.S. Department of Grazing Service, the Soil Conservation Service, and others. Each camp had a letter designation representing the sponsoring agency, such as "F" (Forest Service), "DG" (Department of Grazing), and "SCS" (U.S. Soil Conservation Service). Each camp also had a sequential number indicating the order in which it had been planned. Finally each camp had a trailing letter indicating the state. The camps along NM-44 were SCS-7-N and SCS-8-N.

In New Mexico, over the life of the program, more than 100 camps were established in about 85 general locations, although all were not active at the same time. The number of simultaneously active camps varied greatly but generally ranged from 30 to 50. Each camp had a company commander (who was a regular army or reserve officer), a junior officer, an education advisor, and a camp doctor. A project superintendent represented the technical agency sponsoring the work.

The average camp had about 24 pre-fabricated buildings, including offices, barracks, officers' quarters, mess hall, education and recreation buildings, infirmary, showers, latrine, motor pool, fueling stations, etc. Each camp required a communication and supply link via a main highway or railroad, and a nearby post office. Over the course of the program about 55,000 men served in New Mexico camps, over 32,000 of them state residents. Northern New Mexico was administered by the Albuquerque District, which had a big warehouse in Albuquerque on Mountain Road at the Santa Fe Railroad tracks, and a vehicle repair complex on north 4th Street.

CCC vs. WPA. The CCC should not be confused with another work program called the Works Progress Administration (WPA). The WPA was created in 1935 and focused on adult, often married men who usually worked on municipal projects and generally lived at home rather than in camps. During the late 1930s men from the middle Rio Puerco villages (Chapter 9) "commuted" during the week to camps located first at San Luís and then at Cabezón. With horses and scrapers they constructed the road (now much of CR-279) south from San Luís, to Cabezón, and then west toward San Mateo northwest of Mt. Taylor (García 1994).

CCC camps along old NM-44

During the 1930s the middle and upper reaches of the Rio Puerco watershed was suffering from the combined effect of drought and overgrazing. Erosion was out of control. Tremendous loads of mud were being washed down the river until it emptied into the Rio Grande and was then transported down to Elephant Butte Reservoir. The muddy sedimentation threatened the very existence of the reservoir. Because the reservoir had been constructed in 1916 at a cost of more than $5 million, there was a great deal at stake. It was imperative to stem the destruction of the Rio Puerco basin. The CCC was sent to do what it could. Two camps were established along old highway NM-44 in the mid-1930s under the supervision of the Soil Conservation Service (Figure 101).

Camp SCS-8-N

In July 1935, Camp SCS-8-N was established on the Ojo del Espiritu Santo Land Grant (Chapter 10). The camp site is just off to the east of US-550 at mile 40.6. Although the camp was located in the Rio Salado watershed, vs. that of the Rio Puerco, much of their work was done in the latter. Today the site is on Zia Pueblo land, and only a falling-down adobe building of the old motor pool can be seen from US-550.(Figure 102).

The enrollees at this camp were mainly Hispanic boys from New Mexico. They built roads, fences, flood-control dams and stock tanks, and planted trees and grass. In the spring of 1940 the camp was physically dismantled and loaded onto trucks, leaving only a few foundations behind. The truck caravan drove up the old dirt highway NM-44 to Durango, Colorado, where a new camp was established.

The SCS-8-N camp experience molded the character of the CCC-boys, all members of the "greatest generation." One of those was Rupert López (Figure 103A and 103B). He served at the camp in 1935–1937. Rupert was born in the San José area of downtown Albuquerque in 1916. After high school he joined the CCC, where he served at the SCS-8-N camp as well as one in Santa Fe. When his CCC days were over, in 1939, he married a girl he'd met at a roller-skating rink in Albuquerque's South Valley. In December 1941 he went to work for the Civil Service Department at Kirtland Air Force Base as a supply inspector. He retired from there in 1973 after 32 years. During this time he also joined the New Mexico National Guard and transferred to the Air Force National Guard, which sent him to Korea for a year in 1968. He then transferred to the Army Reserves, from where he retired as E-8 first sergeant. After retirement, as a resident of Corrales he was very active in community affairs and served on several committees. In 1998 received the Corrales Farmer of the Year award. For a while he operated his own travel agency, Rupert's Travel. He and his wife lived about 70 years, until her recent death, in the house he built himself on their small farm on Rupert Lane in Corrales.

A. Northeast view of CCC camp SCS-8-N in 1936. (Photo courtesy of Rupert López, former "CCC-boy.")

B. Inset showing motor-pool garage, part of which survives today.

C. View of remains of camp from US-550. (Photo by author 2008.)

Figure 102. CCC camp SCS-8-N

Another CCC boy at camp SCS-8-N was Alex Gallegos (Figure 103C), who came to the camp after Rupert had left. For generations the Gallegos family had tended sheep on the east side of the Rio Grande north of Tomé. When Alex was a small boy the family moved to the Belen area, where Alex's father worked on the dairy farms. At age 16 or 17 Alex took a job at a mill in Albuquerque. At age 18 he joined the CCC and was assigned to camp SCS-8-N, where he stayed for 18 months during 1939–1940. After the camp was dismantled and moved by convoy to Durango in 1940, Alex was discharged from the CCC. He worked for a while with the Denver and Rio Grande railroad, and then took a job with Boeing in Washington State, where he worked as a welder. He returned to Albuquerque with a pair of bad eyes from the welding. When the war started he enlisted in the military and served in the Pacific Theater, and became an artillery instructor. After the war he returned to Albuquerque and began his career as a master tile worker. He was one of the men who installed the beautiful travertine panels in the interior of the State Capitol building. Today he lives in the South Valley with his wife. The photo (Figure 103C) shows him as a young man posing in front of the motor-pool buildings near the northwest perimeter of the camp—the only buildings that remain today.

A. Group portrait of CCC Company 2836, camp SCS-8-N, 1936 (Rupert López indicated).
(Photo from Civilian Conservation Corps, 1936.)

B. Rupert López at camp SCS-8-N, 2007.
(Photo by author.)

C. CCC-boy Alex Gallegos at camp SCS-8-N,
1939. (Photo courtesy of Alex Gallegos.)

Figure 103. CCC boys.

"Lost" camp SCS-7-N

In January 1936 a second CCC camp, SCS-7-N, was built about seven miles north of SCS-8-N. (Although SCS-7-N was set up after SCS-8-N, it had been the first of the two planned.) Camp SCS-7-N was located about one mile WSW of the present US-550 from MP-48.

For quite some time I was unable to locate the site of this camp. The CCC's 1936 Official Annual (Civilian Conservation Conservation Corps 1936) mentions only that it was located 20 miles south of Cuba. That would place it on US-550 at about MP-44. The essential clue was provided by a tiny (2 x 3.5-inch) photo of the camp in the 1936 CCC Annual mentioned above, showing the profile of a mountain range that clearly is the Sierra Nacimiento (Figure 104A).

I searched in vain in the general area for a matchup of the profile. When enlarged, however, the photo revealed the faint profile of a mesa in front of the Sierra Nacimiento. I immediately recognized it as being that of South La Ventana Mesa. Returning to the area I found a reasonable matchup of the profile at MP-48—at the entry gate to Jémez Pueblo's

Dragonfly Springs Recreation Area on the west side of the highway. However, the low intermediate foreground ridge in the photo (Figure 104A) was nowhere to be seen. The gate was open so I drove west down toward the spring. I saw a suspicious, low gravelly ridge off to the south so I followed a dirt track in that direction. There I found the exact profile (Figure 104B), and there were the camp's foundations. This was the site of SCS-7-N, about 7.5 miles north of SCS-8-N and 16 miles south of Cuba. In the mid-1930s highway NM-44 had followed a different alignment than today's US-550 and passed right by the camp's front gate on its west side (Figure 101). After NM-44 was realigned the old camp became "lost." Today the camp site is on pueblo land and is off-limits to the public. However, its "footprint" is clearly visible from the air (Figure 105).

A. East view of CCC camp SCS-7-N in 1936. (Photo modified from Civilian Conservation Corps 1936.)

B. East view of site of CCC camp SCS-7-N. (Photo by author 2001.)

Figure 104. Search for "lost" CCC camp SCS-7-N.

The legacy of the CCC

During the 9.25 years the CCC was in existence about 3,000,000 young men passed through its ranks nationwide, plus 225,000 WWI veterans. They were fed and clothed, they developed skills, learned discipline, worked hard, built muscle, and gained immense self-confidence. They built much of the infrastructure that is now taken for granted on the nation's public lands. America's involvement in WWII put an end to the CCC as the country's priorities abruptly shifted and the "boys" went off to war. The CCC program was terminated on June 30, 1942, midway through Period 19. During the war ex-CCC members made up about 20% of the armed forces and brought with them work and leadership skills and the maturity essential to the war effort. General George Marshall later stated that these men and the CCC ethic were major factors in the nation's victory in the conflict.

WWII effectively erased the CCC from the public mind. It is no less than astounding that Tom Brokaw in his book, *The Greatest Generation* (Brokaw 1998), never once mentioned the CCC. I'm convinced that he probably didn't know about it! The CCC molded the character of the "greatest generation." This fact is perhaps best expressed by the words of Roy Lemons, an ex-CCC boy who died in 2002 at the age of 83: "In the final years of my life, in assessing the most profound happenings, I find the CCC experience the most rewarding of all. Nothing had more impact on me than that. All that I am was shaped from that 2.5-year interval of my life and the 3.5 years of the Depression that preceded it" (Civilian Conservation Corps Commemoration Committee 2003).

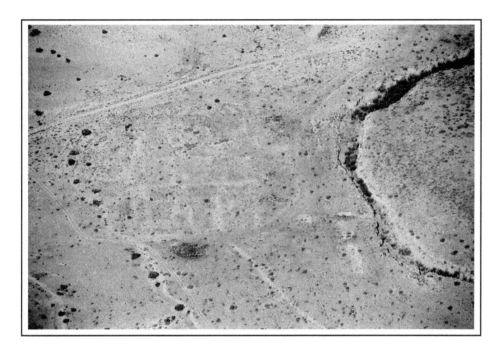

Figure 105. West high-oblique view of former site of CCC camp SCS-7-N. Photo by author 2001.

12

On to La Ventana

At mile 45.5, US-550 crosses a low drainage divide (Figure 106). The divide rises only 12 feet above the opposing drainage courses of a tributary of the Rio Salado on the south and the Rito Olguin, a tributary of the Rio Puerco, on the north. The southern course has a steeper gradient. It allows water to flow along it at a higher velocity and it therefore more quickly erodes headward. In the not too-distant future the Rio Salado tributary will eat its way north through the divide and "pirate," or capture, the upper part of the Rito Olguin drainage. The flow of the Rio Salado will then be significantly enhanced and its rate of erosion will be increased.

As mentioned before, stream piracy is an enormously important geologic phenomenon. As streams cut down and modify a landscape they use the weapon of headward erosion to capture the courses of other, less robust streams. We saw this back in Chapter 8 (Rio Salado anticline), where the streams there have significantly changed their courses as they incised down into the bedrock. This place differs only in that we can see the future "scene of the crime" at close hand. The two steams are very much in competition. One writer poignantly compares the physical process of drainage-system evolution to the Darwinian "survival of the fittest" concept operating in biological systems (Lucchitta 1989). Streams with steep gradients have an advantage over those with lower gradients and therefore have a greater chance of survival.

A stop at MP-41 provides a vivid example of the type of erosional destruction taking place in the two drainage basins (Figure 107). The rounded, tan-colored hills sparsely covered by junipers in the background are made up of the ca. 85 Ma Cretaceous Mancos Shale bedrock (Figures 107A and 107B). The flattish surface covered with snakeweed (gray patches) and cholla cactus is the top of the < 1 Ma valley-fill alluvium. The vertical-walled arroyos are rapidly excavating the alluvium. The story goes like this: First, during the past 1 million years pre-alluvium drainages incised valleys into the Mancos Shale during a period of increased precipitation (Figure 107C). Later, during a period

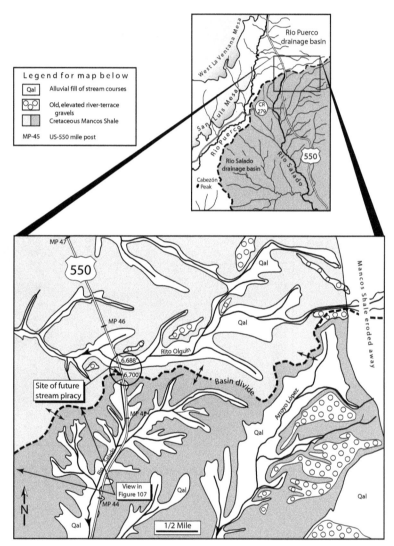

Figure 106. Map showing site of impending stream piracy that will enlarge Rio Salado drainage basin to south.

of reduced precipitation and decreased carrying capacity of the streams, the valleys were backfilled by alluvium that was in effect put into "temporary storage." Now, in front of our eyes, the back-filled alluvium itself is being removed from storage. Eventually all it will be scoured out and flushed downstream. This is a disaster unfolding in slow motion.

La Ventana

On the east side of the highway, at MP-50, is the former La Ventana rest area, behind which is a fin of sandstone cut by a gap (Figure 108). The gap was once bridged completely over, hence the name *La Ventana* (Spanish for "The Window"). The span fell down of its own accord in 1919 (Sherman and Sherman 1975). The site of the coal-mining town of La Ventana is at the single ranch house west of the highway about 0.2 mi ahead (Figure 109).

Today the site is a sleepy, virtually forgotten place, but it was not always so. Miera y Pacheco's 1779 map (Figure 94) shows a mesa with the name "Ventana," but that probably alludes collectively to today's North and South La Ventana Mesas (Figure 101), which in turn were named from the natural bridge that was intact at that time (Kessell 1975). Interestingly, the 1779 map identifies the main, East La Ventana Mesa, on the west side of the Rio Puerco as the *Mesa Grande de La Veta Negra*, or "Big Mesa of the Black Vein." Clearly the area's coal deposits were well known even then.

The La Ventana area occupied a gap between the small agricultural areas of the middle Rio Puerco Valley (Chapter 9) and the community of Nacimiento (today's Cuba area, Chapter 15). Travelers leisurely once picked up loose coal and carried it home, but no one expended serious effort on the gathering (Jenkinson 1965). It is not certain just when the community took form—it was probably a gradual process—but there were certainly people living and being born there as early as 1848 (Romero 1984). However, the little town eventually withered away and sometime before 1910 farming had ceased for good (Widdison 1959).

A. Northwest view from US-550. (Photo by author, 2005.)

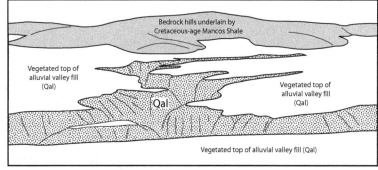

B. Interpretation of photo in "A" above.

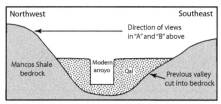

C. Schematic cross section of valley-fill erosion.

Figure 107. Rapid erosion of valley-fill alluvium at mile 44.1 (location in Figure 106).

The first coal mines in the general area were to the north of here—south and southeast of Cuba. They were worked to fuel small, local copper-ore smelting operations in the foothills of the Sierra Nacimiento between 1880 and 1900 (see Chapter 13 below). Serious consideration of the coal deposits in the La Ventana area would have to await the arrival of the railroad.

Sidney Weil and the Santa Fe Northern Railroad

The copper deposits along the west flank of the Sierra Nacimiento and the conveniently-nearby coal deposits, including those at La Ventana, early on caught the eyes of many a speculator, but those of one man in particular—Sidney Weil. (We first met him back in Chapter 4.) He was the driving force behind the construction of the Santa Fe Northern Railroad from its connection with the Santa Fe Northwestern at San Ysidro to the coal fields of La Ventana and, he hoped, beyond to the copper mines in the Cuba area. But first, who was this guy?

Sidney M. Weil was born in Iowa in 1881 of a Jewish family. His grandfather, Moses Weil, was a native of Prague, Bohemia. In the 1840s Moses came to the U.S. and with his five brothers established a successful tanning business in sev-

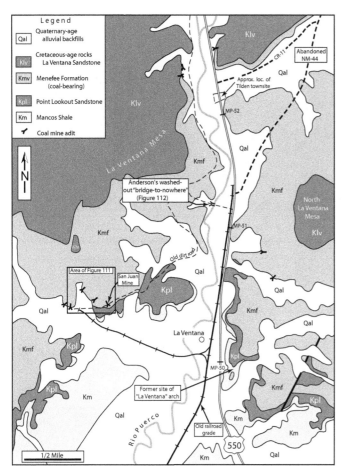

Figure 108. Geology of La Ventana area.
(Modified from Woodward and Schumacher 1973.)

Legend
Qal — Quaternary-age alluvial backfills
Klv — Cretaceous-age rocks La Ventana Sandstone
Kmv — Menefee Formation (coal-bearing)
Kpl — Point Lookout Sandstone
Km — Mancos Shale
⌐ Coal mine adit

eral eastern cities. Moses had eight children, one of whom, Michael, father of Sidney, was born in 1853 in Ann Arbor, Michigan, and lived in Chicago. He was associated with his father's tanning business and assumed management of the firm's branch in Keokuk, a Mississippi River town in the southeastern corner of Iowa. Michael married a local Jewish girl of the Klein family. The couple had two children, and of the two, Sidney was the only survivor (Coen 1925).

Sidney spent his first 11 years in Keokuk before his parents moved to Chicago. He had wanted to attend the University of Chicago, but his father had suffered financial reversals and the young man had to go to work as a newspaper reporter. Later he became a feature writer at the *Milwaukee Sentinel*, and in 1901 he covered the plot to assassinate President McKinley for the Associated Press. He became involved in the advertisement business and coined the phrase, "Ask the man who owns one" for the auto maker Packard (Ripp 1985). But he really wanted to get into mining. To prepare for this new career he took a two-year course on the metallurgy of copper, specifically the electrolytic rendering of copper. In 1916 he came to Albuquerque to investigate copper deposits found in sandstone and quickly focused on the copper occurrences on the west side of the Sierra Nacimiento. He acquired an interest in the Señorito area southeast of Cuba and purchased the San Miguel Mine, which was on-trend with Señorito five miles to the south (Chapter 13). He immediately recognized the economic need to transport heavy tonnage via a railroad that could connect with the main line of the Santa Fe Railroad in Bernalillo (Coen 1925).

In 1920 he married a young woman, Marion Watlington (see Marion Butte in Chapter 13 below). Also in 1920 Weil incorporated the Santa Fe Northwestern (SFNW) Railroad, which ran from Bernalillo to San Ysidro and thence north into the timber country of the Jémez Mountains. The prospectus for this company listed the coal and copper country west of the Sierra Nacimiento as a potential second destination, but this never came to pass. By 1923 Weil was completely separated from the SFNW and he began to promote another railroad to the La Ventana coal beds, the Cuba copper deposits, and perhaps even the Farmington oil fields. That same year he incorporated his railroad as the "Cuba Extension Railway Company."

Between 1924 and 1927 the grade and the standard gauge track were slowly extended along the Rio Salado and into the Rio Puerco drainage basin toward La Ventana. The line was quite easy to construct because it traversed gentle grades and sandy terrain, including innumerable dry washes. The flip side was that the line was just as easily washed away by the region's enthusiastic rainstorms. Almost no one

Figure 109. West low-oblique aerial view of La Ventana townsite up against the Rio Puerco. (Photo by author 2001.)

148

lived along the course of the railroad, so the success of the line depended entirely on the successful development of the area's natural resources. In 1927, for some undisclosed reason, the name Cuba Extension Railway morphed into the "Santa Fe Northern Railroad." Later that year the line was completed to a few miles north of La Ventana, at mile 30.3 at a place called Tilden. For the completion ceremony notables drove from Santa Fe down to Bernalillo and then took the train out to La Ventana. New Mexico Governor Richard C. Dillon commemorated the event at the end of the track (probably at Tilden) by driving a golden spike (Glover 1990).

Tilden was a townsite founded in 1927 (Julyan 1996), and may have been on both sides of the Rio Puerco. On the west side was a depot and construction camp consisting of an office, some frame buildings, and a hotel or bunkhouse (Glover 1990; Nickelson 1988, p. 186). On the other side was a little store and service station owned by Col. Walter Hernández of Cuba (more about him in Chapter 13). Back then, before the paved road was built, only a dirt road wound around the edge of the cliffs (Wiese 2001). The New Mexico State Highway map of 1934 shows a Conoco station at the site.

A trackless railroad grade was eventually extended out into nothingness about two miles north of Tilden. In 1928 the railroad went bankrupt and was sold under court order. In December of that year the operation was reorganized as the San Juan and Northern Railroad Company (Nickelson 1988, p. 173). A rail spur was built west across the Rio Puerco on a wooden trestle to the mines (Glover 1990).

The railroad and the mines had a rather short life due to extremely bad timing. The Great Depression descended on the nation in October 1929, and by the early 1930s economic activity and demand for coal plummeted. In the last two months of 1929 a total of 220 coal cars rumbled down the track to Bernalillo, 853 cars in 1930, and 353 cars in the first four months of 1931. In the spring of 1931 the track was washed out by floods and by the end of 1931 the cash required to repair the track had dried up. That year the company went into receivership. Another big washout occurred in mid-1932. One official acidly remarked that a heavy dew was enough to take out the line. The last train out of La Ventana was in December 1932. The combination of the declining coal market and the undependable service doomed the line (Glover 1990).

For years two locomotives lay abandoned and rusting away under the desert sun at La Ventana. From time to time a mechanic from the Santa Fe Northwestern Railroad would journey out to La Ventana and cannibalize a part or two. In 1941 the rails were torn up and trucked out. Finally, during the scrap-iron drives of WWII, the locomotives were cut up in place and the pieces hauled down to Albuquerque (Glover 1990).

Sidney meanwhile forged on like the human dynamo he was. Over the years he had his finger firmly immersed in many pies. In 1922 he had been elected to the Albuquerque City Commission, and for a while he was director of the Albuquerque Chamber of Commerce (Tobias 1990). He was instrumental in persuading General Hap Arnold that Albuquerque was large enough for an Army Air Corp field, which later became Kirtland Air Force Base (Ripp 1985). In 1930 or 1931 his marriage with Marion fell apart and she moved in with her parents in Albuquerque (more about her below in Chapter 13). Around 1936 Sidney married his second wife, Rhoda Calhoun. Until 1953, the year Rhoda died, he lived in Albuquerque and involved himself in "investments" and real estate. He died at the age of 82 in 1963. Years later his barber, remembered Weil as his most unforgettable character: "Sidney was a little guy, about yay [gesturing to the reporter] high. He was a good promoter, but he was always owing people money. He was a good ducker. When he'd see somebody he owed coming down the street he'd duck into a bar or store and buy himself something. He always came in the shop for a shave five minutes before closing" (Ripp 1985).

La Ventana coal

The coal deposits here occur within the Cretaceous-age Menefee Formation, which has been eroded into a cul de sac around the axis of a north-plunging anticline (Figure 108). The Menefee Formation consists of sandstones, mudstones, and coal beds that were deposited in the low-lying, swampy environment located on the landward side of a shallow sea. The thick and very rugged La Ventana sandstone overlies the Menefee Fm. and forms the imposing cliffs.

La Ventana Sandstone

Only near La Ventana does the La Ventana sandstone occurs in such a massive way. Why only here? The geologic map and a little geologic knowledge provide the interesting explanation. Turn the geologic map about 45 degrees clockwise (Figure 110A), and then generalize the geology by smoothing the formation contacts somewhat (Figure 110B). Because the formation dip direction is to the northwest away from us, our eyeball-view of the geologic map is somewhat akin to looking at the beds in a virtual cross section. With such a cross section in mind we can compare it with a geologic model that shows the development of beach sands (Figure 110C). The thick La Ventana sandstone is revealed as a buildup of beach sands where the shifting shoreline maintained a constant, stable position for a long period of time. The coastal swamps landward from (behind) the La Ventana beach sand were also stalled in this fixed geographic position and therefore, as the land subsided in tandem with continued deposition, developed an unusually thick sequence of coal beds and beach sands. The outcrops of the Menefee Formation, below the La Ventana Sandstone, host the coal-mining beds of the La Ventana area.

San Juan Mine

The approach of the railroad to La Ventana area in the early 1920s caused a flurry of coal leasing and prospecting. The most important mine was the San Juan Mine (Figures 108 and 111). An exploitable bed of coal was identified here in 1923, but activity was limited because no railroad yet existed. In mid-1929 the time had arrived. By then a sizable tipple was in operation and capable of extracting 5,000+ tons per month, and up to 35 men were employed there. As we all know, this was just before the Great Depression was launched on the infamous Black Tuesday, October 29, 1929. By 1931 the railroad was in serious financial trouble and went into receivership. The mines—including the San Juan—were closed. Washouts of the tracks, coupled with reduced demand for coal, had pushed the operation over the brink (Nickleson 1988).

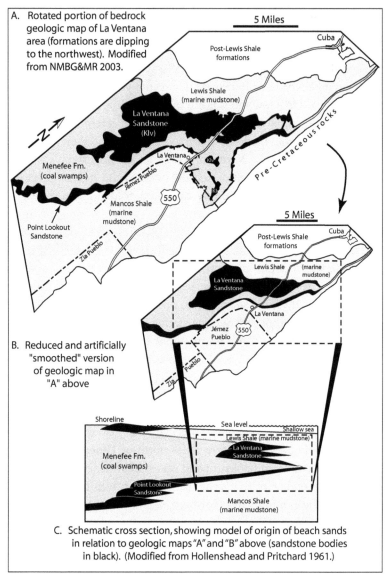

Figure 110. Generalized origin of La Ventana Sandstone (Cretaceous formations shown in grays and black).

In 1935 recommendations were made to plug the mine, but no one had the necessary cash to do even that. Early in 1937 the mine caught fire, probably via spontaneous combustion. The fire spread quickly. Suspicions were voiced by some that the mine had been torched to cover up the theft of rails from the mine. CCC boys from the Soil Conservation Service camp No. 7 were brought in to tear up the rails and part of the tipple in order to prevent them from catching fire, and they also tried to plug the shaft. In 1938 the lease was canceled and the BLM assumed ownership.

In 1939 the mine was still on fire but it was tightly sealed and apparently just simmering away below the surface. The mining company lost a considerable amount of inventory in the form of rails, cars, and other equipment that had been sealed up in the mine. If the mine had been sealed properly right after discovery of the fire the mine might have had time to cool. However, the seals at the mine opening leaked. By 1946 the San Juan Mine was still burning. Legal attempts were made to reopen the mine, but they were rejected. That was the end of the San Juan Mine (Nickleson 1988).

The grim times of the 1930s were also a time of free-wheeling entrepreneurs. Mining regulations were in place but were often poorly applied or enforced. This created an environment that resulted in some interesting stories. In 1936 a man named Nick Luciani illegally began digging a shaft only 1,800 feet west of the San Juan Mine on state land. He was very much in trespass, but he was clever enough to formally request a lease, and at the same time to pay royalties to the state in person, requesting a receipt. The state had been snookered! Because they had accepted royalty payments they had to grant Luciani a lease for his mine, called the No. 1 Peacock. The owners of the San Juan Mine protested and the Peacock Mine was forced to shut down.

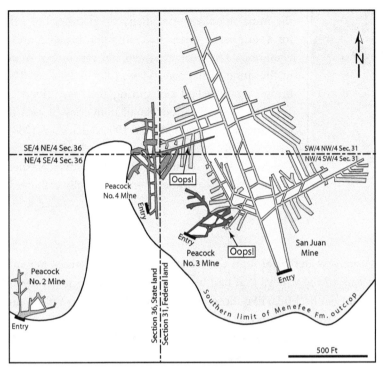

Figure 111. "Dueling" underground coal-mine shafts (gray strips), La Ventana area (location in Figure 108.) (Modified from Nickelson 1988; Woodward and Schumacher 1973.)

In 1939 Luciani moved a few hundred feet to the east, even closer to the San Juan, and began his No. 2 Peacock Mine (Figure 111). The coal was of poor quality and so the No. 2 was shut down too. Later that year Luciani moved east yet again, this time trespassing on federal land, and opened his No. 3 Peacock Mine, only 600 feet west of the main portal for the San Juan Mine. The new shafts broke through into the San Juan Mine in several places (oops!) and caused the reactivation of the fire. In 1940 the No. 3 was forced to shut down. Luciani then moved back west a tad and opened his Peacock No. 4 Mine, a few feet west of the line dividing state land Section 36 and federal land Section 31. In a short time shafts were dug to the east onto federal land and again broke into shafts of the San Juan Mine (oops again). This reactivated the fire in the San Juan and later that year the No. 4 was shut down for good (Nickelson 1988).

Political bridge to nowhere

At mile 51.2, leading off to the west from US-550, is a road to nowhere (Figure 108). A locked gate and a sign stating "Bridge Out" is all that can be seen from the highway. Jump the fence (it's okay—this is federal land) and walk down the road to where it ends abruptly at the bank of the Rio Puerco. On the other side is half of a bridge, dangling over the middle of the deeply-incised river (Figure 112). This bridge once connected old highway NM-44, on the east side of the river, to the Anderson-Sackett Mine, two miles west of La Ventana. Anderson was a well-connected member of a group of investors that acquired a mining permit in 1925. In time he would become one of New Mexico's most influential politicians.

Clinton P. Anderson (1895–1975), born in South Dakota, moved to New Mexico in 1917 seeking a cure for tuberculosis. He recovered and decided to stay. He worked as a newspaper reporter and editor until 1922, and then became involved in the insurance business and dabbled in politics. He served as the treasurer for the state of New Mexico from 1933 to 1934 and participated in various Depression relief agencies until 1940. In November 1940 he won election to the U.S. House of Representatives on the Democratic ticket, but left Congress after one term to serve as Secretary of Agriculture from 1945 to 1948. In 1948 he successfully ran for the U.S. Senate for the state of New Mexico and occupied that post until 1973.

There was only limited development of the mine in 1926 because there was no way to get coal across the river to the highway. The railroad had not yet arrived, but it was on the way. The bridge was constructed in 1929 and, interestingly, this bridge to a private mining interest was financed jointly by (in addition to the permittees) Sandoval

Figure 112. West view of Anderson's bridge-to-nowhere over the Rio Puerco. (Photo by author 2005.)

County and the State Highway Commission (Figure 112). By 1931 a spur of the railroad reached the mine, but in mid-year the line was bankrupt and the road across the bridge provided the only outlet for coal shipments.

Even so, because of the slow economic times the mine remained essentially on standby but idle for a number of years. In 1940, upon his election to Congress, Anderson divested his remaining share in the mine to his son-in-law, John F. Sackett. The bridge was washed out during the mega-flood of 1941, but was repaired. Small deliveries of coal to local markets continued until 1956 (Nickelson 1988). After the Rio Puerco was diverted in 1965–1966 (see Chapter 13) the resulting increased flows washed out the bridge's abutments and it partially collapsed.

Controlling the Rio Puerco

The Rio Puerco has successfully gouged out a narrow defile through the resistive La Ventana sandstone and down into the underlying units. Climate change has caused the valleys that had been cut into the bedrock to be subsequently backfilled with alluvium (Figure 108). The alluvial backfills, like those of the middle Rio Puerco Valley, are today themselves being excavated and sent downstream as sand and mud.

The defile produces a choke point for transportation. For many years the narrow slot was simply too narrow to accommodate both a modern highway and a meandering river. In the 1920s and 1930s the "highway" was essentially a dirt track that worked its way through the defile as best it could on to the east side of the river (Figure 113). Several different alignments were laid out over the years to get past this difficult point (more in Chapter 13). Finally in the mid-1960s the New Mexico Highway Department faced the task of constructing a new, more direct, wider, and safer highway through the defile. To do so they had to first somehow control the Rio Puerco.

A number of years ago I remember a TV ad that was hawking a brand of margarine. "Mother Nature" is seen as being fooled into tasting the margarine instead of the real butter, and when she discovers the deception the clouds gather, the ground trembles, and she admonishes in a quivering voice that "It's not NICE to fool Mother Nature!" An attempt was made to fool Mother Nature along a reach of the Rio Puerco north of La Ventana, and she responded accordingly.

The river naturally meanders from one side to another and crowds the valley such that little room remains for a highway. To negotiate a way through, the Highway Department was faced with the prospect of constructing three expensive bridges. They chose instead to avoid that unpleasant option and to perform an engineering experiment with the river: they would cut off the meanders and divert the river into a straight, man-made channel (Figure 113). In 1965 the construction company dug a 1.1-mile shortcut ditch and in 1966 they built an earthen dam on the north to divert all the river's flow into the new diversion. In the middle of the reach they blocked off the nose of a meander with a pair of berms, and on the south they built another earthen dam over a box culvert to drain the local runoff from the old channel back into the river. The flow distance between the northern dam and the southern reentry was shortened by more than a mile, but of course the drop in elevation between the two points—60 feet—remained the same (Coleman et al. 2001). The gradient (elevation-drop/distance) had been increased from 0.44% to 0.74%. This changed the river dynamics and threw the entire system into a state of disequilibrium. Then Mother Nature spoke.

A stream in equilibrium balances the power of the stream (water volume and gradient) vs. resistance (sediment load in the stream, sediment particle size, and channel roughness). The diversion had almost doubled the gradient. The realigned Rio Puerco, now with an attitude and under the guidance of an angry Mother Nature, began to wreck the new, engineered channel. First it started to cut downward into the soft alluvium until it hit bedrock, and then it began widening the channel and excavating enormous volumes of alluvium. A 15-foot waterfall developed on top of the bedrock and migrated upstream more than three feet per year. The waterfall acts as a regulator on the lateral erosion rate. Downstream of the waterfall the river is widening its channel and developing a meandering pattern. This has caused bank failure as the river pounded against its vertical walls and encroached on the new highway at several points. The channel there has widened more than 12 feet per year to a current width of over 300 feet.

As the regulating waterfall migrates upstream the area of severe erosion and channel widening also migrates upstream with it. Over the course of 30 years the river has excavated over 500,000 cubic yards of sediment and flushed it down the river. In so doing it has created the "Grand Canyon of the Rio Puerco." The volume from this one 1.1-mile reach of the river represents about 20% of the sediment load delivered to the Rio Grande on an annual basis. If allowed to continue the river will eventually gouge out about 1.3 million cubic yards (Coleman et al. 1998) and strip virtually all the alluvial fill off the underlying bedrock. It had to be stopped. Engineering had disturbed the river and engineering was required to correct the blunder.

To meet the challenge, and to deal with overall problems within the entire Rio Puerco watershed, Senator Jeff Bingaman in 1996 introduced legislation to form the Rio Puerco Management Committee. The Committee consists of a diverse group of ranchers, environmentalists, Navajo leaders, and government agencies. The largest of the Committee's projects—the *La Ventana Project*—was a joint venture with the New Mexico Highway and Transportation Department. Its aim was to provide technical input for an effort to return the Rio Puerco to its natural channel in this confused area (Soussan 2001).

In 1999 the Committee and the Highway Department signed an agreement whereby the Department agreed to include the construction of two new bridges over the old channel as part of the widening of NM-44 and the creation of US-550. After completion of the bridges in 2001, work was scheduled to begin on the southern, downstream end. The plan was for construction crews to put in a series of rock steps to allow the restored river to fall 12 feet down into the natural channel, which has been gouged out to a new, deeper level. Then, in the middle of the natural channel where it makes a sharp loop, a soft barrier was to be installed to keep the restored river away from the highway. Crews will then dig out several 15-foot-deep wildlife ponds that would remain wet year-round to provide animal habitats. Next, in the north the river was to be reshaped so that it can enter the channel that it left 40+ years ago—now six feet higher than the present eroded bed. Finally, the earthen diversion dam on the northern end was to be removed and a new earthen berm placed to guide the river back into the old channel (Soussan 2005). Work began in late 2004 and was completed in 2005. Mother Nature must be smiling.

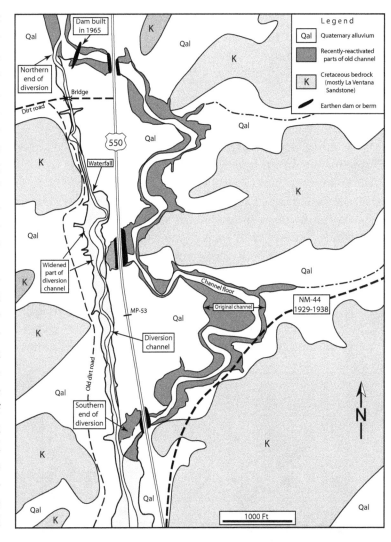

Figure 113. Engineering experiment along the Rio Puerco, 1965–2005. (Modified from New Mexico State Highway and Transportation Dept. 1998; Woodward et al. 1970a; Woodward and Schomacher 1973.)

13

On to Cuba

Mesa Portales

Once north of, and through, the constriction of the Rio Puerco discussed in the previous chapter, another mesa appears to the west from about MP-56 on US-550. This mesa, *Mesa Portales* (Spanish for "Entry Gates"), is actually a north-dipping sandstone bed with a sharp south face. It looks pretty much like every other mesa out here but it has a special significance. The base of the massive Ojo Alamo Sandstone holding up the mesa forms the boundary between two mega-divisions of the geologic record, namely the contact between formations of the Cretaceous Period of the Mesozoic ("middle life") Era below and those of the Tertiary Period of the Cenozoic ("modern life") Era above (Figure 4). The actual time of the change-over was 65 Ma. This is a hugely important point in the history of life at which the dinosaurs, which had dominated life on land for about 160 Ma, suddenly disappeared from the geologic record. We'll expand on this special event in Chapter 14.

Set apart from, and located just east of Mesa Portales is the conspicuous landform of Marion Butte (more below). The term "butte" refers to an isolated, flat-topped hill. As Mesa Portales was eroded back the little butte remains as a survivor of incomplete destruction of the mesa's east flank. Although the Ojo Alamo has been completely removed from its top, a portion of the underlying Cretaceous-age Kirtland Formation has been spared to form Marion Butte.

Multiple-choice roads to Cuba

The present US-550 is the fifth generation of roads connecting the Rio Grande Valley, via the La Ventana area, to Cuba (Figure 114). First, as early as 1879 or perhaps earlier, a wagon road generally followed the east bank of the Rio Puerco from points south of La Ventana to Cuba (Luna 1975). There were no bridges, and traffic had to laboriously work its way across the numerous side drainages that were periodically flooded.

Second, in the 1880s copper and coal deposits were exploited in a number of communities along the Nacimiento Mountain front (see Sierra-Front Towns below). A dirt track from La Ventana linked San Miguel, San Pablo, and Señorito with Cuba (Renick 1931). After the mines were abandoned the communities died and the road fell into disrepair.

A third road, the first graded state route NM-44 built in the late 1920s, led north from La Ventana, struck off to the northeast at the present mile 52.5 of US-550, avoided the meandering bends of the Rio Puerco, snaked its way the best it could through low points in the rugged outcrops of the La Ventana Sandstone, joined the old Rio Puerco track for a short length, and then arched around the east side of the Wiese ranch to Cuba.

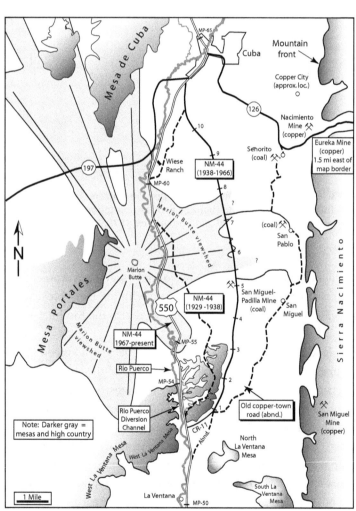

Figure 114. All roads lead to Cuba. (Modified from Renick 1931.)

The fourth, a realigned NM-44 built in the late 1930s, worked its way completely around the east side of the La Ventana Sandstone outcrop and went directly to Cuba. The southern end of this route, branching off from US-550 at mile 51.3, is now abandoned. Sometime in the early 1960s a bridge along the southern end of the route washed out. One source opines that the bridge was set afire by a disgruntled area rancher (Hernández 2000a). Sandoval County Road 11 now branches off US-550 at mile 52.25 and serves as connector to this old NM-44 north of the abandoned stretch and provides a laid-back and nice scenic drive up to Cuba.

The fifth road, the third alignment of NM-44, was laid out in 1967 directly through the defile of the La Ventana Sandstone cliffs along the Rio Puerco. This is the route that led to the diversion of the Rio Puerco discussed above in Chapter 12. Finally, in 2000, the old NM-44 was four-laned, but along the same course, and redesignated as US-550.

Marion Butte

Maps are fascinating. They are rich sources of historical information, sometimes even more so than the written word. Maps not only reveal the shape of land forms and the distribution of geologic formations, but quirks of human history as well. Realignments of roads suggest changing economic and political priorities and the evolution of motor vehicles. And then there are place names. Many of the place names along US-550 are quite unimaginative and simply descriptive, such as La Ventana Mesa, from the landform and the nearby community. But then there is "Marion Butte," a less-than-impressive protuberance due west of US-550, but with an extensive viewshed (Figures 114 and 115).

Figure 115. North view of Marion Butte from US–550 at mile 56.5. (Photo by author 2005.)

Whenever a person's name has been attached to a landform it usually implies a good story. So who was Marion? For the answer we must return to Sidney Weil (Chapter 12). Remember that shortly after his arrival in New Mexico in 1916 he found a distraction in the form of a young lady named Marion Watlington. It is likely no coincidence that Weil was a promoter of railroads and Marion's father worked for the Santa Fe Railroad. In June 1920 Sidney, now pushing 40, and Marion, about 24, were married in her parents' Albuquerque home. She was described as "one of Albuquerque's most beautiful young women" (*Albuquerque Journal* 1920). In 1927, shortly after the completion of the railroad to Tilden, Sidney evidently discovered that the railroad company still had some loose coins in its pocket. He constructed an elaborate "hotel" near the pyramidal landform now called Marion Butte, a few miles north of the end of the railroad grade. The building was used to entertain friends and prospective investors. He may have intended the site to be at the division point on the railroad, halfway between Bernalillo and Farmington (Glover 1990).

Via census records, newspaper announcements, and old telephone directories, this is what we know about our Marion. Her father was Charles Watlington, born in 1866 in North Carolina, where there are many Watlingtons. According to the 1900 census he was a single Santa Fe Railroad man living in Socorro. The 1910 census has him as a married man living in Albuquerque. During the preceding decade he had married a May Andrews from Michigan and in the process had acquired a 14-year-old stepdaughter, Marion Francis Rowe. Then in 1920 Marion married Sidney Weil and entered the busy life of a "blowing-and-going" promoter. We have no details about the quality of their marriage but evidently they had problems. During the 1920s they had a daughter and a son. Around 1927, when Sidney built his hotel, he named the isolated butte visible from the front door after his young wife.

Soon afterwards Sidney found himself in serious financial straits. In 1928 the railroad went belly up and the hotel at Marion Butte was sold off (Glover 1990). The Albuquerque telephone directory for 1931 shows that Marion had by then moved in with her parents. In 1940 her mother died in Albuquerque. Marion, now married to a man named Dudley Long and living in Long Beach, California, attended the funeral (*Albuquerque Journal* 1940b, 1940c). In 1950 her father died in Albuquerque and Marion again rushed to New Mexico. Still married to Dudley Long, she at this time was living in Brookline, Massachusetts (*Albuquerque Journal* 1950). Social Security records list a Marion Long who was born in June 1895 (that checks with the 1910 census age of 14) and died in 1983 in Palo Verde, California, a retirement community along the lower Colorado River. Could this be our Marion?

Today the little butte is located on ranch land owned by the Wiese family of Cuba. Louis Wiese once allowed me to visit the butte to locate the old hotel site, which he insisted was still out there, but I was unable to find a trace. The structure had mostly been torn down long ago and the usable pieces carted off. Interestingly, at the very top of the butte is a sandstone bed containing fossilized palm-tree stumps (Lucas et al. 1992). This geologic unit is the Cretaceous-age Fruitland Formation, the same unit that contains the huge coal fields near Farmington. It is not known if Marion ever again visited the little landmark named after her, or if she even cared.

Sierra-front towns

As we approach Cuba from the south we can see the towering mass of the Sierra Nacimiento off to the east. Up against the mountain is a huge, whitish heap. This is the tailings pile of the inactive Nacimiento Mine. It was the largest of three copper mines in this vicinity, the others being the San Miguel Mine along the mountain front ten miles to the south, and the Eureka Mine three miles east up into the canyon into the Sierra and high up on Eureka Mesa (Figure 114).

People had lived along the foot of the Sierra Nacimiento for a very long time because of the reliable springs issuing from the heights. In time any and all useful or valuable materials in the area were discovered. Particularly attractive were the richly colored copper oxide and silicate minerals—the green "malachite" and lesser amounts of the green "chrysocolla" and the blue "azurite." The Indians and Mexicans had worked the copper deposits in small volumes before 1800, but serious mining didn't start until about 1880 when the copper deposits were rediscovered. It is no coincidence that this is the year that the railroad arrived in central New Mexico and provided a railhead outlet in Bernalillo for copper ore. Conveniently-located coal deposits made it possible to process some of the ore near the mines. Let's talk about these old towns from south to north.

San Miguel

The tiny settlement of San Miguel is located near the mouth of San Miguel Canyon (Figure 114). The canyon is the southernmost occurrence of semi-reliable water along the Sierra Nacimiento front. It was site of the region's third copper smelter and a one-room schoolhouse. The San Miguel Mine itself was located 3 ½ miles to the south-southeast up on the side of the Sierra Nacimiento.

San Pablo

This little, once-coal-mining and trading community is located at the mouth of San Pablo Canyon, about two miles north of San Miguel and two miles south of Señorito (Figure 114). The coal was mined from a steeply-dipping outcrop of the Menefee Formation and supplied to the local smelter (Nickelson 1988). Although San Miguel is the southernmost community, San Pablo Creek is the southernmost perennial stream along the front of the Sierra Nacimiento and was therefore destined to become inhabited. There was an abundance of game, wood, and construction stone.

A historic house, *La Casa Vieja* (Spanish for "The Old House") is located near the mouth of the canyon. Today the house is inaccessible to the public. However, I anticipate that someday it will be opened up and for that reason I include its story in this book. The sole reason the history of this obscure ranch house is known at all is that it attracted the attention of an academic with the improbable name of Bainbridge Bunting (1913–1981).

Bainbridge Bunting received his degree in architectural engineering from the University of Kansas in 1937, and a doctorate from Harvard in 1952. With a new bride on his arm he came to New Mexico in 1948 and joined the staff of the Art Department of the University of New Mexico (he eventually rose to the position of Professor of Art History). Upon his arrival he immediately plunged into the local culture and history, especially so-called "domestic" architecture (i.e., private houses). He believed that the study of architecture revealed much about the society that created it. Here in New Mexico, where much of the historic building were made of perishable adobe, he encountered a curious paradox. He wrote, " . . . most individual buildings are distressingly short lived. Because adobe also lends itself well to remodeling, old structures can be so easily and drastically changed that little trace of their earlier appearance remains . . . " (Reeve and Schalk 2001).

Between 1958 and 1975, with a sense of urgency, Bunting taught the students in his Architecture 261/262 class the basics of describing and mapping buildings, including domestic adobe houses, as term projects. This is not as simple as it sounds because such structures seldom contain right-angles. Mappers armed with T-squares had to be innovative and very attentive to detail. He encouraged them to take as many photographs as possible, and—at the risk of trespassing—he advised them to ask permission only after they had taken the critical photos! Eventually this long project resulted in descriptions of 181 sites in central and northern New Mexico, about half of which were homes, including *La Casa Vieja* in San Pablo (Reeve and Schalk 2001).

The first permanent settlement here was in 1880 built by Santos Cebada and his family, who had come from Spain via Mexico (Bunting papers, undated). During the Spanish-American War (1898) the little town was the site of an armory. After the war the armory and the surrounding land were deeded as a pension to the patriarch of the Cebada family, who had been blinded in the war (Cebada-Córdova 1995).

As Cebada's heirs reached adulthood they each were granted a homestead within San Pablo Canyon. In 1900 the third son, José, took the upper part of the canyon. That same year he built a "temporary" two-room log house perched on a terrace overlooking an orchard. This first house, the *Casa Vieja*, was flat-roofed, set on a foundation of stones removed during planting of the orchard, and oriented west-east. Its walls were of hand-cut timbers laid horizontally on their wide sides, and adobe was used only as mortar between the timbers. It was the beginning of a typical Spanish Colonial house (Figure 116).

The concept of the Spanish Colonial house deserves some elaboration (the following is taken from Cuba historian, Esther Córdova-May 2009a). In this area, as elsewhere in northern New Mexico, very few houses were built all at once due to the area's limited resources, but instead were built in stages. For the family's "first house" or *casa vieja*, a suitable location was first located. Then, if conditions allowed, a foundation was constructed out of stone, either in the form of a filled trench or a platform a foot or two above grade. Usually there was the need to complete a shelter before winter blew in, so the labor-intensive *adobe*

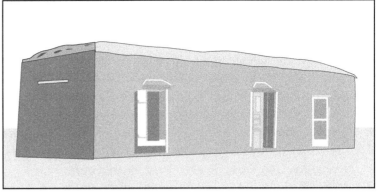

Figure 116. Sketch of typical Spanish colonial house, similar to Cebada's *Casa Vieja* in San Pablo. (Drawing by author.)

stage was postponed and walls of *jacal* (vertical logs set side by side in a trench) erected instead. The tops of the *jacal* logs were trimmed to accommodate an array of *vigas* (horizontal planks chinked with plaster) for the *zotea* (a flat, dirt-packed roof on top of the *vigas*). Once the family was somewhat protected from the elements in their one or two-room house there was time to make a sufficient number of *adobes* for the second stage of construction during the summer months. Often another room or two was added, either in a single row or in an "L" shape, often with a porch enclosed within the "L." The *zotea* naturally tended to leak. It also accumulated a great deal of snow in the winter, which had to be cleared. By the end of the 19th century when milled lumber became available, it became possible to construct a pitched roof directly over the *zotea*. The new roof provided a leak-proof shield against the rain and snow, as well as a *tejaván* (attic) for storage and for bunking down visitors. The next improvement might be the addition of windowed dormers for extra light. The Spanish Colonial house was typically a work in progress.

The massiveness of the foundation and walls of the Cebeda *Casa Vieja* suggests that a second story had been anticipated from the start. One of the original rooms was a combination kitchen/dining room/parlor. A door off the kitchen was carved into the hillside to form a *cueva* (Spanish for "cave") for food storage. The second room was a bedroom with a fireplace for the entire family.

The inside walls were coated with plaster made from *yeso* (Spanish for "gypsum," that was quarried from the vertically-standing Todilto Formation located near the farm), flour for paste, and coal-oil for spreadability. The outside walls were plastered with a mixture of mud, hay, and a deep red sand found nearby. The ceiling consisted of a bed of *vigas* covered with two layers of one-inch-thick decking, and topped with a layer of earth for the *zotea*.

Lighting was provided by windows that were made quite small because of the need to maintain the load-bearing walls and due to the scarcity of window glass. A broad, covered portal ran along the elongated south side of the house. This first house was therefore a fine example of Spanish Colonial architecture characterized by construction via local craftsmanship using local materials (Bunting papers, undated).

In 1918 the growing Cebada family added a second story to the *Casa Vieja*. They removed the roof of the first story, added a new planked floor and ceiling for the second story, timbered walls, and a pitched roof enclosing an attic. Access to the second floor was only via an outside stairway up the portal to a door on the second floor. The *Casa Vieja* remained simple and functional (Bunting papers, undated). Meanwhile José María Cebada operated a trading post and store (Cebada-Córdova 1995).

In the late 1930s the Cebadas sold the house to Richard Wetherill, Jr., son of the discoverer of Mesa Verde, archeologist at Chaco Canyon, and Navajo trader (Chapter 25). (After the murder of Richard Sr. at Chaco Canyon in 1910, his widow Marietta had moved with her children to a small community called Gonzalitos, then located a few miles south of Cuba.) Wetherill, with greater resources than the original owners, the Cebadas, soon began adding a *casa nueva* (Spanish for "new house") on to the original structure. He extended one side of the original house, and included a second floor and attic extensions. He knocked out the wall separating the original two rooms, making it a "hall" to the new addition. He converted the original east window into a door, and extended the portal the length of the entire structure. He then covered the pitched and balcony roofs with tin sheeting in keeping with the style of the times. Finally, in 1953 he extended the portal/balcony around the west end. The house was "finished" (Bunting papers, undated). The house lies there today as private property and it's a pity that this icon of New Mexico architectural history is unavailable the public. Hopefully some day it will become accessible.

Señorito

Another town, *Señorito* (Spanish for either "little gentleman" or "little Christ"), was founded about the same time as Copper City, about two miles to the north (see below), first as a trading area and after 1893 as a coal mining camp (Figure 114). It had a post office (1901–1924), store, sawmill, a few houses, a copper smelter, and a workforce of about 100. The town's purpose was to mine coal from the steeply-dipping outcrop of the Menefee Formation (the same coal-bearing unit mined in the La Ventana area) and to supply the coal to the smelter. The Hispanic miners were recruited locally and lived in tents. The smelter rendered the copper ore into a "matte" (an impure metal concentrate) that was wagon-hauled down to the railhead in Bernalillo. There was no stage line and mail had to be brought in on horseback. The nearest doctor was 60 miles away in Jémez Springs.

Mining activity ceased about 1920 and the post office lingered on until 1924 (Sherman and Sherman 1975). Life in this little camp was nicely captured by Roy Stamm (1999), an Albuquerque businessman, in his autobiography. In 1899 the 23-year-old Stamm received a call from his merchant father to get on up to Señorito from Albuquerque posthaste to take a job with the Jura-Trias Mining Company. He jumped into a buckboard and bounced over the 107-mile post road to the town. There he worked as a "storekeeper, bookkeeper, postmaster, weighmaster, supplyman, and anything else which was loose and not technical." Meals were prepared by local cooks, and consisted of goat meat, beans, and chile twice a day that was "stout enough to hold you up if you stepped into it."

Copper City

The town of Copper City, a few miles southeast of Cuba, was founded in 1882 to service the ores that were wagon-hauled in, mostly from the Eureka Mine to the east (Figure 114). It was a collection of several well-stocked stores, two saloons, a good hotel, a restaurant, and school, and the site of the first copper smelter. In 1883 the town had a population of about 500. Most of the buildings were of timber supplied by a local sawmill, but some were of adobe. The town's principal supporter was Paul Langhammer, who promoted the town in the Albuquerque newspapers during 1882 and 1883. He predicted that Copper City would become the "metropolis of copper producing in New Mexico," and that it would soon rival Leadville, Colorado, in its importance. In 1883 he wrote, "Hundreds of towering chimneys of furnaces and factories will arrive and darken the clear atmosphere with volumes of smoke as sign posts of industry and stepping stones to wealth." He was clearly a man of his time! He continued, "The brand of the Copper City ingots of unrivaled pure metal will be known all over the world and find a ready market" (Bryan 1966). What the town really needed of course was a railroad. At the time there was talk of a Santa Fe line coming northwest from Bernalillo or a Denver and Rio

Grande line coming west from Española, but a railroad was far off in the future. After 1883 little about the town appeared in the newspapers. The town folded in 1890 (Bryan 1966; Sherman and Sherman 1975).

Copper mines

Serious mining in the so-called Nacimiento District began in the 1880s, faded after 1917, and increased again in the late 1960s. The most important deposits are contained within the massive Agua Zarca Sandstone of Triassic age. (This is the same unit that forms the surface of the south-sloping ramp at the southern end of the Sierra Nacimiento west of San Ysidro.) The Agua Zarca comes to the surface from the depths of the San Juan basin and curls and bows up along the western rim of the Sierra. Most of the copper ore is concentrated in ancient, west to northwest-trending river channels, where the sandy sediment is locally coarse, porous, and chock full of ancient tree logs and other organic debris (Figure 117A).

The channel sands are encased in finer-grained, non-porous sands, which are devoid of ore. Mineable ore deposits occur where the linear channel sands fortuitously intersect a favorable structural position on the western, uplifted rim of the Sierra. These two conditions are met at only three locations, i.e., the San Miguel, Eureka, and Nacimiento mines (Figure 114).

The copper minerals occur in two general chemical states: 1) at shallow depths, oxidized ore consisting mainly of the attractive green mineral "malachite," $Cu(CO_3)(OH)_2$; and 2) at greater depths, reduced, sulfide ore consisting mainly of the black mineral "chalcocite," Cu_2S. Most investigators believe that the copper had been leached from older, distant, uplifted copper deposits by oxidizing groundwater during or shortly after the deposition of the Agua Zarca Sandstone (ca. 220 Ma). The copper-bearing groundwater was driven downgradient by gravity, preferentially through porous zones in the Agua Zarca Sandstone (Figure 117B). When the copper solutions encoun-

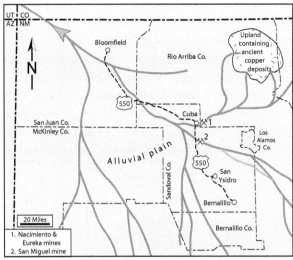

A. Paleogeography during deposition of Agua Zarca channel sandstones (gray lines), ca. 220 Ma. (Modified from Lucas 2004.)

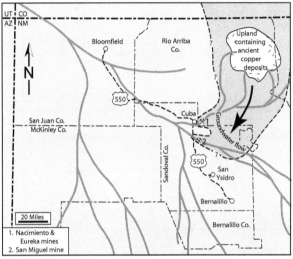

B. Movement of copper-rich groundwater (black arrow and darker gray area) shortly after shallow burial of Agua Zarca channel sandstones (gray lines)

Figure 117. Speculative origin of copper-ore deposits in Triassic-age Agua Zarca channel sandstones. (Modern geographic features shown for reference.)

tered a chemical environment turned acidic by decaying organic matter, some of which consisted of huge logs, the copper precipitated as the sulfide mineral chalcocite, around, onto, and into the organic matter. Much later, when the Sierra Nacimiento was uplifted and exposed to the atmosphere, the upper part of the now-shallow chalcocite was oxidized to the colorful malachite and azurite (Talbott 1984).

Prior to 1900 most of the production came from the San Miguel Mine (Figure 118A) and totaled about 6.3 million pounds of copper and 63,000 pounds of silver. Through 1964 the Nacimiento District produced over 7.5 million pounds of copper and about 75,000 ounces of silver (McLemore 1996). Clearly, from 1900 to 1970 production was minor and sporadic and came mostly from the San Miguel and Eureka Mines (Woodward 1987).

The Nacimiento Pit (Figure 118B) covers the Copper Glance-Cuprite Claims, which were patented in 1929. Until 1917 most of the workings consisted of diggings in the shallow, oxidized zone, although small amounts of high-grade sulfide ore was extracted and hauled by wagon to railheads in Aztec to the north and Bernalillo to the south for shipment to smelters in Durango, Colorado and El Paso, Texas, respectively. Interest dropped off in 1918 after the end of World War I. In 1968 the property was acquired by Vitro Minerals Corp., which merged into Earth Resources Company.

A robust exploration program was followed by startup of the mill in 1971 (Talbott 1984). The mill was designed to handle an estimated reserve of almost 900 million tons of ore containing 0.7% copper (6.4 million tons of copper). The ore was mined by open-pit methods. In 1973 a tailings dam broke and in 1974 the company shut down the mine. The property changed hands several times in the 1970s and 1980s. In the late 1980s, the company Leaching Technology, Inc. purchased the mine. The company conducted a pilot *in-situ* leaching project. Treated water was pumped down special injection wells to the sulfide ore where the water dissolved some of the ore. The copper-rich solution was pumped to the surface via another set of wells, the solution was circulated over electrically charged iron elements, and the copper plated out. However, poor recovery and environmental concerns doomed the operation (McLemore 1996).

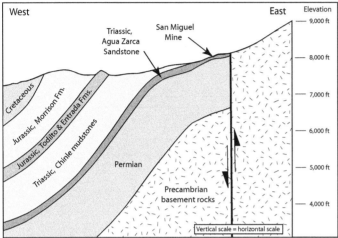

A. San Miguel Mine. (Modified from Woodward and Schumacher 1973; Woodward 1987.)

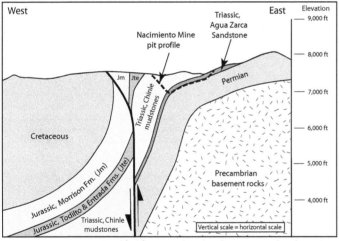

B. Nacimiento Mine (Modified from Woodward et al. 1970a; Woodward 1987.)

Figure 118. Cross sections through Sierra Nacimiento copper mines.

The Señorito Project

Visible from US-550 and clearly visible from the Village of Cuba is the enormous pile of whitish tailings at the Nacimiento Mine, located just north of highway NM-126 at the mouth of Señorito Canyon. The pile was very susceptible to erosion, and an additional source of sediment to the already sediment-choked Rio Puerco was definitely not needed. A group called the Quivira Coalition received a grant from the Environmental Protection Agency to fund what is called the "Señorito Project," and work began in the summer of 1999. The project's aim is to spread seed and hay over the barren, devastated terrain and to run cattle across it. This will create conditions for plant growth and soil development through the tried and true low-tech method of "poop and stomp." The coalition hopes to use this as a demonstration project and then to promote the methods for application elsewhere in the Rio Puerco watershed (Soussan 2000).

14

The Bone War and the Missing K-T Boundary

The so-called great "Bone War" raged during the quarter century from 1869 to 1897. Unlike most battle sites, there are no monuments or commemorative plaques to document desperate infantry charges or artillery duels, no riveting tales of heroism amidst the fog of battle and the acrid smell of cordite. There are no annual re-enactments by weekend warriors and veterans' organizations. In fact the war was fought entirely in print, far away from New Mexico, by two paleontologists. A paleontologist is one who studies the history of past life via their fossilized remains. This war was fought over reputations, and fossils—especially dinosaur fossils—but in significant part over mammalian fossils from the eastern edge of the San Juan basin near Cuba, New Mexico. The opposing principals in the Bone War were O.C. Marsh (1832–1899) and E.D. Cope (1840–1897), two of the most prominent "vertebrate" (=animals with backbones) paleontologists of the 19th century, a time when American science was in its infancy. It is Cope who is of interest to us because he made his most important discoveries near Cuba.

19th century natural science

The Bone War cannot be understood outside the context of the times, so a brief examination of the world of science in which the two men lived is appropriate. The term "scientist" didn't even exist until about 1840. During the second half of the 19th century the entire number of scientists in America numbered somewhere between 300 and 1,000. To become one didn't require specialized schooling. In fact most American "scientists" were medical doctors who used their knowledge of anatomy to study the natural world on the side. Others were self-educated, often wealthy men who followed their research interests as hobbies. There were no specializations like "biologist" or "geologist." More often than not these men considered themselves simply as "naturalists" (Jaffe 2000).

At the time America lacked the prestige that European universities and academies had. Americans in general saw no value in knowledge for knowledge's own sake—they were a more practical people. But by mid-century voices were being raised to make science in America more respectable, and in 1847 the federal government established the first entity in America dedicated entirely to the advancement of scientific knowledge, the Smithsonian Institution in Washington, D.C. (Jaffe 2000).

The middle part of the 19th century was a time of great intellectual upheaval. Early in the century the generally accepted history of the earth was based on the bedrock principle set down in the Book of Genesis (*Genesis*, 1:31), that God created the world in six days (a tenet still faithfully believed by many today). As mentioned earlier in Chapter 1, in 1654 the English Archbishop and biblical scholar James Ussher (1582–1656) performed an intensive study of the Hebrew calendar and the chronology of the Bible. He added up the life spans of the descendants of Adam and determined that the world was created on Monday, October 23, 4004 BCE, at 9:00 AM, and that it had remained unchanged ever since. Thus was born the belief that the world was only 6,000 years old.

Despite observations being accumulated by others, by the end of the 18th century the Biblical version of facts and Ussher's 6,000-year age for the Earth were generally accepted as unshakable articles of faith. But cracks in the dam were appearing. The Industrial Revolution in western Europe led to the construction of many canals, and engineers began to cast a critical eye on the sequences of layered rocks and their contained fossils that they found exposed in the excavations. Fossils had been recognized for many centuries but now careful observers saw that certain sequences of strata contained different kinds of fossils with different shapes and sizes. The observers wondered why these life forms had changed through time, i.e, upward in the sequence of layered sedimentary rock.

Such ideas directly challenged the authority of the Bible, and such a brash action was considered unthinkable at the time. By the early 19th century many concluded that the real purpose of direct observation should be to uncover the evidence needed to reconcile the growing tension between these dangerous new ideas and Biblical teachings. They believed that the rocks and landforms of the modern world were the product of Noah's flood, and they set out to prove that assumption.

The first official dinosaur discovery was made in southern England in the early 1820s by Gideon Mantell (1790?–1852), a medical doctor and avid fossil collector (some said his wife actually made the discovery). The find consisted of several huge teeth. In 1824 another Englishman, the geologist Rev. William Buckland, belatedly described dinosaur fossils that he had found six years before. For the next two decades fossil hunting became the rage in Europe. In 1841 English anatomist Richard Owen (1804–1892), attempting to coin a word to describe these strange creatures, combined the Greek words *deinos*, meaning "terrible," and *sauros*, meaning "lizard." Thus these animals became known as belonging to a group called the *Dinosauria* (Wallace 1987).

In 1859 Charles Darwin (1809–1882) published his *On the Origin of Species by Means of Natural Selection*. He based his theory on observations he had made during his five-year (1831 -36) voyage around the world as a naturalist on the *H.M.S. Beagle*. For years he postponed publication of *Origin* because of worry about the impact his work would have on the established beliefs and on the anticipated backlash. His worries were justified. He proposed that species, including Man, evolved by slow, gradual steps from simple to more complex forms. The public was both astounded and outraged by the implications of this explosive work. If unchallenged this would remove the control of human destiny from the hand of God and place it in the dubious clutch of Nature. Darwin conceded that because of the slow pace of evolution his theory perhaps could not be proven by observations of the modern world. The proof therefore must lie in the distant past, a past revealed by study of the fossil record, i.e., by paleontology.

Meanwhile, on the other side of the Atlantic in the U.S., scientific inquiry took a back seat during the trauma of the Civil War. After the war's end in 1865, however, the victorious North was on a roll. With its abundant pent-up energy, industrial strength and infrastructure, it turned its face westward to the frontier and wondered what was out there. To be sure, the U.S. had already established commercial links to the Spanish Southwest via the Santa Fe Trail, homesteaders had surged to the Pacific Northwest along the Oregon Trail, the Mormons had blazed a trail to the valley of the Great Salt Lake in Utah, prospectors and miners had poured over whatever rocks they could reach by wagon or mule, and in 1849 the California gold rush drew thousands to the Pacific West. All these areas were rather well-known, but what might occupy the vast spaces in between that had largely been skipped over? What mineral wealth could be tapped, what areas could be opened to settlement and farming, what investment opportunities existed? In other words, what was this territory worth?

In 1876 the government responded to this heightening curiosity by launching what later would become known as the "Great Surveys" (Bartlett 1962). Four great surveying efforts were dispatched to map the West. The U.S. Department of the Interior directed two of them: 1) the U.S. Geographical and Geological Survey of the Territo-

ries, led by medical doctor Ferdinand Vandeveer Hayden (1829–1887), and 2) the U.S. Geographical and Geological Survey of the Rocky Mountain Region, led by John Wesley Powell (1834–1902). The War Department supervised the other two: 3) the U.S. Geological Exploration of the 40th Parallel, led by a civilian, Clarence King (1842–1901), and 4) the U.S. Geographical Surveys West of the 100th Meridian, led by Lt. George Montague Wheeler (1842–1909). By 1879 the King and Hayden surveys were finished and the newly-formed U.S. Geological Survey took control of the other two.

The Great Surveys' contributed an enormous amount of knowledge about the geography, topography, geology, and natural history the American West. Because much of the work was supplanted by later researchers, the huge efforts of the men of the Great Surveys are today largely forgotten, and only the daring account of Powell's exciting boat trips down the Colorado in 1869 and 1871 still captures the modern imagination. But the Great Surveys did more than map the West. They took along with them photographers, artists, zoologists, biologists, botanists, and paleontologists in order to construct a truly comprehensive picture of the West and its resources.

The fossil hunters

The first dinosaur fossil in the U.S. was identified in 1856 by Joseph Leidy (1823–1891) based on teeth recovered from Montana. Leidy was Professor of Anatomy at the University of Pennsylvania Medical School. By education he was a medical doctor, but after only two years he abandoned his practice and in 1847 began to teach anatomy at his *alma mater* and to collect vertebrate fossils in his spare time. In 1858 Leidy collected the first dinosaur fossil skeleton in the U.S. from the marls of New Jersey's southern coastal plain on the Atlantic coast—at the time the principal source of large vertebrate fossils. But deep down he remained an anatomist. He published descriptions of his finds but provided no interpretation of their meaning. He can hardly be blamed for that. At the time the available vertebrate fossil record was almost nonexistent. Collection techniques of the time were haphazard and sometimes downright sloppy. The recovered specimens were notoriously incomplete: a thigh bone here, a tooth or skull fragment there. What was needed were some more complete fossils. Enter the two paleontologists Marsh and Cope—the two protagonists of the "Bone War"—onto the stage.

Othniel Charles Marsh

Our first featured paleontologist, Othniel Charles Marsh, was born in 1831 on a farm in western New York state (Jaffe 2000). At the age of 21 he considered a career in carpentry or school teaching, but just in the nick of time his Uncle George intervened. The uncle was George Peabody, a millionaire banker living in London. Marsh inherited the dowry that had been set aside for his late mother by Peabody. With this windfall Marsh entered preparatory school and then, with Peabody's support, Yale College, where he studied geology and paleontology. In 1861 he received his Bachelor of Arts. Due to poor eyesight he avoided military service in the Civil War and instead, with Uncle George's continual financial backing, took graduate studies at Yale and at several universities in Germany.

During the early 1860s the aging George Peabody was planning the donation of his fortune. Marsh persuaded him to consider Yale, and Peabody obliged. Marsh wrote to Yale's science faculty and announced that Peabody would grant $150,000 (a sizable sum in those days) to Yale, and thus in 1866 the Peabody Museum of Natural History was established. In a second letter accompanying this announcement, Marsh inquired about the availability of a teaching position at Yale. Without hesitation Yale formed a chair of paleontology, and in 1866 Marsh occupied it. The position, however, was not funded. However, with an ample inheritance he received after Peabody's death in 1869, Marsh did not have to deal with the messy business of teaching and was allowed to do essentially what he wanted, and that was to collect vertebrate fossils.

Marsh realized that the best place to find large, intact fossils was the vast expanse of the unpopulated American West. The Indian wars were winding down and travel with some degree of safety became possible. He immediately began planning a massive fossil collection expedition out West. With a group of Yalie students in tow he made four expeditions to the West between 1870 and 1873. He amassed a large collection of vertebrate fossils for the Peabody Museum and built a formidable reputation.

Edward Drinker Cope

Edward Drinker Cope was born in 1840 to a Quaker family in the outskirts of Philadelphia, Pennsylvania (Jaffe 2000). His father wanted him to be a farmer, but Cope resisted. In 1860 he convinced his father to send him to the University of Pennsylvania to take Joseph Leidy's anatomy course, pleading that such knowledge would be valuable treating livestock on a farm. Being a Quaker and a pacifist, Cope avoided joining the military at the outbreak of the Civil War, and in 1863 Cope's father sent him to Europe for the grand tour. It was in Berlin that Cope met Marsh, and the two struck up a cautious friendship. In 1864 Cope was back in Philadelphia. His family used its influence to gain Cope a position as professor of zoology at a Haverford College, a small Quaker school. The college immediately granted him an honorary master's degree. Disillusioned by college life he resigned his post in 1867 and the next year moved just across the Delaware River to New Jersey to be close to the fossil localities. He soon realized, like Marsh, that the West held the most promise for major finds and career enhancement.

Seeds of conflict

In 1868 a shipment of fossil fragments arrived in Philadelphia from a huge aquatic dinosaur found in the plains of western Kansas. Cope set to work assembling the disorganized assortment of bones and finished reconstruction the next year and published his results. Marsh visited Cope and with some glee pointed out that Cope had placed the skull on the tail instead of the neck! Cope was deeply embarrassed and offended. The relationship between the two would never be the same and thus began a bitter rivalry that would last until Cope's death in 1897.

Marsh spent only four seasons in the field (1870–1873). Henceforth he became more of an armchair paleontologist and employed crews of professional collectors to provide his specimens. In contrast, Cope, beginning in 1871, spent many seasons in the field but in time also employed professional collectors. Marsh was affiliated with the King and Powell surveys, who sent him fossil specimens, with the Peabody Museum at Yale, and later with the Smithsonian Institution in Washington. Cope received his specimens from the rival Hayden survey and was affiliated with the Academy of Natural Sciences in Philadelphia.

The Marsh and Cope crews operated in great secrecy and were known to spy on each other. When a crew encountered a significant find they would rush to the nearest railroad telegraph station and cable a description back east. Either Marsh or Cope would then, in great haste, rush the description into print in a scientific journal. This led to duplication and error. Often working from incomplete skeletons they gave different names to the same animal. At this time there were no established publication standards. The ensuing mess would have to be sorted out later. Most of the increasingly bitter rivalry was sparked by the fabulous dinosaur quarries from the Jurassic Morrison Formation of Montana, Wyoming, and Colorado. Amazingly they never communicated with each other about their finds.

Bone war in the San Juan Basin

Why is the area near and west of Cuba important to this story? Some additional background is in order. One of the most dramatic events in the history of the Earth's past life was the extinction of the dinosaurs, as well as about 70% of all living species, at the end of the Cretaceous Period of geologic time at 65 Ma. After that fateful date, mammals took over the land and proliferated. This profound shift in the character of life separates the earlier, Mesozoic Era (era of "middle life," also known as the age of reptiles) from the later, Cenozoic Era (era of "modern life," or the age of mammals) (Figure 4). Dinosaur fossils were not only spectacular but enormously popular with the public. Marsh and Cope via their professional collectors scrambled over the areas of Mesozoic outcrops of the West, especially in Wyoming and Colorado, and made some fabulous finds. They found that the overlying, and therefore younger, Cenozoic strata strangely lacked dinosaur fossils and realized that something important had happened between the two sequences. Both sequences are superbly exposed in the San Juan basin, and Cope was the first to hit pay dirt there.

Cope had attached himself to the Wheeler survey for one year (1874) as survey geologist. He was not interested in rocks but rather he really sought to sponge logistical support from the survey for his fossil hunting. While in the area north of today's Cuba he heard reports from the locals about fossil riches in the San Juan basin to the west. But Wheeler was not impressed and intended to move north into Colorado. Cope finagled a split-off from the main party with some men and supplies (Fleck 2001). For two months he blissfully collected fossils in the area centering on Arroyo Blanco between Regina and Llaves, 20 miles north of today's Cuba. On a summer day in 1874 he found fossils of eight different small mammals, much more primitive than anything he, or anyone else, had yet discovered. He assigned these remains to the Eocene ("dawn of the recent") Epoch, the oldest unit of the Cenozoic Era that was then recognized (Figure 4).

During the next 40 days he assembled a large collection of fossil mammals. Cope considered this the most important discovery of his career. But he was nominally still part of the Wheeler Survey and his work was later published as part of the official survey reports. Cope was never invited back to join the survey. That same year Cope searched for fossils in the badlands of the Nacimiento Formation, which makes up the soft slopes underlying the cap of Cuba Mesa (sometimes called the "Mesa de Cuba"), west of Cuba. He found no fossils there but realized that the unit was older than the fossil finds he had made to the north along Arroyo Blanco (Simpson 1981).

During the next decade Cope dispatched professional fossil collectors to continue the search the Nacimiento badlands. One of these collectors was a colorful character named David Baldwin (1835–?). Not much is known of this sturdy field man. He lived much of his life in Abiquiu, on the Rio Chama north of the Jémez Mountains, and the end of his life in Farmington, then the metropolis of the San Juan basin. In 1875 he too worked with the Wheeler Survey. From 1875 to 1880 he collected fossils for Marsh, who was then following up on Cope's discoveries in this area, much to Cope's irritation. He did most of his collecting on the west side of the Continental Divide between today's Lindrith and Gavilan. Marsh was not pleased by what Baldwin was sending him, but Baldwin was more upset by the difficulty he was having getting paid. The two split in 1880 and Baldwin transferred his allegiance to Cope. We can scarcely imagine the hurrumphs that must have issued from the hallowed halls of Yale when Marsh heard the news. To make matters worse, Baldwin went on to make the most important paleontological discovery in the San Juan basin and put the basin firmly on the map (Simpson 1981).

Baldwin must have given an interesting first impression. Fossil collecting takes infinite patience and a determined countenance, and he was an extremely patient and highly skilled collector. One of his colleagues from the Wheeler Survey described him as "equipped like a Mexican with a burro, some corn meal, and a pickaxe." He often worked alone, and often in the winter so that snow found on the ground could be melted for water in this vast arid area (Simpson 1981).

During the years 1881–1886 Baldwin went about his collecting in the basin, especially from that interval between the youngest recognizable Cretaceous and the oldest known Eocene. He shipped a stream of mammalian fossils to Cope. Cope described them as fast as he could and realized that these were the earliest Cenozoic mammalian fossils known. For a while he assigned them—and the beds that contained them—as belonging to the "Lower Eocene." Soon he learned that Baldwin was finding these fossils in beds equivalent to the badland-forming

Nacimiento Formation he had studied at Cuba Mesa. Later this interval of beds was assigned a new age name, the "Paleocene Epoch" ("early dawn of the recent") to accommodate this primitive collection of life forms (Figure 4). The Paleocene, the rocks of which age are nicely exposed in Cuba Mesa, is the lowest division of the Tertiarty Period and thus the crucial epoch that recorded the fundamental shift in the world's terrestrial life forms (Simpson 1981).

K-T boundary

The precise contact between the Cretaceous Period (abbreviated "K") at the end of the Mesozoic Era, and the Tertiary Period (abbreviated "T") at the beginning of the Cenozoic Era at 65 Ma is today known as the "K-T boundary" (Figure 4). Why did the remarkably successful dinosaurs, which had dominated the land for an incredible 160 million years, and many other species to boot vanish in a geologic blink-of-an-eye? (In fact, although the dinosaurs did virtually vanish at this point, their numbers had been dwindling for some three million years prior to 65 Ma due to deterioration of the climate caused by massive volcanic eruptions in India). The mammals, which had existed as tiny, unimpressive creatures underfoot the larger dinosaurs since as far back as 215 Ma, suddenly found whole new habitats totally vacated by their huge competitors. Their numbers and diversity exploded into a new world during the Cenozoic Era. Just what happened at the K-T contact?

Since the dawn of paleontology the demise of the dinosaurs had been explained in terms of changing food supply, volcanic disturbances, and/or climatic change, but no one explanation seemed to work. Then in the late 1970s a research team from the University of California, Berkeley, led by Walter Alvarez, was studying excellent exposures of the K-T boundary in Italy and Denmark, and discovered that a clay unit at the contact contained from 30 to 160 times the normal background content of the rare element iridium (Frankel 1999). No known geologic mechanism could explain this anomalous concentration. The process of elimination led to a question: Could the source of the iridium be extraterrestrial? This question gave rise to what would be called the "impact hypothesis."

In time other iridium anomalies were discovered in many locations around the world, all of the same age. By estimating all the global iridium represented by these deposits (500,000 tons), and comparing that figure to the average iridium content in meteorites, investigators determined that the meteor or comet that impacted the earth was about six miles in diameter and would have left an impact crater about 110 miles in diameter. But if such an extraterrestrial body slammed into the earth, where was the crater? A number of ancient impact craters are known from the geologic record, but none had the proper age or size.

In the late 1980s studies of unusual sedimentary deposits in the Texas Gulf Coast and in Haiti pointed to a nearby source of an impact, perhaps near Mexico's Yucatan Peninsula. A Canadian geologist, Alan Hildebrand, was intensely studying the problem. He consulted with petroleum geologists from the Mexican national oil company Pemex, which had drilled a number of exploratory wells in the 1950s in the Yucatan Peninsula (but had found no oil) and in 1978 had conducted an intense geophysical survey of the area. Their surveys found a huge crater, now largely-filled-in, half on land in the northwestern part of the peninsula and part in the shallow waters to the north in the Gulf of Mexico (Figure 119A). By 1990 Hildebrand and his colleagues had pulled all the clues together and fingered the Yucatan feature as the impact event responsible for the great K-T die-off, and in 1991 he finally convinced the scientific community (Frankel 1999).

The Yucatan crater was dubbed "Chicxulub" (pronounced Chick-shoe-LUBE), from a small fishing village on the Yucatan coast at the crater's center. The crater measures about 110 miles in diameter. As the figure shows, at the time of the impact, 65 Ma, the area was occupied by a shallow sea. From the estimated size of the impactor (six miles in diameter) and the average velocity of most asteroids and comets crossing Earth's orbit (about 12.5 miles per second) the intensity of the impact was calculated to be equivalent to 100,000,000 megatons of TNT (Frankel 1999). This equates to about 6.6 billion Hiroshima-size atomic bombs, detonated in an instant at a single locality! The planet must have rung like a bell!

All above-ground animals and most land plants were incinerated during this awful episode. Those marine plants and animals living in the shallow, oxygenated waters were killed. Only small, burrowing animals such as mammals, fresh water creatures, some lucky land plants, and marine life that lived in deeper waters tended to survive. Eventually the atmosphere recovered and the survivors entered a new age and found a host of unoccupied niches to exploit (Frankel 1999). The dinosaurs were gone.

Or were they? Here is an interesting little postscript. For years there have been questions about whether perhaps a few dinosaurs might have survived the event. Recently the partial remains of a single Cretaceous-type dinosaur was found near Farmington in a unit called the Ojo Alamo Sandstone, dated at about 64 Ma (oldest Paleocene age). This is the youngest dinosaur ever found. It is believed that of the thousands of dinosaur eggs laid just before the impact, a few had somehow survived. The buried eggs apparently provided a safe haven for maybe a year or two after the event. When the critters hatched a small community of dinosaurs was able to eke out a life in a new world with new competitors. However, they lasted only for a million years after the impact and then they indeed disappeared from the earth forever (Fassett et al. 2002).

Via an exquisite quirk of geologic irony, the distinctive iridium bed at the exact K-T boundary has never been found in outcrop in the San Juan basin (Figure 119B). The K-T boundary is a physical thing that would have been preserved if deposition had been continuous between the latest Cretaceous-age rocks below to the earliest Tertiary rocks above at the precise time of 65 Ma. It turns out the age of the youngest Cretaceous unit in the basin, the top of the Kirtland Formation, is about 71 Ma, and that of the oldest Tertiary unit, the base of the Ojo Alamo Sandstone, is about 64 Ma (Cather 2004, Figures 4 and 119B). Therefore, for a short period of time, only about seven million years (71 to 64 Ma) the San Juan basin was a low-lying alluvial plain that was essentially a surface of non-deposition. The iridium dust that fell out of the sky onto this land surface was washed away before it could be buried, and therefore it was not preserved. In contrast, to the east near Raton, New Mexico, where the landscape at that time was closer to sea level, the K-T boundary with its distinctive iridium bed has been preserved (Orth et al. 1987).

The whole story is therefore much more interesting that the one that was available to Marsh and Cope. They would have been astounded — and utterly fascinated. As it was, the Bone War between the two fizzled out with the death of Cope in 1897. Today the once-bitter conflict is at best an esoteric footnote in the history of science.

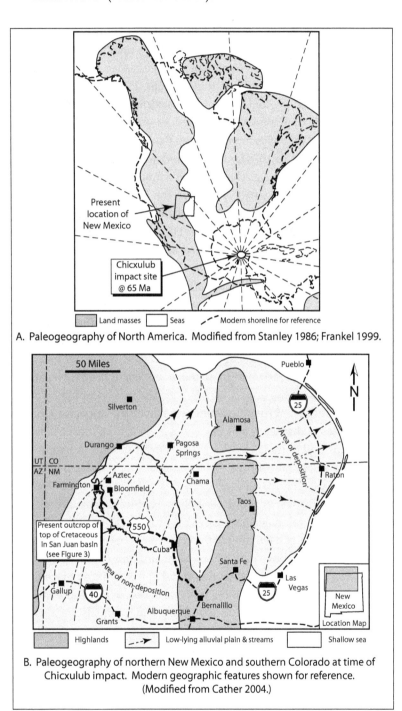

A. Paleogeography of North America. Modified from Stanley 1986; Frankel 1999.

B. Paleogeography of northern New Mexico and southern Colorado at time of Chicxulub impact. Modern geographic features shown for reference. (Modified from Cather 2004.)

Figure 119. Northern New Mexico and southern Colorado at end of Cretaceous Period, ca. 65 Ma.

15

Cuba — Midpoint on Old NM-44/US-550

The Village of Cuba, located about midway between Bernalillo and Bloomfield on highway US-550, is the only watering hole and overnight rest stop along the lonely 151-mile stretch between the two end points. Downtown metropolitan Cuba occurs at a lofty elevation of 6,900 feet and therefore receives a hefty dose of winter weather. The community incorporated as a village only in 1964. Village government consists of a mayor, a mayor pro tem, and three occasionally combative councilors. The population in the incorporated area is about 800 (2000 Census), but the town serves an area population of nearly 12,500 (*Cuba News* 2001a). The merchants depend heavily on the Navajo trade from the "Checkerboard Area" to the west. The community is truly tri-cultural, with a population that is about 60% Hispanic, 25% Navajo, and 12% Anglo. Cuba is not an old community — only a century and a few decades old — and it therefore lacks the depth of history of a place like Santa Fe, Bernalillo, or even Albuquerque. However, what history it has is appreciated by its residents. The village is very much a gateway, both to the vast San Juan basin country to the west and to the Santa Fe National Forest and San Pedro Parks Wilderness areas to the east.

Because of its central location I have placed considerable emphasis on this place and its essential character. Many people, especially native Cubanos, of course know a great deal more about this area than I ever will, so I'm not posing as an expert. However, I have compiled the area's history and geology in such a way designed to create a different and interesting way of looking at the place. I hope Cubanos approve.

Pre-settlement

The site of present-day Cuba was always destined to become a viable community because of its location at the western foot of the Sierra Nacimiento "water tower" (Figure 15A). The northern and loftiest part of the Sierra, known as the San Pedro Parks area, is a huge, flat-topped platform, much of it above an elevation of 10,000 feet. This elevated mass sweeps moisture out of the atmosphere and sends it gushing down in streams in all directions. Three of these streams — the Rio Puerco, the Rito Leche, and the Rio Nacimiento — water the Cuba area. The Sierra is also the main source of the area's groundwater, some of which discharges from springs at the lower elevations. As we will see later in this section, an unusual set of circumstances allowed the Rio Puerco to pirate the headwaters of several west-flowing streams and funnel a disproportionate amount of the flow of water into the Cuba region. The word *nacimiento* is Spanish for "birth," or birthplace of the waters. It was an appropriate name. It was a good place to settle.

Long prior to settlement the rich surface water resources provided an ideal habitat for abundant game. Stone-age hunters, from at least as early as 2000 BCE to as late as about 400 CE, established bases in the hilly terrain around present-day Cuba. Several "scatters" of obsidian (a dark, volcanic glass) and chert (a microcrystalline form of the mineral quartz), both ideal for the manufacture of stone tools, have been found along the present course of US-550 generally between miles 67 and 69. These sites indicate that the people worked the lithic material into projectile points and likely staged hunting forays from there (Post 1994). Much later the presence of water and game attracted the Navajo, along with their rivals and archenemies the Utes, after 1500, and especially by the 18th century. Bands of these people considered the place theirs, although they never established permanent settlements.

Nacimiento Land Grant

In the second half of the 18th century population pressures in the valley of the Rio Grande and its main tributaries forced Spanish settlers to look elsewhere for land and opportunities. Some naturally cast their eyes on the lush upper reaches of the Rio Puerco Valley. In 1769, 36 families petitioned the Spanish governor for a grant of land in the area. That same year an Act of Possession was issued for the San Joaquin del Nacimiento Grant (Figure 120).

The act defined the grant limits as:

On the north: a small arroyo commonly called the Arroyo Tortuga;

On the south: a point of a mesa which points to the south and to the west toward a small hill under the large mesa, and past which runs an arroyo commonly called El Arroyo;

On the east: the east side of the mountain extending from Jémez to the Piedra Lumbre (a valley north of the Sierra Nacimiento and the Jémez Mountains);

On the west: an arroyo commonly called El Arroyo en el Medio.

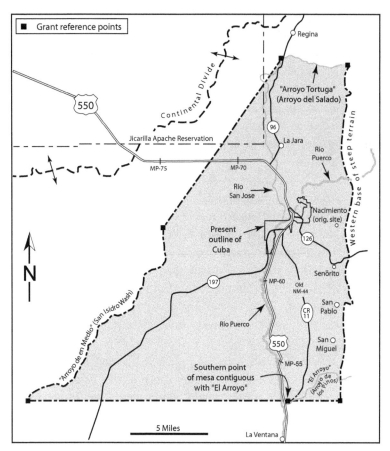

Figure 120. San Joaquin del Nacimiento Grant (gray). (Modern geographic features shown for reference; modified from Luna 1975.)

To acquire ownership of the land the Spanish custom required that the settlers build houses on the land and protect and occupy them for four years. They were also required to divide, clear, work, and stock the land. However, this region was often ravaged by raids from the Navajo, Utes, and even the Comanches. In view of these threats the settlers chose a defendable townsite located about two miles east of the present town on an uneven mesa top overlooking their fields in the valley below (Figure 120). The colonial government, also recognizing the danger, had specified in the grant document that the place should be constructed "with battlements and in compact form." The document was written in provincial Spanish, vs. modern Spanish. Later this apparently esoteric detail would lead to ignoble misunderstandings, as we will see. The tiny community became known as "Nacimiento" (Luna 1975).

Indian raids worsened in 1779. That same year the Spanish governor ordered the preparation of Miera y Pacheco's map of the New Mexico territory (Chapter 9). The map clearly showed the existence of the community of Nacimiento at that early date. However, in 1780 the governor ordered the settlers to relocate for their own safety to such places as Cañon de Jémez (north of the present Jémez Pueblo) and Abiquiu, on the north flank of the Jémez Mountains. In 1782 the observer Padre Morfi reported that all the middle and upper Rio Puerco villages, including Nacimiento and La Ventana, and the middle Rio Puerco villages of Cabezón de los Montoyas (later Cabezón), Rancho de los Mestas (later San Luís), and Guadalupe de los Garcías, were abandoned. For almost the next century there were no permanent residents at Nacimiento, but descendants of the original 36 families would from time to time check on their lands, only to be chased out by the Indians in turn. (Luna 1975).

In 1848 the Treaty of Guadalupe Hidalgo was signed by the United States and the Republic of Mexico. The treaty specified that the inhabitants of the ceded territory of New Mexico would have their liberty and property protected. In 1854 the U.S. created the office of Surveyor General to unravel the virtual "Gordian Knot" of New Mexico land-grant claims, and the next year the first Surveyor General put out a request for the claimants' original documents. In 1870 the heirs and legal representatives of the San Joaquin del Nacimiento Grant submitted their documents to the Surveyor General's office, written of course in the original provincial Spanish. In 1871 the documents were officially filed in an English translation, and in 1872 the sitting Surveyor General recommended that the grant be approved, pending a survey. In 1879 the claimants contracted for a survey, which found that the grant as previously defined contained 131,726 acres. Confusing the issue was the fact that some 60 homestead claims, under the 1862 Homestead Act, had been filed and that about 238 families, mostly heirs to the original 36 families, were already living on the grant (Luna 1975).

But then things quickly began to unravel. In 1885 a new Surveyor General had arrived on the scene. George Washington Julian (1817–1899) was a former congressman from Indiana who had made it his mission to prevent plunder of the public domain by speculators. His philosophy was that no one person or persons in the United States should be able to acquire huge swaths of the public domain to build personal kingdoms. The essential point that Julian overlooked, or chose to ignore, was that the San Joaquin del Nacimiento Grant was never a part of the public domain, and that the U.S. was treaty-bound to honor all genuine titles. Nevertheless, in 1886 Julian rejected the grant (Luna 1975).

Later that year the claimants filed a protest, so Julian ordered one of his men, William M. Tipton, to visit the townsite of Nacimiento and report back to him. Tipton focused his attention on the Spanish requirements to build a town and occupy it for a definite time. He walked over the site and was struck by the lack of remains of a well-defined town with streets and a plaza, unaware of the mistranslation of the original Spanish that had specified a settlement "with battlements and in compact form." He even wondered whether this was really the correct site of the grant. The original grant document had indicated that the northern boundary was the *Arroyo Tortuga* (Spanish for "Turtle Arroyo"). In the late 19th century no one knew of such a feature because over the years its name had mutated to *Arroyo del Salado* ("Salty Arroyo"). The southern boundary had been supposedly defined as a mesa near *El Arroyo* ("The Arroyo"). The translation into modern Spanish had the name as *Arroyo Hondo* ("Deep Arroyo"). At least as early as 1812 this drainage was merely a shallow depression in the ground, and only after the mid-19th century did this drainage begin incising its channel that today is a vertical-walled arroyo some 30 feet deep. To complicate matters, sometime before the late 19th century the Arroyo Hondo became known as the *Cañada de los Pinos* ("Gully of the Pines"). So a combination of mistranslation and geographic name changes over the course of a century had fatally muddied the waters — to the claimants' detriment. Surveyor General Julian rejected the grant a second time (Luna 1975). To this day a legacy of deep bitterness lingers in the hearts of the area inhabitants because of the perceived injustice.

Town of Laguna

Now we move to the middle of the 19th century. Over the decades descendants of the original 36 families had hung onto their land despite the threats from the Indians. They visited the area and ran their cattle in the area whenever they could get away with it. By the mid-19th century the Cuba area was still a grassy wilderness, and the rivers hadn't yet cut their deep arroyos. With the lack of well-defined channels, the area was poorly drained and surface runoff lost itself in innumerable lakes and swamps full of cattails (Calkins 1937; Wetherill 1978).

In the late 1860s and early 1870s, after the threat of the Navajo had been ended, livestock barons Mariano S.

Otero, Sr. and Mariano Perea (son of Bernalillo *patrón* José Leandro Perea) controlled the entire upper Rio Puerco basin. The two of them carved out spheres of influence: Otero controlled the region west of the Rio Puerco and Perea the region east of the river. Together they ran about 240,000 sheep and 9,000 cattle. Beginning in the late 1870s settlers moved in to establish homes and farms. By 1878 there were 20 families in the area (Calkins 1937). A few resettled the old site of Nacimiento but most preferred a drier site two miles to the west, on an elongated drainage divide between the unentrenched Rio Puerco on the west and a swampy area on the east — the site of modern-day Cuba. The new site was variously called *Laguna* ("Lagoon") but many continued for a while to call it "Nacimiento." When the Post Office arrived in 1887 an official name was needed. In the early days people were spread out and lived in tiny communities surrounding Cuba with such names as Lagunitas and Gonzalitos to the south along the Rio Puerco, and Vallecitos del los Pinos and Vallecitos del Rio Puerco up against the Sierra. Another of these was a small cluster of habitations on the west side of the Rio Puerco called *Cubita*, ("Little Water Tank"), and the name was likely borrowed from there. The main town — today's Cuba — was where the stores were located and where the *ricos* ("rich folk") lived (Wetherill 1978).

Geology sets the stage

The Village of Cuba, like the entire stretch of US-550 on the west side of the Sierra Nacimiento, is located on the eastern flank of a huge geologic feature known as the San Juan basin. All the geologic formations dip — i.e., are inclined downward — gently to the northwest toward the center of the basin (Figure 3A). Where the rock layers abut the very western edge of the Sierra they curl sharply upward at very high angles. The deformation of the rock layers occurred during a protracted period of compression mentioned earlier called the Laramide orogeny, which affected this area roughly from 75 to 40 million years ago (Ma). During the following 40 Ma erosion worked its way down into the sequence of dipping formations and sculpted the bedrock landscape (Figures 121, 122A, and 122B).

By about 3 Ma the landscape had been beveled off to a low-relief, stable bedrock surface (Figure 122B). About 1 Ma this part of North America entered a period of great Ice Ages during the Pleistocene Epoch, which after about 800 ka came with regularity about every 100,000 years. Around 600 ka a fluctuation in the rate of precipitation caused the bedrock landscape to be buried by a thin sheet of gravel washed off the eastern heights by streams (Figures 123 and 122C). Remnants of this gravel sheet today form the high, gently-sloping surfaces perched up against the mountains northeast of the village.

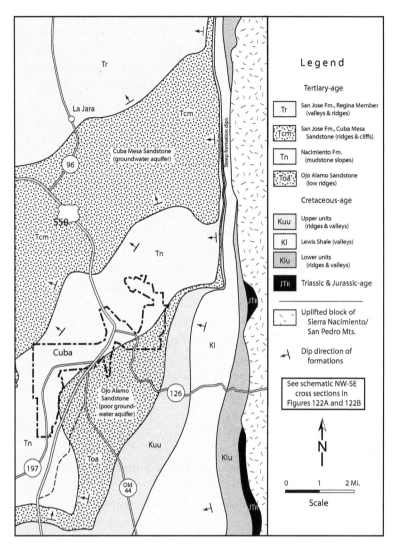

Figure 121. Geohistory of Cuba area–I: Distribution of bedrock units (now partly buried). (Modern geographic features shown for reference; modified from Woodward et al. 1970a, 1970b.)

Figure 122. Highly schematic, unscaled sequential cross sections across Cuba area (see legend in Figure 121).

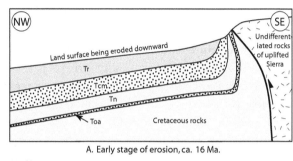

A. Early stage of erosion, ca. 16 Ma.

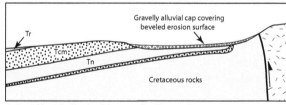

B. Continued erosion (see Figure 121).

C. Partial burial of bedrock by gravelly alluvium from east (see Figure 123).

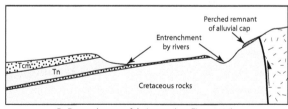

D. Entrenchment of drainages (see Figure 124).

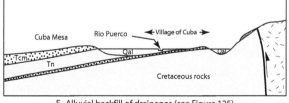

E. Alluvial backfill of drainages (see Figure 125).

After this depositional episode the landscape was once more deeply eroded, this time through the gravel and into the underlying bedrock (Figures 124 and 122D). The dip of the formations ensured that rock units of variable hardness would occur at the surface in this area and produce a rugged landscape of sandstone cliffs alternating with mudstone valleys. A final period of deposition filled in the latest bedrock valleys with up to 90 feet of fine alluvium (Figures 125 and 122E; Bryan and McCann 1936). The low-lying alluvial surfaces would eventually become the preferred sites for human settlement.

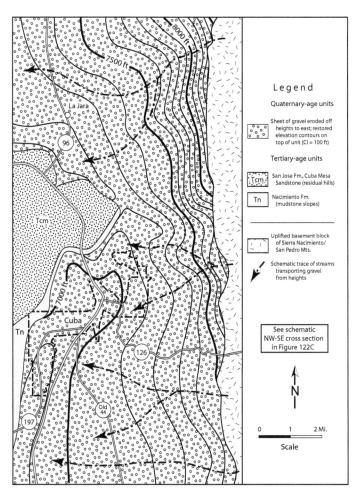

Figure 123. Geohistory of Cuba area–II: Partial burial of bedrock by gravelly alluvium from east. (Modern geographic features shown for reference; modified from Woodward et al. 1970a, 1970b.)

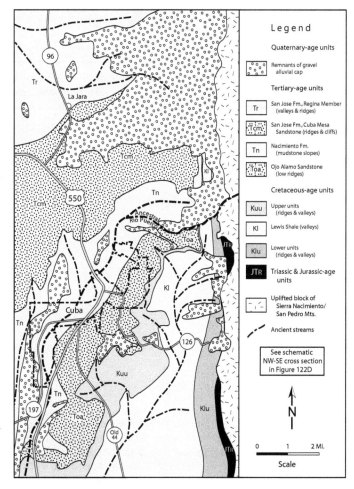

Figure 124. Geohistory of Cuba area–III: Entrenchment of drainages. (Modern geographic features shown for reference; modified from Woodward et al. 1970a, 1970b.)

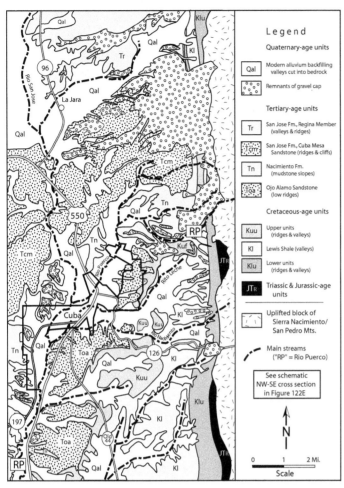

Figure 125. Geohistory of Cuba area-IV: Realignment and alluvial backfill of streams to form modern landscape. (Modern geographic features shown for reference; modified from Woodward et al. 1970a, 1970b.)

Water—the staff of life

For a time the upper reaches of the Rio Puerco were the best watered in the area, and the farmers tended to favor the relatively flat alluvial surfaces in those areas. At the site of modern Cuba the villagers concentrated their activities on one such patch of alluvial land along the main road coming up from the south, lying between the then-unentrenched Rio Puerco on the west and a low-lying swampy area to the east (Figure 126).

The Rio Puerco carried the highest flow because of a fluke of geology. Long ago, sometime after the deposition of the gravel pediment gravel cap, the ancestral Rio Puerco cut deeply in the soft mudstones of the Nacimiento Formation and progressively undermined the gravel cap from the south (Figure 124). The steeper gradient of this southwest-flowing stream allowed its drainage basin to advance headward into those areas in the Sierra once drained by west-flowing streams. The result was the progressive capture, or piracy, of the headwaters of those other streams and the gradual augmentation of the flow of the Rio Puerco at their expense (Figure 127).

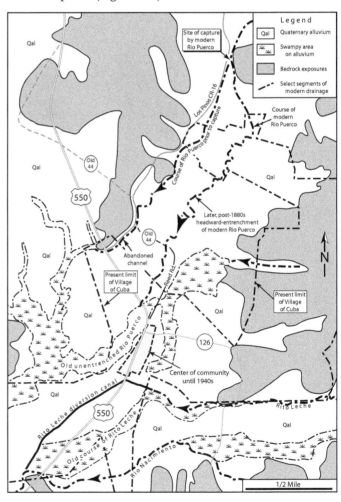

Figure 126. Interpreted pre-entrenchment drainage in Cuba area, late 19th century. (Modern geographic features shown for reference; based on 1935 Soil Conservation Service aerial photographs #2640 and #2641.)

172

Even via the increased flow due to stream capture, the supply of water from the upper reaches of the Rio Puerco and its main tributaries, the Rio Nacimiento and Rito Leche, was neither steady nor reliable enough for predictable irrigation agriculture. The discharge water from the springs along the mountain front was certainly by itself inadequate. Therefore, agriculture was either intermittent, flood-irrigated, or dry-farmed. Most of the snowmelt and runoff from the lofty San Pedro Parks bypassed the Rio Puerco drainage basin and instead flowed to the south down the Rio de las Vacas and its main tributary, Clear Creek, to the valley of the Rio Jémez (Figure 127). One of the first orders of business for the settlers was to augment their surface water supply. In 1882 they organized themselves and via a huge effort dug a 3,000-foot-long ditch to connect the channel of Clear Creek to the head of the Rio Nacimiento. The ditch in effect pirated (or beheaded) Clear Creak and diverted its significant flow down the Rio Nacimiento to the waiting fields with their crops of alfalfa for livestock, beans, and corn.

In 1890 several Hispanic settlers, for $800, deeded a right-of-way to the Rio Puerco Irrigation and Improvement Company. The company had been founded in 1889 by several men, including none other than the powerful politicians Mariano S. Otero Sr. and Thomas B. Catron. The right-of-way was to contain a 14,000-foot-long ditch con-

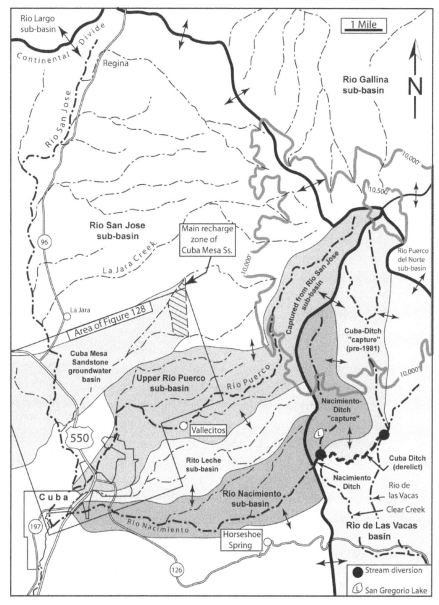

Figure 127. Water in the upper Rio Puerco basin and its subbasins.

necting the existing diversion of Clear Creek to a higher diversion point to the east on the main Rio de las Vaca. This was part of a grand scheme to artificially capture runoff water from the Rio de las Vacas and transport it down the Rio Nacimiento to the Rio Puerco, and from there downstream to vast irrigated lands in the middle Rio Puerco Valley south of Casa Salazar (Dortignac 1962). The hubris displayed by this project was monumental because the company fully intended to completely bypass the existing settlements along the middle Rio Puerco (López 1982). However, the rapid incision of arroyos made it impossible to construct a durable retention dam on the middle Rio Puerco and in 1895 the company went belly up. Eventually the proposed new ditch was constructed and it became known as the Cuba Ditch.

In 1958 San Gregorio Dam was constructed across Clear Creek to store water for the Nacimiento Ditch, and San Gregorio Reservoir became a fishing lake for the New Mexico Department of Fish and Game. In the summer the water is released for diversion downstream to the ditch. In 1964 the reservoir was included within the boundaries of the San Pedro Wilderness. Sometime before 1981 the Cuba Ditch fell into disrepair, although the water rights are still valid (Fischer and Borland 1983).

In the 1980s the Pueblos of Jémez, Zia, and Santa Ana filed suit to acquire an accurate accounting, and possible acquisition, of water flowing out of the wilderness area and in particular the waters from the beheaded Rio de las Vacas and Clear Creek. Today the Nacimiento Community Ditch Association (NCDA) irrigates fields of alfalfa for livestock. The lawsuit put the fate of the NCDA and the valley in severe jeopardy. The NCDA was able to retain their water but the legal settlement required that the flow from Clear Creek and San Gregorio Reservoir be accurately monitored and regulated. This required construction of a concrete diversion dam and monitoring gate. The problem was that because the ditch was within the wilderness area, no motorized equipment was permitted. In 2002 the NCDA had to haul, by horse, ten tons of sand, 45 tons of rock, 100 sacks of cement, and 120 vigas up to the site in order to build the dam (Linthicum 2002).

At this time the mayordomo of the NCDA was Carlos Atencio. The Atencio family of Cuba, including fathers, sons, cousins, and uncles, has been involved with the ditch from its very beginning. The name Atencio also appears on records from the first settlement in the late 18th century. One of Carlos' earliest memories is going with his dad up to the ditch's headwaters to build a floodgate. In 1954, at the age of 17, Carlos left Cuba to find work. As a migrant worker he followed the crops from Arizona to Washington. He eventually married a girl named Rita and settled down in California. Twenty years later, in 1975, he moved back to Cuba, and rejoined his family and its involvement with the ditch.

Carlos and Rita made their home in a unpretentious, cosy house near the base of the Sierra Nacimiento, not far from the old Copper City. In January 2006 their peaceful home was "invaded" by two local thugs. The pair severely beat the Atencios and stole their vehicle (*Cuba News* 2006a). The case sent shock waves through the close-knit community. The punks were caught and sent to prison, but the elderly Atencios might never again feel safe in their rural home.

A second man-made ditch has had an important controlling effect on the pattern of development for the community of Cuba. A main tributary of the Rio Puerco is the *Rito Leche* ("Milk Creek"), which today flows subparallel to and north of the Rio Nacimiento and enters the Rio Puerco on the south edge of the town (Figure 126). Early on, possibly contemporaneously with the Nacimiento Ditch, this second ditch was dug to direct flow from the original Rito Leche and to short-cut it west to some fields.

Prior to construction of this ditch the old Rito Leche had flowed to the southwest and a lazy tributary from the north had probably added to the flow. The new ditch and its berm disrupted this original flow and caused the tributary drainage to back up north of the ditch, resulting in a swath of swampy, low-value real estate that has resisted development to this day. When the Rio Puerco entrenched its channel after the 1880s the flow from the old swampy area eventually found its way into the main channel on the north end of the town, but the swampy south of that point area remained. This explains why the present incorporated area of Cuba has a dumbbell shape, with the neck in the middle being the old swamp. It also became the unfortunate site for the town's first public school (see below).

Since the early days the area's inhabitants acquired their drinking water from springs in or near the mountains. One of these was Horseshoe Spring (Figure 127). Between 1995 and 2004 the spring revealed traces of coliform bacteria. The Ranger District of Santa Fe National Forest risked being fined up to $32,000 per day by the New Mexico Environment Department if it continued to pipe unsafe water to the spout. The system had been completely rebuilt as late as 2001 but vandals had complicated matters by shooting up the spigot. The Cuba Ranger District disconnected the spout in August 2004, although the spring itself was left open (*Cuba News* 2004).

Shortly after incorporation in 1964, Cuba formed the Cuba Water Users Association, with Col. Walter Hernández as its chairman and 81 members. The group acquired a government loan to pipe spring water to the village from the Vallecitos area east of town (Figure 127). Then politics intervened. In 1972 the New Mexico State Health Department condemned the water in part because it came from an open source (*Albuquerque Journal* 1972a). The case went to court but the village argued that if filtered properly the water could be rendered safe. However, Cuba lost its bid to continue operating the system. A state of emergency was declared because the village then had no adequate supply of drinking water. For a time potable water had to be trucked in (*Albuquerque Journal* 1972b), as late as 2005 by the New Mexico National Guard (MRCOG 2004). The problem of inadequate volumes of drinking water continues to this day.

In the 1970s the village turned to the Tertiary-age Cuba Mesa Sandstone aquifer north of town as a source of groundwater (Figure 128). Over the years the town had drilled a number of wells into the sandstone to a depth of about 600 to 700 feet. The wells produced adequate water volumes, but with high concentrations of magnesium, manganese, and iron that build up and plug the casing and pipe. When that point is reached an expensive replacement well is needed. The water is of such poor quality that it cannot even be used in car radiators or for watering lawns (Hernández 2000a). The water also contains sulfate, which combines with the magnesium to form epsom salt. This gives the water a bitter taste and renders it generally unpalatable (Anderholm 1979). An interesting aside is that epsom salt serves as a purgative in the human body because of the unpleasant effect that magnesium has in the digestive tract!

Streams flowing west out of the Sierra Nacimiento cross a remnant of the alluvial gravel cap. At first the water remains in the streams and within the cap because most of the rocks underlying the cap are non-permeable and therefore form a floor. However, as the stream flow moves farther downslope to the west it crosses the beveled edges of the underlying, permeable Cuba Mesa Sandstone. In that narrow zone much of the water permeates downward into the pore spaces between the sand grains of the waiting Cuba Mesa Sandstone aquifer. The water percolates downward in the aquifer until it enters the zone that is fully saturated, and then flows gently to the southwest within the aquifer.

Today three wells pump groundwater to the surface from a small well field north of the village in Santa Fe National Forest Land (Figure 128). The water is piped to a nearby treatment plant and then to two storage tanks, one next to US-550 at

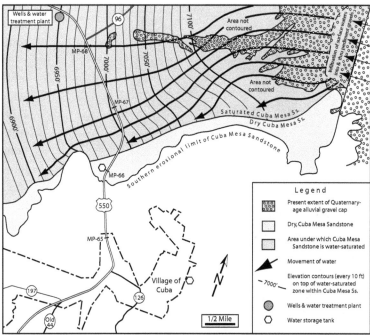

A. Movement of surface water and groundwater in Cuba Mesa Sandstone (alluvium "artificially removed" for clarity. Location in Figure 127.)

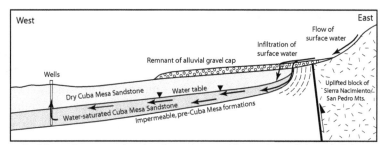

B. Schematic cross section showing movement of water (arrows) through Cuba Mesa Sandstone (section drawn approximately across northern edge of "A" above)

Figure 128. Generalized portrayal of groundwater system in the Cuba Mesa Sandstone. (Modified from Woodward et al. 1970b; Anderholm 1979.)

MP-66, and one down in the village near the school complex. As if other problems were needed, the distribution system was plagued by low pressure due to small-diameter and aging piping (MRGOG 2004). The problem was remedied by upgrading with six-inch piping.

The sheep barons

The mid-19th century cattle empires of Otero and Perea gave way to sheep empires controlled by four men. One was J.E. Matthews, who arrived at Cuba in 1883. Three years later he had opened a store and was dealing in lambs and wool. By the 1890s he was shipping large quantities of lambs and wool to railheads in Chama and Bernalillo. In time though he yearned to return back east. He left Cuba in 1897 with considerable wealth and moved to New York City, where he opened another store. His business failed and in 1900 he committed suicide (Calkins 1937).

The second man was Epimenio A. Miera (Figure 129). This fascinating and enigmatic character deserves more than just passing notice. However, it is difficult to assemble an accurate biography for him because the available historical sources are inconsistent. One source states that he was born in 1865 in Algodones, New Mexico, and he worked as a clerk

Figure 129. Epimenio A. Miera.
(Photo from Peterson 1912.)

in his home town until 1879, when he moved to Cuba and got into the mercantile business (Peterson 1912). However, it seems unlikely that he was only 14 years when he got into that tough business.

It is clear that Miera had intended to get into the sheep business because that was the way to wealth. He was well-educated and spoke English, Spanish, French, and German (Peterson 1912). He also had a sharp eye for opportunities. He knew of Matthew's desire to get out of the sheep business. One source reports that Miera learned of a widow in Bernalillo, who just happened to own 1,000 sheep. He rushed to her side and married her soon afterwards. He must have been extremely persuasive. In 1897 he took his newly-acquired sheep and used them to buy out Matthews (Calkins 1937).

However, oral tradition doesn't quite square with the marriage story above. A second line of evidence has been provided by Marietta Wetherill, who earlier, with her husband Richard, had made an indelible mark on the history of Chaco Canyon (Chapter 25). After Chaco Canyon she lived for several years during the early 1900s in the Cuba area. Many years later, in 1953, she dictated her life story to a reporter from Silver City. The transcripts of the

tapes are now housed in the University of New Mexico's Center for Southwest Research. The material that follows comes from those transcripts.

Epimenio had an older brother named Venceslado and they both moved to Cuba at about the same time. Venceslado met two beautiful and wealthy sisters in Santa Fe and became engaged to one of them. In those days it was the groom's duty to buy the "trousseau" (the bride's household items), so while Ven was away on this important errand Epimenio stepped in and married her! Could she be the "widow" mentioned above? Epimenio built his new bride a beautiful house in Cuba and the couple moved there (Wetherill 1953).

The new Mrs. Miera loved to ride horseback. One day the stableman saddled up a horse for her morning ride. Just out of the gate a dog ran up and frightened the horse. It reared up (she was riding sidesaddle), and she was thrown off and seriously injured. The sad result was that she was paralyzed from the waist down. The ever-attentive Venceslado then ran off and married the sister! He moved her to Cuba too and built a nice home. Family get-togethers must have been fascinating! For some reason Ven turned to drink and lost everything that he and his wife had. Meanwhile, Epimenio never got over the shock of his wife's injury. For years until her death she required constant nursing care. Epimenio assuaged his pain by stabling a mistress in the little community of Los Gonzalitos a few miles south of Cuba (Wetherill 1953).

With his increasing wealth and influence Epimenio became heavily involved in Bernalillo County politics (at this time Cuba was part of a much larger Bernalillo County). When Sandoval County was formed in 1903 Miera suddenly found himself thrown out of his county commissioner's office (Anderson 1907). Given his volatile nature he must have been enormously annoyed by this turn of events. However, he then became a member of the new Sandoval County Commission (1903–1905) and later (1907) was a New Mexico State Senator (Peterson 1912). Last, but not least, he was one of the 100 men who served on the Constitutional Convention (October 3–November 21, 1910) who drew up New Mexico's state constitution.

Miera was without a doubt a complex individual. He could be kind to those he liked and very much a scourge to those he didn't. As an example of his good side, he "adopted" two black orphan boys from Texas, Andy Jackson and John Clark, when they were very young. The women in his household raised the boys to adulthood, after which they continued to work for him as hired hands. According to one source, in words characteristic of her time, "they were two great, big, husky, black-as-ink negroes and their hearts were the biggest and the whitest hearts that's ever been made" (Wetherill 1953).

There is no question that Miera ruled the territory he perceived as his roost with an iron fist. For example, if somebody didn't vote as he wished, Miera might just drive the ingrate out of town, or worse (Wetherill 1953). However, Marietta Wetherill generally had nothing but good things to say about Miera. She was the widow of amateur archeologist Richard Wetherill, Sr., who in 1910 had been murdered at their home in Chaco Canyon (Chapter 25). After his death she had to move out of the canyon. She deeded her homestead there to Miera, who had known her husband. Miera in turn offered to sell her a piece of land in the Sierra Nacimiento above Cuba for the price of eight mules. This was too good a deal to turn down so in 1912 Marietta moved with her five children to Cuba. For the rest of his life Miera was very kind to Marietta, and she never forgot it (Wetherill 1953).

Through clever trading and dealing, Miera by 1904 had amassed 32,000 sheep and was one of the two wealthiest men in Cuba. Upon his death in 1913, he had assets of $100,000 in the bank—a huge sum for the times—and 15,000 sheep (Calkins 1937). He was a remarkable man in a remarkable era.

The third sheep baron was Augustin(e) Aaron Eichwald (1860–1926). Born in Wesphalia, in what would later become Germany, in 1882 he emigrated to the U.S. and arrived in Cuba in 1885 as a peddler. Soon he was able to open a small store. In 1890 he began dealing in livestock, wool, and hides, and by 1904 owned 20,000 sheep. Between 1900 and 1910 Augustin married a New Mexico girl, Isabel Córdova, and sometime during that same decade built an imposing house on the west side of the main street, across from the Catholic church. Relatively speaking the house was a mansion, and certainly the most impressive residence in town. Upon his death in 1927 Augustin had $150,000 in the bank and 15,000 sheep (Calkins 1937). The old house on the main street burned down long gone and was replaced by a single-story building that exists today.

The fourth baron was Frank Bond (1863–1945; mentioned earlier in Chapter 9). Although not a resident of the Cuba area, he exerted a powerful influence on the region's economy and is very relevant to this story. Franklin Bond was born on his parents' farm in Quebec, Canada. In 1882 he followed his older brother George to New Mexico. Earlier, brother George had been given a loan of $125 from his father to travel to Toronto to take a test for his supposed law career. Instead he bypassed Toronto and took the train down to Colorado. For a time the brothers operated a wool-processing business in Trinidad, Colorado, before purchasing a mercantile business in Española in 1883 and naming it the G.W. Bond and Brother Mercantile Company. This was a few years after Española had been connected with Colorado and the outside world by the Denver and Rio Grande Railroad (D&RG). The brothers were quick to recognize the lucrative connection of wool and rails. Their business boomed. The Bonds opened franchises in Wagon Mound, Roy, and Albuquerque, and then got into the sheep-raising business (Bond House marker, Española).

Meanwhile, in 1887 Frank had married a girl from Trinidad, Colorado, May Anna Caffal. U.S. Census records show that in 1900 the Frank Bond family lived in Española, but in 1910 they lived in Pueblo, Colorado. Between 1900 and 1910 brother George left New Mexico and moved his family to Trinidad, Colorado, and in the 1910s Frank moved back to Española for good (Bond House marker, Española; U.S. Census records). Frank acquired extensive holdings of land, including, in 1917, the enormous Baca Location No. 1 in the Valle Grande of the Jémez Mountains. From this land base he gained control of the lucrative wool market, and his businesses expanded to include wool storage and marketing (Martin 2003).

By the time Bond first visited the Cuba area in 1905 he was a well-established sheepman in Rio Arriba County. In 1912 he opened a store in Cuba and dealt in lambs, wool, and hides. In 1920 he began offering the smaller sheepmen a number of sheep on shares—the so-called "partido" system. He moved his store out of Cuba in 1922 but his sheep holdings continued to increase. After the cruel winter of 1928 killed about 50% of the sheep in the area, the Bond Company found itself fully in control of the sheep industry (Chapter 9; Calkins 1937).

In 1925, Bond relocated to Albuquerque for family-health reasons and set up other ventures with his son R. Franklin, such as the Wool Warehouse Company. During his first few years in Albuquerque he lost his parents, wife, and two daughters. He moved to California and son R. Franklin took over operations. In 1945 Frank died from chronic heart disease in Los Angeles, California, and the young Bond took over as president of Frank Bond and Son, Ltd. He maintained the vast family interests until his sudden death in 1954. Lacking a family member able to manage operations, in 1959 the family ceded control of the Baca Location No. 1 to the King family of Stanley, New Mexico. The Frank Bond empire had ceased to exist.

16

An Evolving Cuba

The first settlers in the Cuba area (sometimes called the Cuba Valley) in the late 18th century found a lush land at their disposal. Pristine streams issued from the headwaters of the Rio Puerco and its tributaries and delivered abundant water to the land via willow-lined courses. These conditions were essentially the same when the second wave of settlers returned to the Valley in the late 19th century. These settlers were mainly farmers. Shortly after their arrival the land began to deteriorate, but despite that by 1900 Cuba had become a wheat and corn-growing center.

The nasty winter of 1928 marked a major shrinkage of the area's resource base, and more reductions were to follow. A declining resource base has been an ongoing challenge to people in the Cuba area for quite some time. It is not a stretch to compare the process to the party game of musical chairs, where a number of people circle a number of chairs, one fewer than the number of players, while the music plays. When the music stops the participants scramble over each other to find an open seat. The frustrated one who fails to find a seat is out of the game. The chairs are in effect a decreasing resource base, analogous to grazing range, irrigatable land, copper ore, and timber.

The outlying open range was grazed by the large livestock operators. At first they were mainly Hispanic, but later they were replaced by Anglos. With the arrival of the Anglo operators the herds of sheep and cattle grew to huge numbers and reached a peak around 1910. The herds placed an enormous burden on the land. Eventually some of the villagers also acquired herds of sheep and cattle and grazed them near their homes to supplement their farming livelihood (Oberg 1940).

The only markets for crops were mainly the large stock owners and less so for local consumption. Therefore almost immediately there began a slow transition from farming to livestock raising. Until that time there was little or no cash available and virtually the only medium for the exchange of value was livestock. One exception to this general trend was during World War I when large shipments of grain were made to the railhead in Bernalillo to supply outside markets. After the war agricultural production continually decreased until by the mid-1920s it no longer could even supply local needs (Calkins 1937). Upon the arrivals of the railroad in central New Mexico in 1880 and the Anglo stockholders to the Cuba area, the isolation of the Hispanic inhabitants of the Cuba valley slowly came to an end. The availability of cash produced a demand for wage labor—a new concept. But the transition to a cash economy was a very slow one.

The evolving land use, and several pieces of pivotal legislation, have left their indelible mark on the land. The Cuba area—especially its hinterland to the west—is a complex checkerboard of private, state, federal, and Indian land (Figure 130). This fractured—and largely non-tax-paying—ownership has not been conducive to economic development.

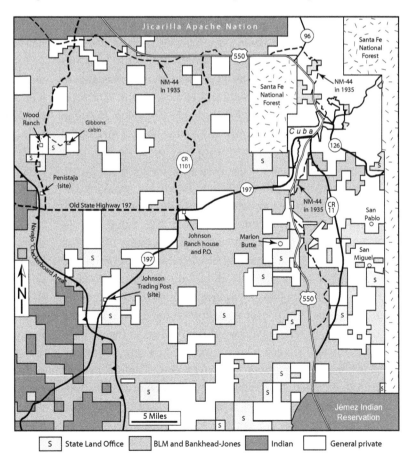

Figure 130. Surface land-management status and general geography, Cuba area. (From BLM 1997, 1999a, 1999b, and 2004.)

Pivotal legislation

The end of World War I ushered in a major change in the economy and society of Cuba. Much of it was due to three big changes to the federal government's homestead policy.

Homestead Act of 1862

Way back in 1862 Congress passed the Homestead Act, which granted 160 acres of public land in the West as a homestead to "any person who is the head of a family, or who has arrived at the age of 21 years, and is a citizen of the United States, or who shall have filed his declaration of intention to become such." The homesteader was required to pay a small filing fee, live on the land for five years, and make certain improvements to the land in order to receive clear title—a "patent"—to both the surface and mineral estates. This act was the culmination of years of controversy over the disposal of the public lands. Beginning in the 1830s some groups had proposed the free distribution of public lands. This demand had been adopted by the Free Soil party, which saw such distribution as a means to eliminate the spread of slavery in the territories, and it was later adopted by the Republican party in its 1860 platform. The Southern states had been vehement opponents of the policy, and their secession cleared the way for its enactment.

Although a homesteader could live quite well on 160 acres of well-watered land in the Midwest, such was not the case in the arid West. Many tried and failed to make a go at dry-farming when stock raising was really a more viable option. The policy also opened up doors for shenanigans of dubious legality. Using proxies, powerful individuals could file on multiple parcels containing water and thus control vast expanses of adjacent rangeland. Many of the best lands came under the control of the railroads and speculators. Realizing the folly of the policy, the government finally passed the Enlarged Homestead Act of 1909, which increased the homestead size to 320 acres. Eventually Congress passed the Stockraising Homestead Act of 1916, increasing the size to 640 acres, but—significantly—severed the mineral rights from the land and reserved them for the U.S. The homestead act was by implication repealed by the Taylor Grazing Act of 1934 (see below, and Chapter 9).

By the end of World War I there began a slow migration of settlers, mainly from the southern plains of Texas and less so from Oklahoma, to the dry, so-called "Penistaja" (counter-intuitively pronounced Pen-ish-tah-HAH) district lands west of Cuba. The region near the Rio Puerco had been pre-empted by strong-man Epimenio A. Miera, so they were obliged to settle farther to the west out in the dry country.

These settlers did their best to dry-farm but quickly learned that it was not feasible. The "dry-nesters" were forced to either move away or shift their emphasis from farming to livestock operations. The two groups—the newcomers and the people from the village area—formed two very distinct societies. The Penistaja district was peopled mainly by English-speaking Anglo, Protestant, Democratic-leaning, cattle-raising families. In sharp contrast the Cuba community was made up of mainly Spanish-speaking Hispanic, Catholic, Republican-leaning, cattle-raising and irrigation-farming families. The Anglos were economically tied to Albuquerque for their needs, while the villagers were more focused on the village itself for theirs. Because of the 640-acre homestead size the Anglos were spread out thinly, and the establishment of a more centralized community, like Cuba, was not practical (Oberg 1940). The two societies tended to stay separate and not mix. In fact in 1930 there were only three Anglo families living in Cuba (Smith 1965).

Meanwhile, the irrigation farmers of the Cuba area were facing pressures of their own from the deteriorating land base. If able, they tended to migrate upstream to the headwaters of the rivers where the water was more plentiful. But as the water table continued to fall, many springs became wells. The rivers entrenched their courses and became arroyos. Erosion and destruction of the land accelerated. All the while agricultural markets remained limited and intermittent. It was tough to make a living as a farmer. With all these woes the inhabitants of the Cuba Valley had to adjust their lifestyles. The changes were slow and barely perceptible, but the economy continually shifted its emphasis from farming to livestock raising, and gradually to a mixture of livestock raising and wage work (Calkins 1937).

By the mid-1930s the Great Depression was full blown. Markets were soft for livestock as well as for everything else. The need for wage-labor accordingly relatively increased, and in the Cuba Valley that often meant wages earned via the federal relief agencies such as the Civilian Conservation Corps (CCC) and the Works Progress Administration (WPA). At the same time much grazing land was taken over by the federal government and placed off-limits to ranchers.

Taylor Grazing Act of 1934

For many years the land had been terribly over-used. As mentioned earlier in Chapter 9, in 1934 Congress passed the Taylor Grazing Act (TGA). The act's stated purpose was to protect the public domain by preventing overgrazing and soil erosion, to provide for its orderly use, improvement, and development, and to stabilize the livestock industry that depends on it. The act mandated the establishment of grazing districts to be administered by the U.S. Grazing Service under the U.S. Department of Agriculture. The act also mandated the issuing of grazing permits that were preferentially sold at "reasonable" rates to local individuals for terms of up to ten years. One effect of the TGA was that it inserted the federal government squarely into the affairs of the ranchers, often to the great irritation of both.

Bankhead-Jones Tenant Act of 1937

The third piece of pivotal legislation was the Bankhead-Jones Farm Tenant Act of 1937, which was aimed at farmers. Congress recognized that, largely because of the Depression, farmers had great difficulty in raising the necessary capital to operate a family farm, and that operating on submarginal land inflicted great environmental damage. The act intended to secure stable, economically viable ownership of farms through loan programs and the retirement of submarginal land. The retired land was to become rehabilitated and entered into the public domain. In 1941 many of the Anglo homesteaders were forced to sell out to the government and move away (Smith 1965), and as a result most of the federal land west of Cuba was acquired via the Bankhead-Jones Farm Tenant Act (Figure 130).

Continuing change

World War II drew many of the inhabitants away from the area into the military or to defense jobs. After the war many did not return. In 1946 the Bureau of Land Management was established to take over the function of the Grazing Service. The new agency maintained the policy requiring permits to graze on the public domain, thus shutting out many cash-strapped ranchers.

Lumber industry

Lumbering was for a while an important activity in the Cuba area. Several sawmills processed trees harvested from the nearby national forest land and the finished lumber was trucked to places like Bernalillo and Albuquerque. By the mid-1960s the industry was at its peak. But trouble was in the wings. First, the supply of large-diameter trees was declining. Then in 1973 President Nixon signed into law the Endangered Species Act, and in 1993 the Mexican Spotted Owl was put on the list of endangered species. For these two reasons, thenceforth the public lands ceased to supply large logs. Unfortunately the existing mills had been designed to handle large-diameter logs.

The following year Duke City Lumber Company, Cuba's large sawmill (two miles south of town on the old NM-44, now CR-11) shut down. Fifty eight employees lost their jobs and up to 200 support people went out of business (Thompson 1994). In 1988 New Mexico had 17 mills but only two in 2000, both in the northern part of the state, due principally to the lack of the large logs from the public lands (Romo 2000). The state's last large mill—on land leased from San Juan Pueblo near Española—shut down in 2003 (*Albuquerque Journal* 2003c). The only lumbermill that lingered in Cuba was a small operation operated sporadically by local resident John Hernández that had been founded many years ago by his parents Walter Hernández and Bessie Young.

Because the closed mills had been designed to process large trees, new, replacement mills must be designed to handle smaller-diameter trees. Most today would agree that our forests could use some thinning of small trees for firewood and wood products. A number of pilot projects are afoot to determine the economic viability of harvesting small-diameter trees and reducing them to wood chips to be used as fuel in space-heating systems. An official from the state's Energy, Minerals and Natural Resources Department has stated that, "This is a giant first step in creating an economic demand for small diameter wood in unhealthy forests . . . and it is an excellent source of new jobs in rural New Mexico communities" (*Cuba News* 2005a).

Electricity at last

Electricity did not come to Cuba until the late 1940s. Today it is difficult to imagine life without this "essential" utility, but prior to this time its wasn't essential at all because it just didn't exist in many of New Mexico's rural communities. Rather, coal-oil or kerosene for lamps was a staple item. The history below is from the Jémez Mountains Electrical Cooperative, Inc.'s website.

The tale of Cuba's electrification begins just before World War II in the small Village of Jémez Springs. Fred Abousleman and his brothers had installed a small hydroelectric generating plant on the Rio Jémez to provide electricity for themselves and two other families. The primitive system was an improvement over nothing. Then the brothers went into the military. In Germany they were profoundly impressed by the fact that every small town had hydroelectric power generated from a nearby creek. When they returned home they found the old hydroelectric generator in bad shape. They were determined to bring the town into the 20th century.

Abousleman approached the Public Service Company of New Mexico (PNM) and quickly learned that large power companies had no desire to invest in rural areas where the profit-to-cost ratio was so low. However, an interesting model was available to them. Following the federal Rural Electrification Act (REA) of 1936, the town of Grants in west-central New Mexico had formed the Continental Divide Electric Co-Op. The act made funds available to rural organizations so that they could form cooperatives to provide power for their areas. Abousleman teamed up with a guy from San Ysidro and attempted to persuade Continental Divide to extend electrical service, but they got nowhere. Lobbying the New Mexico congressional delegation was also futile. They realized that they needed to organize.

On Easter Sunday, 1947, a group of interested men from Jémez Springs, San Ysidro, and Cuba joined forces to form their own cooperative, the Jémez Mountains Electric Cooperative (JMEC). In April 1948 they incorporated and hit the pavement in search of federal funding. The REA approved a loan for $930,000, most of which was designated to purchase the Inland Utility Company, which had been supplying the power needs of the Española area. The service area of the spanking-new JMEC was thus significantly enlarged. They hired an engineer to draw up plans and a crew to begin installation of poles and lines.

The first co-op board members included Cuba citizens Eric L. Freelove, Walter R. Hernández, and Rudy Velarde, as well as representatives from Jémez Springs, San Ysidro, and Española. Headquarters were relocated to the Española area (later moved to the present location at Hernández, six miles north of Española). Eventually other communities in southern Rio Arriba County were linked into the system. Soon however, the old diesel generator in Española was maxed out. In 1957, JMEC joined 12 other New Mexico electric co-ops as a part of the Plains Generation and Transmission Cooperative, whose gas-fired plant at Algodones plant, north of Bernalillo, supplied all of JMEC's needs. JMEC gradually grew into the largest coop in the state. In 2000 JMEC and the other members of Plains agreed to merge with the Tri-State Generation and Transmission Association. Today Cuba receives its electrical power via a Tri-State substation at about mile 75.5 on US-550. From there JMEC's 69-kilovolt transmission line delivers the power to JMEC's substation on NM-126 on the east side of town.

Santa Fe National Forest

The U.S. Forest Service lands near Cuba are broken up into two non-contiguous blocks: a main body to the east sprawled across the Sierra Nacimiento, and a strange isolated "outlier" to the west covering Cuba Mesa. To the student of maps, such isolated pieces of real estate are provocative puzzles that always have a story to tell and usually involve politics. Let's take a look at the evolution of the National Forest.

After the Civil War the development of the West really took off. The philosophy of the time was that the resources on the public lands were inexhaustible and that they should be used. By the 1880s unrestricted timbering and grazing had taken a heavy toll. In 1891 Congress was immersed in a debate about abuses and fraud involved with the Homestead Act and about grazing on the public domain. In a bill intended to revise some land laws, it attached a (long) one-sentence rider:

> "The President of the United States may, from time to time, set apart and reserve, in any state or territory having public land bearing forests, in any part of the public lands, wholly or in part covered with timber or undergrowth, whether of commercial value or not, as public reservations, and the President shall, by public proclamation, declare the establishment of such reservations and the limits thereof" (Williams 2000).

This innocuous rider to the Forest Reserve Act of 1891 gave the President the authority to establish forest reserves from the public domain on his (or her) own volition. The nation's first forest reserve was created in Wyoming the following year by President William Henry Harrison. In 1894, grazing was completely banned on the forests. In 1897 President William McKinley signed the Organic Act, which (as amended) mandated protection and management of the forest reserves. Regulated grazing was permitted in the late 1890s when it was recognized that the complete banning of grazing was not practical.

And then along came Teddy. The sturdy President Theodore Roosevelt (1859–1919) was personally acquainted with many of the western lands and had become an ardent preservationist. During his two terms (1901–1909) he would create 150 national forests, 18 national monuments, and five national parks. In 1905 Congress organized the U.S. Forest Service, and that same year Roosevelt created the Jémez Forest Reserve. The original western edge of this reserve included and ended at the western foot of the Sierra Nacimiento (Figure 131A).

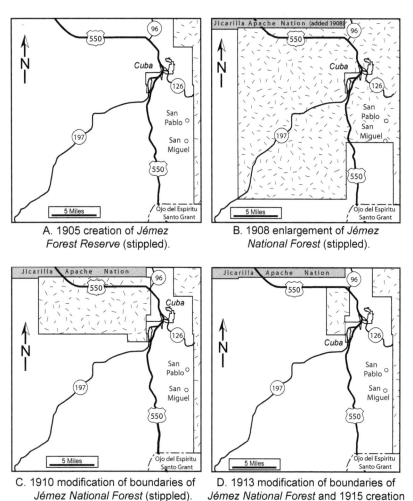

A. 1905 creation of *Jémez Forest Reserve* (stippled).

B. 1908 enlargement of *Jémez National Forest* (stippled).

C. 1910 modification of boundaries of *Jémez National Forest* (stippled).

D. 1913 modification of boundaries of *Jémez National Forest* and 1915 creation of *Santa Fe National Forest* (stippled).

Figure 131. Evolution of western "outlier" of Santa Fe National Forest, 1905–1915. (Modern geographic features shown for reference.)

By 1907 opposition to the president's authority to create forest reserves reached a crescendo. Congress, faced with an agricultural appropriations bill, added an amendment to it that became known as the Fulton Amendment. The rider nullified the president's authority to form forest reserves given by the 1891 Act, and gave that function to Congress. That same year forest reserves became known as national forests.

In 1908 Roosevelt signed a proclamation (PR No. 806) to enlarge the western side of the Jémez National Forest to—strangely enough—include Cuba (Figure 131B). Later that year he signed an executive order (EO No. 849) that cleaved off the northeastern half of the forest to form the separate Carson National Forest. In 1910 President William Taft issued a proclamation (PR No. 1081) that removed a large block of acreage from the west side of the Jémez National Forest, including the community of Cuba, and returned it to the public domain, thus creating an outlier of forest separated from the main block (Figure 131C). The removed acreage was then opened to homesteaders.

In 1913 President Woodrow Wilson (terms 1913–1921) issued a proclamation (PR No. 1250) that whittled the outlier down in size to include only Cuba Mesa, thus creating the outline of the forest's western outline (Figure 131D), and in 1914 he issued an executive order (EO No. 2069) to throw open the newly-removed land to homesteaders. All of these boundary changes must have involved a considerable amount of political arm-twisting and deal-making, but those delicious details are virtually lost to history. Finally, in 1915, Wilson issued another executive order (EO No. 2160) to combine the Jémez National Forest (lying completely on the western side of the Rio Grande) with the Pecos National Forest (formed in 1892 and lying on the eastern side of the Rio Grande) to form what is now Santa Fe National Forest.

A major internal administrative change was the creation of a wilderness area in the San Pedro Parks area above Cuba. The wilderness was designated as a "Primitive Area" by the Forest Service in 1931. In 1941 the Secretary of Agriculture reclassified it as a "Wild Area," and set its size as 41,132 acres, or about 64.25 square miles. In 1965 it became the San Pedro Parks Wilderness. It is closed to all motorized transport.

The incorporated Village of Cuba

The Village of Cuba is and will likely always remain small. Growth is constrained by a limited, dependable water supply (discussed earlier) and supply of annexable private land. Until the 1940s the community consisted of only a small block of businesses and homes along the main road through town (then NM-44, today's US-550) and ended on the south at the Rito Leche (Figure 132). At the time the main street through town was a 30-foot-wide dirt track. The only automotive-service garage on the 190-mile dusty road between Albuquerque and Farmington was the G.R. Hoche Garage in Cuba (*Cuba News* 1964). The road to Albuquerque was only paved in 1938.

In 1964 the community incorporated as a village. The first and second orders of business were to establish a water system (discussed above) and a sewer system. A third was to widen the main drag through the village because of the fear that some future highway might completely bypass the town. Many of the streets were given names at this time. Sometime before 1979 the village annexed a swath of land on the sandstone hills east of town to accommodate new homes, and, for some reason, a spike of land to the south. Finally, in 1985 the village had assumed its present strange "dumbbell" outline via

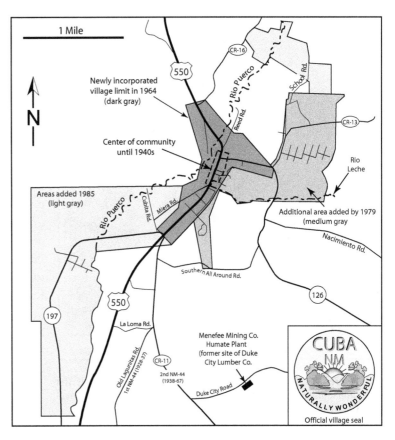

Figure 132. Evolution of incorporated Village of Cuba, 1964 to present.

annexations of additional residential areas to the west and a donated site for a waste-water plant on the southwest.

A succession of Catholic churches

The collective soul of a town, especially a small one like Cuba, consists of its churches, schools, and—to a lesser extent—its stores. For a dominantly Hispanic population, the town's Catholic church is of central importance. The present church, Our Lady of the Immaculate Conception, sits on the east side of US-550 on the town's main street in the center of town. It was built in 1967 and is the last in a succession of three Catholic churches. The first, the church of *Nuestra Señora del Cobre*, was built in the 1880s on a site across the street from the old jail (Chapter 17). Because there were no benches, the people had to bring their own rugs on which to kneel on the dirt floor. The priest came up from Jémez to say mass (Cuba Visitor Center, undated pamphlet). This first church was a simple affair, perhaps only a *morada* (Spanish for "dwelling," more commonly for "meeting place"), consisting of posts and beams. Its name, the "church of our lady of copper," derives from a curious tale taken from the cult of the Virgin Mary on the Caribbean island of Cuba (Rivero undated).

In 1509 a Spanish soldier fell seriously ill on the island of Cuba and was forced to remain behind as his companions continued on. The local Indians nursed him back to health. In gratitude he learned the local language and began to preach Christianity, emphasizing the importance of the Virgin Mary. The cult of the virgin persisted on the island during the 16th century as Spanish settlement expanded. At the same time the Spanish settlers greatly increased their herds of cattle, and in 1598 the mining of copper began in the mountains of the eastern part of the island. The Spanish government established the ranch of Varajagua, which had a large number of cattle. Of course, salt was necessary to prevent spoilage of the meat.

183

Figure 133. East view of Catholic Church of Our Lady of the Immaculate Conception, late 1940s. (Photo courtesy Sandoval County Historical Society, cat. #78.003/002.)

One day in 1607 or 1608 two Indian brothers, Juan and Rodrigo de Hoyos, and a ten-year-old black slave named Juan Moreno, were sent by the administrator of Varajagua to the coast of the northeast corner of the island to gather up some salt. They arrived in a fierce storm and had to wait it out for three days. Returning to their task at hand they took their small canoe into the surf. Soon they encountered a boat-like object. It turned out to be an exquisite image of the Virgin Mary holding a small tablet that said, in Spanish, "Yo soy la Virgen de la Caridad, ("I am the Virgin of Charity"). The three rushed their discovery back to the ranch. The people quickly erected a small alter in the town of *Cobre* (Spanish for "Copper") to hold the object. The royal administrator of *Minas de Cobre* (a copper-mining province) was notified of the find. He ordered that the image be housed in a sanctuary, continually illuminated by a copper lamp. The devotion of this Cuban town for the *Virgin de la Caridad del Cobre* grew over the centuries. In 1915 Pope Benedict XV proclaimed the Virgin de la Caridad del Cobre the Patron of Cuba. Finally, during his 1998 visit to Santiago de Cuba, Pope John Paul II elevated the Virgin to Queen and Patron of Cuba.

Meanwhile, back in little Cuba, New Mexico, in the 1880s the Catholic clergy must have been quite aware of both the Virgin of Cuba and their own local copper deposits. The name *Nuestra Señora del Cobre* must have seemed a natural choice for the name of the new church. It burned down in 1905 (Cuba Visitor's Center undated).

The second church, Our Lady of Immaculate Conception, a quaint, steepled affair, was built in 1915 (Córdova-May, 2009) of adobe and painted blue inside (Figure 133). A stone statue of the Virgin Mary had been hauled by wagon from Guadalajara, Mexico in the late 19th century, and was displayed over the door of the church. In about 1922 a huge, three-story adobe convent and school was built on the parcel of land just south of the church. In 1948 the church's buttresses were added to shore up the walls because the adobe structure was crumbling. By the 1960s both buildings were in bad shape.

As mentioned above, the village incorporated in 1964. Local leaders felt that the old church was "outdated" and that Cuba needed a modern one. In 1965 the church was leveled, much to the dismay of the many Cubanos who had grown up with it. The third church, the second Our Lady of Immaculate Conception, was built in 1967 on the site of its predecessor. The site of the grand old convent and school, demolished that same year, is today part derelict empty lot, part playground, just south of the church (see Cuba Area Schools below).

Public schools in New Mexico

Schools symbolize a town's future and are always of A-number-one importance. The history of education in New Mexico though has traveled a rocky road. In 1850, only two years after New Mexico became a territory, the dynamic personage of Jean Baptiste (a.k.a., John B.) Lamy (1814–1888) arrived in Santa Fe from France to attend to the religious needs of New Mexicans. He found the education system—what there was of it—in shambles and he set out to fix it. He appealed to Catholic religious orders to come to New Mexico and establish schools. The Sisters of Loretto responded first in 1853, and during the next several decades other orders followed.

From 1850 to 1891 public debate raged between Hispanic Catholics and Anglo Protestants over control of the territory's schools. After decades of running the schools the Catholic orders had blended in and become part of the cultural landscape, especially in rural Hispanic areas of central and northern New Mexico. However, the intimate intertwining of religion and education undoubtably was a major factor in the territory's long-delayed success in attaining statehood. Most small communities were unincorporated and therefore the church and school occupied central positions in social and political life (Baca 2005). The Public Schools Law of 1891 for the first time formally established a territorial public school system. However, the religious tensions simmered for decades, until they exploded with the Dixon Case in 1948.

The infamous Dixon Case

In 1947, Lydia Zellers, daughter of a former Protestant pastor in the little northern New Mexico community of Dixon (about midway between Española and Taos), led a group to file a complaint with the Rio Arriba Board of Education. In 1921 Dixon's school had opened as a parochial school by an order of Catholic sisters. During the Depression the school had been converted to a "public" school in order to secure funding. This was not an altogether unusual arrangement. In fact, New Mexico in 1938 was one of 42 states in which public funds were used to subsidize religious schools. The sticking points were that classes were taught by Catholic clergy, and that part of the curriculum consisted of Catholic catechism, despite the fact that some students were not Catholic (Baca 2005).

The Zellers group resented this and maintained that it violated the constitutional principle of separation of church and state. They requested public funds for an alternative, truly public school, but they were told that funds were lacking. They collected donations and built their own school, and optimistically donated it to the county as an alternative institution. Then, via an egregious thumb-in-the-eye, the county school board appointed a Catholic sister as principal, who then turned around and hired other sisters as teachers. The Zellers group was outraged. However, even as their complaints made it up the administrative ladder they failed to achieve their goals (Baca 2005).

Finally, thoroughly frustrated, in 1948 they filed a class action lawsuit, Zellers vs. Huff, in State District Court in Santa Fe that would become infamously known as the Dixon Case. A colossal, passionate struggle followed in the court, in the media, and in the communities themselves. The Zellers group insisted that the issue was a secular conflict, not one pitting religion against religion, but the Church and a great many of its adherents saw it as a religious and very personal one. A ruling was reached in 1949 in favor of Zellers. The trial judge declared that Catholic clergy would be removed from tax-supported schools, and the distribution of state-owned textbooks and the providing of tax-supported transportation to parochial schools would be stopped (Baca 2005).

The Cuba school and a dozen others were declared to be "public schools with no separation of church and state" in place. Surprisingly, in the ruling the judge opined that public payment to religious orders did not constitute endorsement of religion. The Zellers group would have none of it and appealed the case to the State Supreme Court. However, even before the Supreme Court in 1951 issued its ruling in favor of Zellers, the Catholic Church threw in the towel and announced that it would cease teaching in public schools. The Church, freed from its local responsibilities, dug in its heels and devoted its considerable energy in strengthening its parochial school system. At the same time the state vastly improved its public school system, especially in rural areas that had previously been cared for by the religious orders (Baca 2005).

The wounds that resulted between Catholics and Protestants ran deep. Full reconciliation was not reached until 1999, when "An Affirmation of Hope, Reconciliation and Unity" was signed by representatives of the two sides (Baca 2005).

Schools in Cuba Area

Meanwhile, little Cuba did not escape the maelstrom. In the early days schooling was a communal affair because true public education lay far in the future. Each family took turns providing the largest room in their home as a school room for a year, and the teacher would board at the home. The teachers were sent to Cuba from Bernalillo or Albuquerque. They were typically young girls themselves and eighth-grade graduates at best. This was often their first time away from home, they were desperately homesick, and they were genuinely afraid of their students. They usually lasted one year (Arango 1980b).

In the early part of the 20th century the Franciscans fathers had returned to New Mexico for missionary work. They called on their auxiliary order, the Sisters of St. Francis of the Perpetual Adoration, to attend to educational matters. The sisters founded six "public" schools in northern New Mexico between 1904 and 1923. The Sisters started their first "public" elementary school in Cuba in 1917 (Reeve 1961). It was a rock, one-room building (Córdova-May 1973) located on the site of the present Cuba Visitor's Center (Velarde 2006). It was understood that once the school got on its feet it would be funded by Sandoval County and that it would provide instruction for grades one through eight. An interesting and sad Cuba story has it that at the celebration for the opening of the school, a certain Rev. William Grave played the fiddle all night with a bad cold. He caught double pneumonia and died at the age of 33 (Sauter 1994).

By 1923 the town had a spanking-new, quite imposing, three-story adobe, combination convent/school building located just south of the church (Figure 134). In 1932 the Sisters added the first two grades of a high school and three years later had a full four-year high school (Ruíz-Esparza 1993). Kids from the outlying villages and ranches boarded at the convent. Also a number of smaller, more local schools were scattered about the region. It must be remembered that back then distances were formidable obstacles and many kids were simply unable to get into town every day to attend school. An ominous turn of events occurred in 1927 when all the Anglo teachers were told to get out of the schools. Some of these Anglo teachers returned when groups of Anglo homesteaders amassed enough kids to open their own school to the west of Cuba (Fisher undated).

A. Southwest view of Convent School (left) and Catholic Church of Our Lady of the Immaculate Conception (center), 1939. (Photographer Thomas F. Ball.)

B. Southeast view of Convent School, 1939. (Photographer E. Cordova-May.)

Figure 134. Franciscan Convent School. (Photos courtesy Sandoval County Historical Society.)

Figure 135. Penistaja schoolhouse reverting to dust. (Photo by author 2003.)

Remember that during this time what were considered "public" schools were actually being run and taught by members of the Catholic clergy. The dry-nesters living west of Cuba were overwhelmingly Protestant and never bought into the Cuba "public" schools. They had their own one-room school houses spread over the countryside. One of these was the Penistaja school built by the federal government's Works Progress Administration (WPA) in the late 1930s. The tiny community near the school took shape around Penistaja Spring.(Figure 135).

Penistaja was a short-lived, dry-farming and cattle-raising area. It was settled about 1920 by farmers from the southern plains, many of them veterans of World War I. A trading post opened about the same time, and a post office operated from 1930 to 1943. The name seems to be a corruption of the Navajo word *Bíniishdáhí*, meaning variously, "I forced him to sit," "forced to sit," or "where I sat leaning against it" (Julyan 1996; Linford 2000). One can only wonder if the school kids, in their hard seats, mused at all about the irony of the town's name and their current situation.

One of the students at the Penistaja school was Lovena Wood. She grew up in the area and enrolled in the one-room school in 1940. She and her brother rode the three miles to school on horseback. She was the only one in the first grade, so when they graduated her to the second grade she was again the only one. Her teacher was from Mountainair, New Mexico (southeast of Albuquerque). After the teacher was attacked by some of the older boys he brought a gun to school. He was a crack shot and would send a round right over the students' little hunkered-down heads. After one such disciplinary action the kids crawled under the school and refused to come out. (Reader, can you just imagine what would happen if a teacher tried that today?) After the eighth grade students either ended their education or were forced to attend high school outside the area. Lovena, for example, attended the Menaul High School in Albuquerque, where she graduated (Sandoval County Historical Society 1992). Curiously, the name Penistaja appears on a list of "offensive" (to someone) names found on U.S. maps that are recommended for purging because of the word "penis" embedded in the name!

As mentioned above, the Dixon Case of 1948–1949 turned the educational system of the state, including Cuba's, completely on its head. Sandoval County was forced to organize a secular school system from scratch. The community constructed a K-12 facility 0.5 mile east of Main Street (US-550) in a low-lying area on NM-126. While the school was under construction the kids had to attend school in two make-shift quarters on Main Street. One of these was in a laundromat near the new El Bruno's Restaurante y Cantina. While there the kids would play on the banks of the Rio Puerco (Velarde 2006). Finally the new school opened in 1949. As one would expect, the first

187

school year, 1949–1950, was one of chaos, but over the course of two to three years the county had worked the kinks out.

The location selected for the new school turned out to be less than ideal. In the late 1950s the school was forced to install a succession of septic tanks because the impermeable, clayey soil was incapable of providing an effective drain field (*Albuquerque Journal* 1970). To make things even worse—much worse—it was soon realized that the walls of the new school contained asbestos. This material is a known health hazard and its presence in the school was clearly unacceptable. At about this time John F. Young, owner of Young's Hotel and chairman of the Cuba Independent School Board, circulated a petition to build a new junior high and high school (Hernández 2000a). The school, the second public one, was built on the site of the present school complex on CR-13, and opened in 1957.

In 1951 the far-off metropolis of Bernalillo had formed its own independent school district. Impressed by this example, Cuba decided to follow suit and break away from the county system. It took a while but the new school district was up and running for the 1959–1960 school year (Ruíz-Esparza 1993).

Meanwhile, in 1967 the old adobe, parochial convent/school on Main Street was torn down (Sauter 1994). In 1974 the old public school on NM-126 was abandoned due to new restrictions on asbestos in public buildings. The building was remodeled in the early 1980s and used as a youth center until 1988 (Velarde 2006). However, before the structure could be demolished the asbestos had to be encapsulated and transported to a sanctioned disposal site. A cost estimate for this procedure was $37,000 (*Cuba News* 2002a). For years the school complex sat as a derelict eyesore while it slowly rotted away because the cash-strapped School Board had more pressing priorities. Finally, in 2008 the old school was razed.

The federal government provides funds for the new school to accommodate and educate the Navajo students from the Checkerboard Area. The Navajo families resisted sending their kids to Indian schools because of teacher-quality issues, and the U.S. Bureau of Indian Affairs (BIA) wanted to integrate them into mainstream society. The infusion of Navajo kids changed the character of the Cuba public schools forever. Suddenly the environment advanced from bi-cultural to tri-cultural. Such an environment can be enriching but it can also create major problems. Today the Navajo kids attend BIA schools through the 9th grade, and then move on to the Cuba High School. They often arrive poorly prepared to perform adequately due to their poor grasp of English—a core problem. In 2005 the student ethnic make-up was 77% Navajo, 18% Hispanic, and about 5% everything else including Anglo.

In the best of conditions, to teach non-English-speaking students can be a real challenge, but when the first language is Navajo the task can be daunting. There is a paucity of bilingual English/Navajo teachers. By the late 1990s the New Mexico State Department of Education (SDE) took notice of the Navajo students' low test scores, poor attendance, and high dropout rates. The District soon found itself on the "needing improvement" list. In early 2003 the SDE fired the District's superintendent and took over financial control of the schools due both to the test score problem and inadequate record keeping. This traumatized the community, to put it mildly. The District just had to get the Navajo kids' academic achievement up to acceptable levels. After almost two years of herculean effort the District had made major improvements and had regained financial control.

Complicating the situation even more is the fact that the Navajo kids are bused in daily from as far away as 53 miles. The first bus company had to drop their contract because they could not make a profit given the low student-number to miles-driven ratio. The District was forced to organize and finance a bus system itself, with state financial help, and was also charged with providing free meals to all the students (*Cuba News* 2001b). There seems to be no end to Cuba's problems.

17

Cuba Today

Much about the character of a town is reflected by the stories of some of its families. The Putney/Wiese, Young/Hernández, and Eichwald families provide three of these.

Three Cuba families

Putney/Wiese

Two prominent merchant families in Cuba were Putney and Wiese (the latter pronounced WEE-See). The Putney name has today disappeared from Cuba. The two families were linked by marriage.

In 1872, Lyman Beecher Putney, Sr. and his brother, having become dissatisfied with their lives in Lawrence, Kansas, came down the Santa Fe Trail by covered wagon to make their fortunes. They intended to sell supplies to the contractors building the advancing Santa Fe Railroad. In 1876 they set up a temporary store in Mora, New Mexico, and six months later moved to Trinidad, Colorado. In 1877 the brothers packed up and moved to Las Vegas, New Mexico. L.B. decided to stay and establish a mercantile business. His brother, though, decided to move on to California and was never heard from again. L.B.'s wife Hanna and their five or six-year-old son, Robert Earl, joined him in Las Vegas. After the railroad arrived in 1879, and after the Putney store got shot up a few times, his wife announced that she had had enough of that. She returned to Kansas, where son R.E. attended school. In 1880, L.B. Sr. moved to Albuquerque. He opened up a store and livery stable, and then a wholesale grocery store called the L.B. Putney Wholesale House (Sandoval County Historical Society undated).

In 1888 he again sent for Hanna and his son back in Kansas. When they arrived in Albuquerque they spent exactly one miserable, stormy night, and then hightailed it back to Kansas. When they eventually returned in 1890, Albuquerque had grown and improved considerably. Hanna stayed in Albuquerque for the rest of her life. In 1897 the young Robert Earl returned to Kansas and came back with a new wife, Margaret Love. L.B. Sr., died in1900 and left everything to his widow and son. R.E. went on to establish six retail stores in several outlying towns, including Bernalillo, Jémez Pueblo, and Cuba (Wiese 1978). An interesting sidebar is that the post office at Jémez Pueblo used to be called "Putney" (1907–1908) in honor of the family. That post office had been called "Jémes" from 1879 to 1907 and from 1908 to 1950, after which it became known as "Jémez Pueblo" (Julyan 1996). The Cuba store was located at the site of the present Town and Country Store on the west side of the main street (Wiese 2005; Figure 136).

Figure 136. L.B. Putney store in Cuba at present site of the Town and Country store. (Photo courtesy Sandoval County Historical Society.)

R.E. and his wife had three sons, one of whom was L.B. Putney Jr. After World War I, L.B. Jr. and his brother went into business with their father. In 1923 L.B. Jr. and his first wife had a daughter, Alice Francis, but his wife died during childbirth. He and his second wife went on to have a son and two daughters. In 1937, in the throes of the Great Depression, L.B. Putney, Inc. went bankrupt. L.B. Jr. took over the store in Cuba for wages owed. The next year he moved his family to Cuba (Wiese 1978).

At this time there were three competing stores in Cuba: Young's, E.W. Halpin's (who had run the Cuba Putney store before the bankruptcy), and the H.L. Wiese General Merchandise (Wiese 1978). L.B. Putney Jr., died in 1940 at the young age of 43 (*Albuquerque Journal* 1940a). After his death the family moved back to Albuquerque, and Alice Francis Putney eloped with Henry Louis Wiese Jr. (Wiese 1978).

Henry Louis Wiese Sr. was born in 1881 in Bremen, Germany. He came to the U.S. in 1900 and worked at his uncle's store in Watrous, New Mexico. His uncle was also from Germany but the two didn't get along, so he set up a butcher shop in nearby Wagon Mound. Henry Louis Sr. married Mary Alice Paddock, a girl from Wisconsin whose family had moved west for health reasons and had also moved to the Watrous area. The couple had five kids, including Henry Louis Jr., born about 1914. Watrous was the head office for the Frank Bond mercantile operation. Bond was always looking for good people to run one of his stores on a percentage basis (Sandoval County Historical Society undated manuscript).

In 1917 Henry Louis Sr. moved the family to Cuba to manage Bond's store there. The store was known as the Cuba Mercantile ("Cuba Merc") and was on the site of the present C.C. Paisano's Pizzeria on the east side of Main Street (Wiese 2005). They rented a house from local strong-man E.A. Miera. The move wasn't an easy one for a German family during World War I and they never felt entirely welcome. After a few years the family moved to the Gallup area to manage another store. During the Depression they went to Los Angeles, where Mrs. Wiese bought and operated a laundry (Wiese 2001). In 1935 they returned to Cuba and Henry Louis Sr. started the H.L. Wiese General Mercantile Store (Sandoval County Historical Society undated manuscript). He died in 1961.

Henry Louis Wiese Jr. began ranching in the Cuba area in 1937. As mentioned above, in 1940 he had eloped with Alice Francis Putney. They had two daughters and two sons, including Henry Louis III in 1942. Louis grew up in a house in town then located in the area behind the present McDonald's Restaurant. In the mid-1960s Louis and his new wife Carol built their house three miles south of town, at the junction of the old 1930s NM-44 and the then-proposed new alignment of the highway. Today the ranch is a prominent feature on the east side of US-550 at mile 60.25 on the approach to Cuba, tucked into an alcove eroded into the Ojo Alamo Sandstone (Figure 114). By 1980 the spread-out Wiese ranch consisted of 2,481 acres owned by the family and an 4,861-ac grazing allotment (Ripp 1980). Henry Louis Jr. died in 1987 at the age of 72. Alice died in 2005 at the age of 81 (*Cuba News* 2005b).

Young/Hernández

John Floyd Young was for many years the proprietor of the Young's Hotel, Cuba's most eye-catching landmark. He was born in 1883 to a Mormon family in Fairview, Utah, the oldest of 12 children. After finishing business courses at Brigham Young University in 1904 he moved to the Red River Valley area of northern New Mexico where his father was involved in sawmills and mining. While working at the curio shop owned by Frank Staplin, an artist, he met his future wife, Bessie Lee Watson, who had moved to New Mexico from Wisconsin for her health. They married in 1905 at Raton. The two lived for a short time in Tacoma, Washington, and there their only child, Harriet Elizabeth, was born in 1907. They were drawn back to New Mexico, however, and John worked at the Staplin store in Taos. When the store moved to Farmington he formed a partnership with his former employer. Farmington at that time was very isolated. Staplin heard that business was booming in the Cuba area and in 1912 he moved the store there, and the Youngs followed. The footloose Staplin went bankrupt and John worked for a time for Augustin Eichwald.

Tired of this bouncing around Young decided to go into business for himself, and in 1918 opened his own general store at the location of the present McDonald's Restaurant (Figure 137). Often his customers would arrive with a wagon-load of kids and the Youngs felt obliged to put them up for the night in the store's two adjoining rooms. They began losing money. Eventually the Youngs began to rent the rooms out and then they added a few more. Such was the beginning of the Young's Hotel. They operated the store/hotel until the building burned down in 1927 (Córdova-May 1973). Besides the store/hotel business, Bessie in her own right took an active in community affairs and served as Cuba's postmistress from 1915 to 1934 (Córdova-May 2009b).

Figure 137. First Young's store and hotel, ca. 1918. (Photo courtesy Palace of the Governors Photo Archives (NMHM/DCA), negative #57283.)

Figure 138. Second Young's store and hotel, pre-1937 (located just east of present hotel). (Notice mud chains on vehicle and more reliable transportation at right; photo courtesy Palace of the Governors Photo Archives (NMHM/DCA), negative #57282.

At this time the northern edge of town was low-lying and prone to floods. It was also available for purchase. John bought a piece of property there and built a store just west of the present Young's Hotel, smack in the middle of what is now US-550 (Hernández 2004; Figure 138). In the early 1930s he phased out of the retail business and converted the building into a full-time hotel (Reeve 1961). The building burned down in 1937 and it was immediately replaced by the present structure (Figure 139), a short distance to the east and now set back on the east side of US-550.

Both hotels served the cattle-buyers, timber people, construction people from the copper and coal mines, and the occasional traveler between Albuquerque and Farmington. Young also focused on the sawmill business and provided lumber for a large regional market. He went on to live a long life characterized by public service. He served as a U.S. Commissioner to deal with homestead and other federal issues, and as Chairman of the Cuba Independent School District (Reeve 1961). John Young died in 1960.

Daughter Harriet Young married Walter R. Hernández. Walter was descended from an old Spanish family with deep roots in Mexico dating back to the 16th century. The family had acquired a land grant in southwestern Chihuahua State, and had lived there for many years in a valley that became known as *Valle de los Hernández* (Spanish for "Hernández Valley"). The families in the area were staunchly Catholic, but in about the 1880s Walter's family converted to Protestantism, causing a terrific row within the extended family. Walter's father carried a horrible scar on his face where his uncle had hit him with an axe because he had had the audacity to become a Protestant (Hernández 2000b and 2004)!

Walter's father came to the U.S. and prepared for the Congregationalist ministry at a seminary in El Paso, Texas. Walter was born in 1905 and attended elementary school in New Mexico. He graduated from Albuquerque High in 1920 and earned his B.S. in geology in 1925 from the University of New Mexico. He became a rancher. In Cuba he met Harriet Young and in 1927 they were married in Albuquerque. Their

Photo 139. Third and present Young's
(No. 13 in Figure 142). (Photo by author 2002.)

only child, John Sebastian, was born a year later (Reeve 1961). For about two years they ran a little store and gas station in the little roadside stop of Tilden about three miles north of La Ventana (Chapter 12). In 1932 four-year-old John was sent to Cuba to live with his grandparents, the Youngs, so he could attend school. John remained with his grandparents in Cuba and was raised by them (Hernández 2000b and 2004).

In 1938 Walter became involved in the sawmill business in Cuba, and he also ran a trucking operation called the Hernández Truck Line. He hauled Ford auto parts from Detroit to Albuquerque and canned meat from Chicago. The main line, though, was from Durango to El Paso. Like his father-in-law, Walter became involved in community affairs. He served in World War II and reached the rank of major and later became a Lt. Colonel in the New Mexico National Guard. He died in 1991 (Hernández 2004).

In 1953 John S. Hernández married Eleanor Law. Eleanor was born in 1928 in Los Angeles, California. John and Eleanor had four children, all raised and schooled in Cuba. The couple ran the Young's Hotel until about 1980 when the other motels that exist today along the main drag were just getting started. For years they never had a lock on the front door of the hotel. However, many times they'd get up in the morning and find a drunk, who'd come in to escape the cold, passed out on the floor. They had to put a stop to it, so they closed the hotel. The first floor of the building for a time served as a bus station, and

the second floor provides space for publication of the village newspaper, the *Cuba News*.

At the age of 46, Eleanor enrolled in college and earned degrees in library and media education. She became a teacher and taught school at Pueblo Pintado, in the Checkerboard Area, for several years (*Cuba News* 2008a). In 2003 their house, located just east of the hotel, burned to the ground with all their belongings. Eleanor died in April 2008, after 55 years of marriage to John.

Eichwald

The Eichwald family has been prominent in Cuba for over a century. The family possesses an unpublished biography of the sheep baron Augustin Eichwald (Chapter 15), appropriately titled *Don Augustin*. Augustin died ca. 1926, and the manuscript was written by one of his sons, Alec. (Unfortunately I was unable to pry a copy loose from the family.) Another son, also named Augustin, acquired the funding for the town's community center located across the street from the old abandoned school on NM-126. In appreciation, Cuba named the building the "Eichwald Center." Augustin, Jr. served as Cuba's mayor 1978–1982. *Don* Augustin's grandchildren include Mr. F. Kenneth Eichwald, who in 2002 was elected to serve as Sandoval County Magistrate Judge, and Ethel Maharg (nee Eichwald), whom in that same year was elected to serve a term as village mayor.

Some "dry nesters"

The people who settled in the Penistaja district west of Cuba affectionately referred to themselves as "nesters" or "dry-nesters." A favorite expression at the time, referring to the area, was, "It's purty good country in a way. But nothin'll grow, though, and it's shore hard on hogs and women" (Johnson 1953). The stories of three families—Johnson, Wood, and Gibbons—nicely bestow flavor to the history of this region, which had been thrown open to stock-raising or grazing homesteads (640 acres) in the years following World War I.

Johnson

One of the early arrivals was the Johnson family. Mrs. Carla May Johnson, widow of Eugene E. Johnson, lives on the Johnson Ranch at MP-9 on NM-197, west of Cuba. Carla May, nee Aker, was born in Caddo County, Oklahoma. Aker is a German name, and her great grandparents spoke only German. In the 1930s her father and uncle were on their way to Arizona to look for a place to settle. When they got to Albuquerque the car broke down and they had to stay there a few days waiting for parts. They liked the place so well they stayed. She was three months old on the day the rest of the Aker family got to Albuquerque. Her dad homesteaded on the mesa west of the city.

Her father-in-law, William E. Johnson, was from Haskell County, Texas, about 140 miles west of Ft. Worth. He had sent his two older boys out here from Texas in 1920 to find a place to settle. They wrote back and said, "Poppa, come on out. It's a wonderful place. It's warm in the winter. We have grass, everything out here. It's a good country—come on out" (Johnson 2003). The Johnson family consisted of four sisters and seven brothers. In 1924 the parents, two of the sons and their families, and Carla May's four-year-old future husband, came out in two covered wagons. They decided to stay and homestead. The father returned to Texas and brought several family members out by car, but some of the girls stayed behind to complete their university studies at Baylor. When they finished their studies they joined their family in New Mexico. All three girls became teachers. When the boys reached their teens they had to stay with the oldest sister in Albuquerque to attend Albuquerque High and they came home only for vacations (Johnson 2003).

Will Johnson, the father, with his two sons who were then of age, filed for three homesteads on three 640-acre sections of land near Penistaja, and by the early 1930s they had proved them up. In the early 1940s the government offered to buy out all those homesteaders willing to sell. One son sold out to the government and the other to his father. The youngest son, Eugene E., later bought out his father and his sister-in-law's (Smith) homestead (Johnson 2005). It's now the present Johnson Ranch marked by a sign saying "Carla's Trading Post."

Eugene and Carla May were married in 1950. Their ranch was stocked with rabbits, chickens, a milk cow, and a pig to provide meat and lard for soap and meat. The oldest Johnson son, Plemen, bought a section of private land and around 1950 built the Johnson Trading Post on NM-197 at about mile 15.6, southwest of the Johnson Ranch. The post is closed today. Another Johnson son, B.B., married Naomi

Gibbons, the sister of the writer Euell Gibbons (see below). She taught school at Penistaja, which is now just a ruin. The last of the Johnson siblings—two of the sisters—died in their middle 90s within two months of each other in 2002. For 50+ years Carla May has operated the Johnson Ranch Weather Station that gathers vital precipitation data for this portion of the state (*Cuba News* 2002b).

Wood

While my wife and I were visiting Carla May Johnson in her home in 2003, I asked her about the location of the Gibbons homestead (see Gibbon family below). She suggested we drive up to the Wood Ranch, about eight miles northwest as the crow flies. We found Mrs. Wood's rustic ranch house snuggled back in a protected alcove eroded into the Cuba Mesa Sandstone, miles from any other soul (Figure 140). Lovena Wood lives alone at the ranch with three horses, four dogs, and a cat. She runs 40 head of cattle on 640 acres. She explained that her father bought the homestead from a Reverend Payne and moved the family onto the property when she was three years old (Wood 2003).

Figure 140. Lovena Wood at her ranch house up near the Continental Divide. (Photo by author 2003.)

Lovena graduated from college in 1955 and began a career as a teacher. For a while she taught high school in Cuba, but hated it and returned to her family's ranch. Later, in 1965, she was persuaded to teach at a Bureau of Indian Affairs school at Encino, a small Navajo town in the Checkerboard Area. She taught kindergarten and first grade there for 22 years until she retired and returned to ranching full time (Sandoval County Historical Society 1992).

She knows the Gibbons place well and offered to take us there in her pickup. Mrs. Wood is not very nimble. She carefully placed a step-stool at the front door of her truck, gracefully stepped up in to the cab, and somehow reached down and retrieved the stool. We jumped in. Carefully she drove us north up the main road fronting her house, and then east for more than a mile down a steep and deeply rutted track. Finally we came to a two-room cabin partly dug into the hill. A fireplace was tucked into a corner of the main room. This was where the writer Euell Gibbons spent his early teenage years in the middle 1920s. It must have been a very small place indeed, with two adults, four kids, and the specter of unrequited hunger.

Figure 141. Gibbons family semi-dugout (location in Figure 130). (Photo by author 2003.)

Gibbons

If you've heard of Euell Gibbons you are dating yourself. He died in 1975. During his time he became a very popular cultural icon of the wild-food movement. But let's back up. Euell Theophilus Gibbons was born in 1911 in Clarksville, Texas, in the Red River Valley just south of the Oklahoma line. While there he learned from his mother how to gather wild foods. In 1922 his parents moved with their four children to a 400-ac homestead with a spring, 14 miles due west of Cuba. The father, Eli Joseph Gibbons, built a humble semi-dugout (Figure 141). This was the time of the Dust Bowl, and Eli had to leave home from time to time in a desperate search for work. One time the family's food supply had dwindled down to a few beans and an egg. Finally one day, in despair, Euell took a knapsack and went out to scavenge the surrounding countryside. He returned with puffball mushrooms, piñon nuts, and yellow prickly pear fruit. For a month the family lived on this trove of wild food. He saved their lives (McPhee 1962).

At the age of 15 Gibbons left home and worked his way to California and the Pacific Northwest. He cowboyed, picked cotton, lived in hobo camps, and foraged for wild foods to stay alive. He spent a stint in the army (1933–1936) and while there he married. The couple had two sons. During World War II he worked in a shipyard in Hawaii. After the war, now a divorcee, he became a beachcomber. For almost five years he lived exclusively on wild foods and supported himself by composing crossword puzzles in Hawaiian (McPhee 1962)! His level of formal education is unclear. It is likely that as an adult he completed high school and in the late 1940s for a time attended the University of Hawaii (Texas State Historical Association 2002).

In 1949 he married for the second time and he and his new wife taught on the island of Maui until 1953. They joined the Quaker movement and moved to the eastern United States. While employed at a school in Philadelphia in 1955 he began writing about edible plants. His very-understanding wife agreed to support him while he wrote full time. His ambition was to become a novelist. He wrote short stories and completed a novel, some of which contained items about wild foods. Nothing worked. He passed his 50th birthday without a single publication and he increasingly saw himself as a failure. His editors wisely advised him to skip the novels and emphasize the wild foods. And so he did. In 1962 Gibbons published his first book, *Stalking the Wild Asparagus*, and suddenly established himself as the master of his field. The book's timing was exquisite in that it appeared at the beginning of the back-to-nature movement, and it quickly became a best-seller (McPhee 1962; Texas State Historical Association 2002).

In 1963 he and his wife moved to a farm in Pennsylvania that he dubbed "It Wonders Me." He taught foraging, and wrote five more books about wild foods with such whimsical titles as *Stalking the Blue-Eyed Scallop* (1964), *Stalking the Healthful Herbs* (1966), *Stalking the Good*

Life (1966), *Beachcombers Handbook* (1967), and *Stalking the Far Away Places* (1973.) In 1972 and 1973 he published two articles about wild foods in *National Geographic*.

In 1974 he appeared twice on the Johnny Carson show and in 1975 on the Sonny and Cher Show. During the latter he slyly picked up a wooden plaque awarded to him and began eating it (it was a prop made out of chocolate). He made television commercials for Post Grape Nuts cereal. To this day I still remember his trademark phrase, "it tastes just like wild hickory nuts." But these commercials got him into trouble with his hard-core supporters. Many of them thought he'd "sold out to big business," and that he had become a mere mouthpiece for a cereal company (Kallas 1998). Despite the setbacks, the kid from the semi-dugout near Cuba had done well.

Gibbons spent the last of his days in Pennsylvania with his wife. For years he had cooked his wild healthy foods in such unhealthy things as bacon grease. He smoked, as did most people back then. In his later years a worsening arthritis prevented him from exercising. In 1975 he died at the not-so-advanced age of 64. It is sadly ironic that despite eating wild and supposedly healthful foods, he died of heart disease. (Kallas 1998).

I often wondered if Euell, while in northwestern New Mexico working on his 1973 article for *National Geographic*, ever took the occasion to visit his boyhood home. Perhaps he associated the place with the unpleasant memories of hunger and despair. After he left home, the other members of the Gibbons family moved out of their semi-dugout cabin as soon as they could. The father had died early, in 1947, also of a heart attack, but his mother lived to a ripe old age and died at 92 in 1980 in a Clayton, New Mexico nursing home. Euell's sister Naomi (d. 1965), brothers Hoyt (d. 1981) and John Bowers (d. 1984) spent the rest of their lives in New Mexico.

Select places of interest in Cuba (numbers keyed to Figure 142)

It must be admitted that the Village of Cuba is not a tourist destination. It's a place where people live, earn a living, and call home. It serves as a pit-stop and occasional overnight spot along the long haul between Bernalillo and Bloomfield. Yet, like any place with a soul, Cuba does some sites that warrant a second look or evoke an interesting story.

No. 1. Main street, Old NM Route 44/ US-550

Most of Cuba's businesses and points of interest are arrayed along the main street—US-550, or Old 44. The main drag has changed mightily over the years. Until perhaps as recently as the 1940s the southern edge of town ended at the Rito Leche acequia. South of the acequia was all irrigated wheat and alfalfa (Wiese 2001). The northern end of town ended at a swampy area just north of the present Young's Hotel (No. 13 below). Between the two end points was a boulevard of dirt (Figure 143).

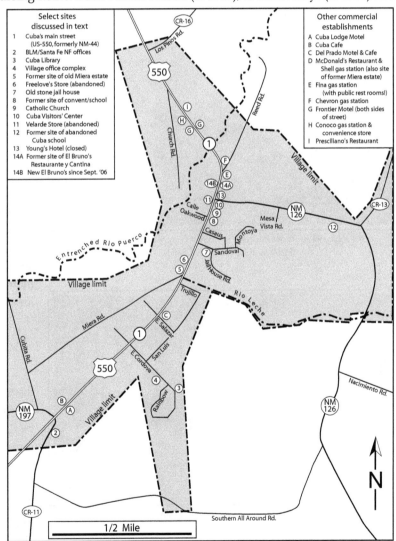

Figure 142. Village of Cuba and select points of interest.

Figure 143. South view along Cuba's main street, now US–550, late 1910s. (Second Catholic church on left; Eichwald mansion on right, shadow in foreground from first "public" school where Visitor Center is today; end of town at trees in distance; photo courtesy Sandoval County Historical Society, photographer Thomas F. Ball.)

A. View ca. 1920s. Photo courtesy Sandoval County Historical Society. (Photographer Thomas F. Ball.)

B. Same view as in "A" above. (Photo by author 2005.)

Figure 144. North view along Cuba's main street, now US–550.

By 1929 the main street had become a segment of state highway NM-44 and it remained so until 2000 when it was redesignated US-550. In 1970 the street was widened, the fronts of a number of old abandoned buildings were hacked off, and bridges were built over the Rito Leche on the south and the Rio Puerco on the north. Finally, in 1979, the main street was paved (Figure 144A and 144B).

No. 2. Bureau of Land Management and Santa Fe National Forest offices

Cuba's portion of US-550 is anchored on the south by the joint offices of these two federal-government entities. They are major employers in the Cuba area. Information and maps are available there.

No. 3. Cuba library

Another symbol of Cuba's community pride is the Cuba Public Library. It is the fifth and hopefully last in a series of facilities. The first was sponsored by the Nacimiento Women's Club in 1966. The group acquired a 10 x 10-foot, dirt-floored tool shed, with no heating or power. The dump was remodeled and it officially opened the next year, with 200 donated books. By 1969 the library had outgrown the building and the Village Council gave its okay to relocate and occupy a portion of the old Fire Station on Main Street. In 1981, when the Eichwald Community Center was opened, the library moved into it (*Cuba News* 2000). Over the years the library filled up the entire Eichwald Center's gym. In 2003 an opportunity presented itself. The Cornerstone Cleaners, on Main Street, was being foreclosed on by the bank. The village elected to lease the building as a temporary library. Finally, in the summer of 2008, the village completed its spanking-new facility across from the Village Center (*Cuba News* 2003, 2008b).

No. 4. Village complex

The village municipal complex was built in 1977. It contains the village offices, fire station, police station, and jail. The senior center was added later.

No. 5. Former site of E.A. Miera's estate

Not a trace of this sprawling estate — once located on the northwest corner of the intersection of US-550 and Miera Road — remains today. Place names, however, are conservative things and tend to last, so the strong man's name at least will survive on the village's road map. From the early 1900s until the 1940s the Miera buildings anchored the southern end of the community. The McDonald's Restaurant was built on the site in 2003, over stiff local resistance from members of the community who felt that it would wipe out other businesses.

No. 6. Abandoned Freelove's store

Roy Freelove and his wife moved to the Cuba area sometime prior to 1912 and bought a homestead in the mountains (Wetherill 1953). Their son, Eric Leslie Freelove, was born in 1908. He began his business career in 1923 (at the age of 15?) as manager of the El Ojo Verde Lumber Company in Cuba and ran that enterprise until 1933. In 1930 he married a Texas girl named Lena Mae Randall, and they went on to have seven children. From 1934 to 1941 Eric was a rancher in the La Jara area. In 1941 he embarked on a public service career that culminated with his four-year term as State Senator (1956–1960). In 1945 he opened and operated a store in La Jara and in the late 1950s opened a branch store in Cuba, with his son as manager. Eric died in 1987, and his wife died in 1991 (Reeve 1961). The business eventually failed and the building was abandoned in the late 1990s.

No. 7. Old jailhouse

In 1918, stone reclaimed from a nearby copper mine and smelter was used to build Cuba's first lockup (Cuba Visitor's Center undated; Figure 145).

No. 8. Former site of Catholic Convent School

This desolate parcel of land once contained the handsome three-story adobe convent (Figure 134). It was torn down in 1967 and never replaced.

No. 9. Catholic Church of the Immaculate Conception

This is the second church of this name built on the site. The stone statue of Virgin Mary in front of the church was originally above the door of the old church. As mentioned earlier, the statue had been hauled to Cuba by wagon from Guadalajara, Mexico.

Figure 145. Old Cuba jailhouse (No. 7 in Figure 142).
(Photo by author 2002.)

No. 10. Cuba Regional Visitor Center

Until quite recently the Nacimiento Heritage Committee, a community-driven committee, maintained the Visitor's Center at the intersection of US-550 and NM-126. It was built in the mid-1990s at the site of the first public school and a younger, vacant gas-station building. Pamphlets, maps, and information were available. The center had been open all year except for the three winter months. Sadly, in December 2008 the facility was closed down (*Cuba News* 2008c). Fortunately in May 2009 it was reopened, with posted hours for Fridays and weekends.

No. 11. Abandoned Velarde's store

This now-abandoned Velarde Groceries store was built by Cuba natives Rudy and Mary Velarde in about the early 1940s (*Cuba News* 1964) on the site of the house owned by Gerald and Leonor Hoch (Córdova-May 1973). Leonor Hoch had been the town's first postmistress. At that time she was a very pretty girl and was considered the "belle of the town" (Wetherill 1978). Velarde Groceries sold general merchandise and bought wool from the local ranchers. In 1948 Rudy became one of the first 11 members of the Jémez Mountains Electric Cooperative, Inc.(along with Eric L. Freelove and Walter R. Hernandez of Cuba), that was successful in bringing electricity to Cuba for the first time. The Velarde Store went out of business in the 1950s (Cuba Visitor's Center brochure undated). Recently the old store has fallen on hard times. The village now owns the derelict old structure and for a time has rented it out in return for its upkeep and renovation. There have been plans afoot to use the building as a gallery for local artists, a temporary site for the Village Library, and a site for a veterans memorial park (*Cuba News* 2002d). Meanwhile it sits there awaiting its fate.

No. 12. Abandoned Cuba school (site)

The unfortunate eyesore that once occupied this site, the old Cuba school built in the 1950s, was abandoned in 1974 because of the presence of asbestos. For years it sat pending abatement for the asbestos, a highly expensive procedure. Some frustrated soul even suggested that the National Guard blow up the building as a training exercise! Finally, in June 2008 the old structure was torn down (*Cuba News* 2008b).

No. 13. Young's Hotel

The existing Young's Hotel is actually the third Young's commercial structure in Cuba. The first was built in 1918 and was located south on the main street where the McDonald's Restaurant is today (Figure 137). The second was built in 1927, immediately west of the present building where US-550 now runs (Figure 138), and burned down in 1937. The third, present hotel was built in 1937 (Figure 139). Each wall of the massive hotel consists of two parallel wooden walls, and the space between them is filled with rubble. The exterior is of stucco and hand-hewn pine. For many years the hotel was a landmark along NM-44. It had 16 rooms upstairs for guests who would let themselves in and pay the next morning. The hotel business was discontinued in about 1982 (*Time* 1986).

For a while the building provided space downstairs for a bus station (the bus service has since been discontinued) and space upstairs for publication of the town's newspaper, the *Cuba News*. This monthly paper had been inaugurated in 1964 when the community was first incorporated. The paper was originally a skinny four-page affair, but over the years it has grown to 20+ pages. From the beginning the typist and editor of the paper was Betty Jane Curry. In about 1995 she, her daughter and a friend bought the paper and incorporated as COM (Curry-Ohler-Meeks), Inc. Press. Each month since the mid-1980s the polished manuscript had been dispatched to Farmington where it was printed for distribution by the *Farmington Daily Times* (*Cuba News* 2006b). With the demise of the bus service to Farmington the paper was forced to move into the 21st century. Finally, in 2006, instead of the tedious "cut and paste" method the process has been finally computerized. Almost until her death in November 2010 Betty Curry penned a no-nonsense opinion about local and national events in her *Speaking Out* section.

No. 14. El Bruno's Restaurante y Cantina

For many years El Bruno's has been a must-stop for travelers along NM-44 and now US-550 (No. 14A). For the 14 years leading up to 1975 the restaurant was run by Renee Jaramillo as the Silver Star Bar and Café, before being purchased by her brother and sister-in-law, Bruno and Hazel Herrera. The Herreras were born and raised in Cuba and were high school sweethearts. They, with their children, have developed their own special New Mexican recipes that are named for the surrounding communities, such as the Gallina, the Regina, and the Cuba (Moore

1999). Hazel learned many of the recipes from her grandmother. The Silver Star had only a small dining room, but the new restaurant had four dining rooms, an outdoor courtyard, and a bar (*Cuba News* 2005c).

On June 9, 2006 Bruno's, with virtually all of its collectables and ambience went up in flames (*Cuba News* 2006c). After an agonizing period of introspection, the Herreras' new El Bruno's—like a phoenix rising from the ashes—opened on September 1, 2006 across the street in a remodeled Tastee Freeze (No. 14B). The sole survivor of the June fire was a mural of a Hispanic woman in a chile field. This sole survivor now graces the wall of the new place (*Cuba News* 2006d).

Some sites of interest west of Cuba

Mesa Portales Ancestral Puebloan ruin

A well-worthwhile side trip west from the Village of Cuba is to the obscure Indian pueblo ruin at Mesa Portales (Figure 146). The pueblo is located on public, Bureau of Land Management (BLM) land. It is utterly unheralded and unmarked, but easily accessible from State Route 197 (Figure 147A). At about MP-14 there is an earthen dam to the east. A dirt road (I recommend high clearance, 4WD vehicle, although it might not always be necessary) skirts the north side of the dam and leads east for about four miles into a big reentrant cut into Mesa Portales, which itself is carved out of the Ojo Alamo Sandstone. At about 3.25 miles is a small pond, and about 0.75 mile beyond that is a dirt track that goes to the south toward the sandstone bluffs. This track circles around to the south of a small, unimpressive sandstone knob.

The knob is a defensible, isolated projection of the main mesa (Figure 147B). The large masonry pueblo atop the knob was occupied during the Ancestral Puebloan, Pueblo III period (1100–1300 CE), a period of upheaval characterized by regional site abandonment and major migrations. The site is officially known as LA-4568, and is the subject of the Mesa Portales Archeologic Project jointly conducted by the Albuquerque District of BLM and Eastern New Mexico University. Field work began in 2002 and is now complete. Ongoing study is focused on precise dating of artifacts. Fortunately, there is little of interest to potential looters (Figures 147C). However, the area does provide a pleasant day-trip destination and an excellent picnic site.

Lonely homesteader's cabin

An interesting and scenic bypass loop west and northwest from Cuba is State Route 197 and Sandoval Country Road 1101, the "Chuilla Road." CR-1101 connects NM-197 on the south to US-550 to the north at MP-73 (Figure 146). The road is well-graded dirt, dusty in summer and to be avoided when wet. It traverses badland country cut into the mudstones of the Nacimiento Formation and then gradually rises onto the somewhat cliffy Cuba Mesa

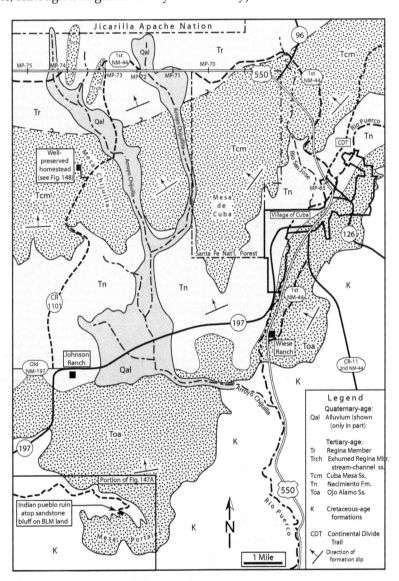

Figure 146. General geology and select points of interest west of Cuba. (Modified from Woodward et al. 1970a, 1970b; NMBG&MR 2003.)

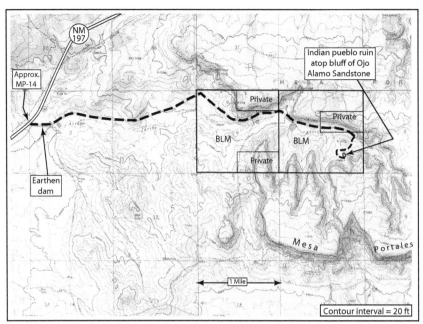

A. Topographic map showing access route to Mesa Portales ruin.

B. Northwest view of small mesa (middle distance) on which ruin is located.

C. Cryptic traces of pueblo wall.

Figure 147. Ancestral Puebloan ruin at Mesa Portales (general location in Figure 146).

Sandstone. After six miles a side road to the left leads 0.55 miles to a rather well-preserved homestead on BLM land. The site contains a main log cabin, two other less impressive log cabins, and a root cellar (Figure 148). The cabins are partly charred by fires, which possibly took down the roofs. The stories these walls could tell! A visit to this old homestead is well worth the effort. The sense of solitude is palpable. These long-gone folks certainly had their privacy!

A few parting words

The Village of Cuba, by virtue of its scenic setting and its strategic location at the midpoint on a major four-lane highway, could and should be a prosperous site. Instead it is a troubled place with a less than vibrant economy. The demise of the lumber and copper-mining operations did not help. Unemployment hovers around 17.5%, about three times the national average.

Statistics gathered for the decade 1990–2000 reveal much. The population is in decline. It is also aging: during that same decade the percentage of people over the age of 50 has increased from 23% to 29%, while the median age has increased from 28 to 32. The trend is for more retirees and a smaller workforce, and fewer people in their child-bearing years means slower growth. Best-guess projections for the next 25 years suggest a measly annual population growth of less than 1%. Forty percent of the houses are mobile homes, and an astounding one quarter of all homes are vacant. Many businesses are vacant or up for sale, as is starkly evident along the main street. Many other streets remain unpaved or below standard, and loitering transients from the Checkerboard Area and litter are constant problems (MRCOG 2004).

Fortunately, many Cubanos realize that the Cuba's future depends on the ability to attract and retain residents. This will require providing amenities and upgrading the appearance of the town (MRCOG 2004). The community and the Forest Service have joined forces and in late 1992 organized a so-called Future Search Conference (FSC). About 100 participants gathered to discuss Cuba's past, present, and future. The FSC created a number of community-driven committees, including the Cuba Regional Economic Development Organization (CREDO), the Rio Puerco Watershed Committee, and the Nacimiento Heritage Committee.

Limited funds and conflicting priorities among Village officials and citizens severely restrict what can be done. In 2000 the town received a cash grant from the Rural Economic Development through Tourism agency to spruce up the place to make it more attractive to tourists by erecting banners and wooden welcome signs, and to hire a part-time staffer for the Cuba Visitor's Center (*Albuquerque Journal* 2000). Cuba indeed has its work cut out for it, but it is trying mightily.

Figure 148. Homesteader's lonely cabin off CR–1101 west of Cuba. (Photo by author 2005.)

Perhaps Cuba's proudest hour in some time occurred on Veterans' Day weekend in November 2005. On that sunny Sunday the town hosted a super parade to launch the annual Capitol Holiday Tree on its way on a special flatbed truck trailer to Washington, D.C. The 80-foot-tall Engelmann Spruce tree was harvested in the Santa Fe National Forest above the town to the east, but the Cubanos considered the tree their gift to the nation (*Cuba News* 2005e). For the tree's send-off they prettied up the main street (US-550) and decorated it in Christmas holiday regalia. Then followed the three-mile-long parade with whoops and hollers.

Perhaps indicative of a Cuban renaissance is the effort launched by local historian Esther V. Córdova-May in the *Cuba News* in November 2007. Since 1972, when she received a grant from California's Mills College to do an oral history project, she has researched and collected material about Cuba. Her monthly column entitled *Antes* (Spanish for "before" or "a time before") tells stories about her favorite town and makes the area's rich history available to its people. Esther recognizes that only by knowing your home's history can you really be at home.

18

Up into the "Inner" San Juan Basin

Highway US-550 north from Cuba climbs up through the cliffy ramparts of the Cuba Mesa Sandstone (more below), takes a sharp swing to the west, and takes us into the central portion of the San Juan basin, sometimes called the "inner" basin. Today this vast region is almost empty of people. Up until about a half century ago, however, it was the center of a busy livestock industry. The grass was tall, the water plentiful, and the life strenuous but good.

First, just what is the San Juan basin? It is both a geologic and geographic feature (Figure 3). The Cretaceous-age (95 - 70 Ma) rock formations were folded downward during the regional Laramide compressive event in three big pulses: 75 - 70 Ma, 65 - 60 Ma, and 55 - 40 Ma (Cather 2004). As the Cretaceous rocks were progressively pushed downward in the two pulses after 65 Ma, topographic lows or "basins" were produced into which younger sediments were deposited. The second pulse provided accommodation space for the Paleocene-age, muddy, Nacimiento Formation, and the third for the Eocene-age, sandy, San José Formation (Figure 4). Today the rim of the "outer" San Juan basin is generally defined as the outcrop of the base of the Cretaceous, and the rim of the "inner" basin usually as the outcrop of the base of the San José Formation (Figures 149 and 150). The buried Cretaceous rocks in the inner basin now contain the huge deposits of natural gas, and the Cretaceous rocks cropping out in the outer basin contain enormous deposits of extractable coal.

The San José Formation consists of three subdivisions, or "members." Only the first two are exposed along US-550. From bottom to top these are the Cuba Mesa Sandstone Member and the Regina Member (Figure 150).

The southern and eastern outcrops of the Cuba Mesa Sandstone form a cliffy rim that defines the perimeter of the inner basin, and the Village of Cuba lies just south of it. The rim has acquired some colorful names, as one might expect for such an annoying topographic barrier. West of Cuba Mesa and Chujuilla Mesa (Figure 146) the name changes to "Sisnathyel Mesa," from the Navajo word, *Sis Naateel*, meaning "descending wide belt" (Linford 2000). Farther west, near Lybrook, the rim is instead formed by channel sandstones of the Regina Member, and it takes the more innocuous name "Crow Mesa" (Figure 149).

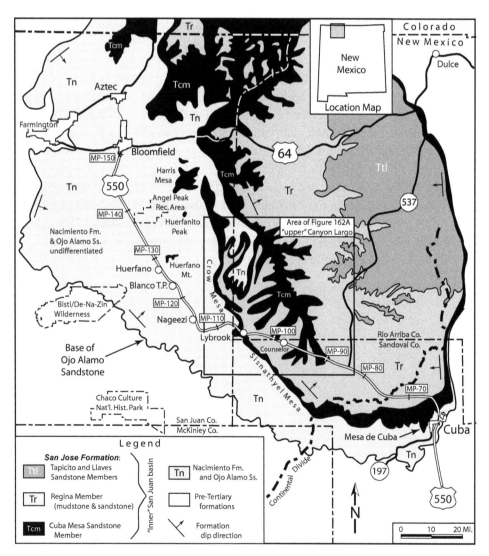

Figure 149. Generalized Tertiary bedrock geology of San Juan basin. (Modified from NMBG&MR 2003; Smith and Lucas 1991.)

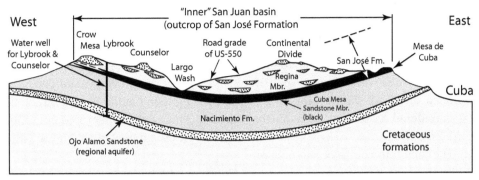

Figure 150. Schematic cross section (not to scale) across US–550, showing "inner" San Juan basin.

Both members of the San José Formation consist of sediments deposited by rivers, but the two differ in one important respect. The Cuba Mesa Sandstone, as its name implies, consists almost entirely of sandstone. Ancient braided rivers (rivers with multiple, "braided" channels) flowing downslope from uplands to the northwest deposited a pile of river sand as the land gently subsided to accommodate it. Later, during the deposition of the overlying Regina Member, the supply of sand had diminished. As a result, the Regina Member consists of a mixture of ancient river-channel sands—exactly like those of the Cuba Mesa—but with a great volume of intervening mudstones deposited in the wide flood plains that once surrounded the rivers (Figure 151).

Erosion has eaten away at the original outer limits of the San José Formation and reduced its original extent to what we see today. A prominent mesa capped by an outlier of Cuba Mesa Sandstone, *Huerfano* (Spanish for "Orphan") Mountain to the west, clearly shows that the formation had once extended at least that far west (Figure 149). Continued erosion has even carved up the Regina Member itself. The result is that some of the old river channels have been exhumed to form north-trending ridges or resistant mesa tops surrounded by low-lying flats cut into the softer mudstones of the old flood plains. A few of these exhumed channels occur at MP-73 and MP-74 (top left of Figure 146), and another group forms the mesa caps in the Counselor area up ahead.

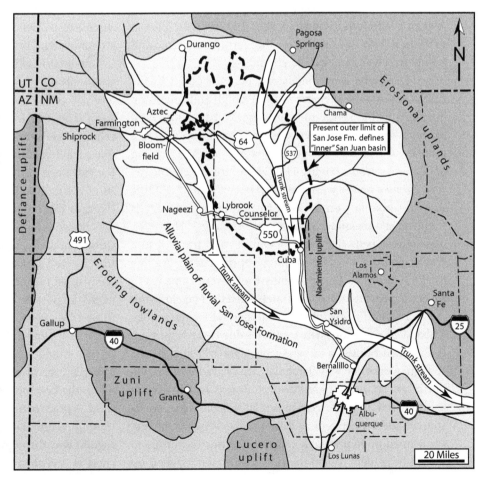

Figure 151. Schematic paleogeography during deposition of San José Formation (lightest gray and white), ca. 55–51 Ma. (Modern geographic features shown for reference; modified from Smith 1992; Cather 1992, 2003.)

Through the ramparts of the Cuba Mesa Sandstone

The Cuba Mesa Sandstone north of Cuba presented a significant obstacle for highway engineers. The original highway NM-44 north from Cuba was built in 1926–1927 as a 16-foot-wide, graded dirt road (Figure 146). At the time this route was considered a "trail road," with the emphasis on the word "trail" (*New Mexico Highway Journal* 1926a, 1926b, 1926c). The circuitous course of the route reflects the construction equipment of the time and of course the vehicles that were expected to use the road. Because it was dirt, maintenance was difficult and was required often. The highway was finally oiled in 1938, reducing the dust, and generally improving life for the traveler (*New Mexico Magazine* 1938). The present, straighter alignment from MP-65 to MP-70 north of Cuba was constructed in 1958 to reduce the grade and turns in order to accommodate higher-speed and heavier vehicles, and to improve safety (New Mexico State Highway Department 1959). This grade has since been widened and improved to form today's US-550.

Once up onto the Cuba Mesa Sandstone the highway makes its big swing to the west. Until this point the route of our highway has been largely governed by the ever-present south-to-north spine of the Sierra Nacimiento to the east. Now we say *adiós* to the Sierra Nacimiento and point our wheels toward Bloomfield.

The Cuba Mesa Sandstone plunges down to the northwest into the subsurface, from a high at its mesa-top positions to the south and east (e.g., Cuba Mesa). West of Cuba Mesa this massive sandstone has been cut into a number of separate low mesas by a pair of south-flowing ephemeral streams (Figure 146). These streams are the *Arroyo Chijuilla* on the west and the *Arroyo Chijuillita* ("Little Arroyo Chijuilla") on the east. My Spanish dictionary is mum on the meaning of the word "chijuilla." However, one source spells the name "Chihuili" and traces it to a Navajo word, *Tsé _itso*, meaning appropriately, "Yellow Rock" (Linford 2000). Apparently the present name was a futile Spanish attempt to accurately spell the Navajo word.

What little precipitation that does fall on the porous Cuba Mesa Sandstone here infiltrates downwards through it until it encounters a impermeable layer, a "tight" streak. It then works its way laterally "downgradient" to the north on top of the tight streak until it is locally able to seep out at the surface as a spring. These local water point-sources and the convenient passageways between the mesas teamed up to present a marginally attractive terrain to would-be settlers. When the U.S. government opened up the area to homesteaders after World War I, all the pieces came into place. Settlers rushed to the area, mainly after 1920. This was the northern end of the Penistaja District mentioned earlier in Chapter 16. Like in the Penistaja area, these settlers first attempted to dry-farm, but they soon saw the light and either moved out or switched to stock-raising. Almost all of them sold out to the government in the late 1930s and moved on (Figure 148).

Continental Divide

The term "Continental Divide" invokes an image of a rugged mountain pass sharply dividing two drainage basins. Sometimes that is the case, but here the divide is a subtle, rather nondescript feature, snaking here and there apparently with little reason. On US-550 the Divide is located at mile 76.75, on a little unimpressive rise of only 7,379 feet elevation. The Divide is simply another drainage divide, but one with a special significance: to the east all streams flow down toward the Rio Grande and ultimately to the Gulf of Mexico, and to the west they run down toward the Rio San Juan, the Colorado River, and ultimately to the Gulf of California and the Pacific Ocean. However, remembering the process of stream piracy discussed earlier, as the landscape is eroded downward, stream courses and their divides are transient things. The Continental Divide likewise is a moving target and shifts about through time. Its position here on US-550 is not fixed and will certainly shift somewhere else in the future.

The roadside stop on the Continental Divide is an excellent place to take stock of an astounding geologic story. Since leaving Cuba, highway US-550 has been basking in the viewshed of the lofty Sierra Nacimiento and especially its northern culmination, the flat-topped San Pedro Parks (Figures 152A and 153). However, about 25 Ma the Sierra was not so lofty. The patch of ground now occupying the Continental Divide and the top of San Pedro Parks were then located at nearly the same elevation. If so there obviously must have been a great deal of uplift and distortion of the landscape since that time. How do we know this to be true? Because of compelling forensic evidence left behind by a deposit of a peculiar and very distinctive type of rock called "chert" on part of the Continental Divide. Chert is a form of the mineral *quartz* (chemical formula SiO_2), the common stuff that forms most of the world's sand. A grain of quartz sand is typically an abraded, single crystalline unit, i.e., the atoms in the unit are arranged in a regular way throughout the grain. Unlike common quartz, chert is "cryptocrystalline," meaning that the individual crystalline units are cryptic, or "hidden" because they are too small to be detected by the naked eye. The material is extremely hard and breaks with a sharp cutting edge that made it a darling of stone-age tool makers.

A famous deposit of chert occurs at *Cerro Pedernal* (Spanish for "Flint Hill"), located about ten miles west of the Village of Abiquiu on the north flank of the Jémez Mountains (Figures 152A, 152B, and 154). This is the same distinctive landform (elevation 9,862 feet) that so attracted the attention of the artist Georgia O'Keefe in many of her paintings. A few hundred feet beneath the lava cap on the top of the hill lies a distinctive bed of chert called the "Pedernal Chert" dated at 25 Ma. Careful geologic work has determined that the sediments immediately above and below the chert were transported by streams flowing down westward from source areas to the east. Therefore the regional land surface at that time sloped down from east to west. Fifteen miles to the west of Pedernal Peak, on portions of the very gently-rolling summit of San Pedro Parks at an elevation of more than 10,000 feet, a thin veneer consisting of transported pieces of Pedernal Chert has been preserved. This elevation is on the order of about 500 feet higher than the chert at Cerro Pedernal (Church and Hack 1939). Of course it should be lower, not higher, because streams dependably flow downhill. Therefore, at 25 Ma the essentially flat summit of San Pedro Parks must have been laterally continuous with, and at a lower elevation than, the chert layer at Cerro Pedernal.

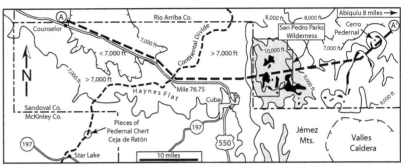

A. Generalized topographic map and distribution of Pedernal Chert (black). Elevations > 9,000 ft in gray. (Modified from Church and Hack 1939.)

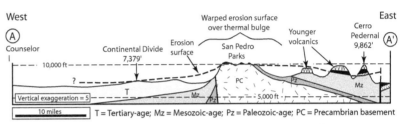

B. Generalized cross section from Counselor to Cerro Pedernal, showing Pedernal Chert (black) and 10 Ma post-chert volcanics ("V"s). (Modified from Church and Hack 1939.)

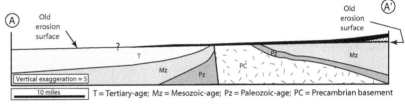

C. Ancient topography at time of deposition of Pedernal Chert (black), ca. 25 Ma.

Figure 152. Continental Divide and warping of erosion surface beneath Pedernal Chert.

Figure 153. East view of San Pedro Parks (skyline) from Continental Divide. (Photo by author 2005.)

205

Figure 154. South view of distinctive Cerro Pedernal (right horizon). (Photo by author 1992.)

The Continental Divide is located some 35 miles to the west of San Pedro Parks on a relatively flat area called "Haynes Flat" (Bryan and McCann 1936; Figure 152A). Pieces of Pedernal Chert have been found on a portion of the divide called *Ceja de Ratón* (Spanish for "Eyebrow of the Big Rat") at an elevation of 6,900 feet, more than 3,000 feet below the towering flat summit of San Pedro Parks (Figure 152A and 152B). What could have caused such a enormous distortion of the landscape? The prevailing theory is that the upward bulge of the old land surface (the base of the Pedernal Chert) was caused mainly during the past three million years or so due to the thermal expansion of the earth's crust attendant to the volcanism of the nearby Jémez Mountains. High heat flow from deep below the earth's surface causes volcanism. Heat causes expansion, or thermal swelling, of the rocks. This explains why a torch is summoned to free a stubborn metal nut that is "locked" onto a bolt. Heat applied to the nut causes the nut to expand more than the inner bolt, thereby making it possible to work it loose from the bolt. This anomalous heat flow caused the San Pedro Parks at the northern culmination of the Sierra Nacimiento to bulge upwards more than 3,000 feet in only the past few million years (Figures 152B and 152C; Formento-Trigilio and Pazzaglia 1998).

Continental Divide Trail

Despite the quirky and erratic path of the Continental Divide, it is the proposed locator for the Continental Divide Trail. This hiking-path-in-progress has an interesting history. In 1968 Congress passed the National Trails System Act. Public land agencies were directed to scout a north-to-south, border-to-border trail that would compliment the Pacific Crest Trail on the West Coast and the Appalachian Trail on the East Coast. In the 1970s, Jim Wolf, of the Continental Divide Society in Maryland, began pushing the idea of a trail along the backbone of the continent—a Continental Divide trail. He scouted the terrain and in 1973 wrote a guidebook on the proposed trail. In 1978 he testified before Congress about his idea and was instrumental in gaining congressional authorization that year of the construction of the Continental Divide National Scenic Trail (CDT) to span the 3,100 miles between the Canadian and Mexican borders (McKee 2001). Serious work on the trail began only in the late 1990s. The effort is being coordinated by the non-profit Continental Divide Trail Alliance, based in Colorado. Most of the work is being done by volunteers, clubs, and corporate sponsors.

Making the proposal was the easy part. Putting the CDT on the ground was a very different thing. The law dictates that the trail be located within 13.5 miles (20 km) on either side of the Continental Divide and as much as possible on public land. Unfortunately, in addition to public land, the Continental Divide crosses State land, Jicarilla Apache tribal land, private ranches, and Hispanic land grants. Owners of the non-public lands often want little to do with legions of hikers and horsemen trespassing on their turf. Trail advocates encountered their stiffest resistance in the area of the Hispanic land grants in northern New Mexico. The first 50-mile stretch of the trail just south of the Colorado state line runs through the Carson National Forest. This is precisely that land most bitterly contested by descendants of the land grant holders who have never recognized the validity of the forest's creation out of what they consider their community lands.

Part of the CDT was proposed to run through the forest above the little town of Tierra Amarilla, the tiny county seat of Rio Arriba County (McKee 2001). Moises Morales, a longtime activist from Tierra Amarilla, has compared the planned trail with efforts by environmentalists to restrict wood-gathering or logging in national forests long used by Hispanic villagers. "It's another example of people not even from our part of the country wanting to come into our community and destroy our privacy," said Morales. He says that trail advocates want to "destroy more of the forest so these guys can come and play" and that the area residents will fight this to the bitter end. In short, to some in the area it's an issue of the rich against the poor. Location of the CDT therefore will likely involve a host of compromises. The result is—or will be—a trail that often diverges significantly from the actual Continental Divide.

A second gap is an 18-mile stretch that includes the Village of Cuba. From the north the trail traverses the San Pedro Parks Wilderness, and descends the Los Pinos Trail to the village (Figure 146). From here the trail must somehow get to the Rio Puerco country to the south and reach the El Malpaís National Monument south of Grants in west-central New Mexico. However, the land ownership status south and west of Cuba is an awkward quiltwork of BLM, State, and private lands (Figure 130). Accordingly the trail is forced bypass this area. Hikers coming down from the San Pedro Parks Wilderness will have to come into Cuba and hike down Cuba's main drag (US-550). Perhaps a long stop for enchiladas and a few cold brews, combined with a night in a local motel, might seem quite compelling. The hikers will then have to follow US-550 south for ten miles to mile 53. From there they will finally be able to leave the din of the highway and enter the Rio Puerco country to the west. Today much of the trail is complete—except for these two gaps. The trail is therefore not continuous and possibly never will be.

Apache Nugget II Casino

At MP-78 highway US-550 begins its traverse of the southwest corner of the vast Jicarilla Apache Nation. Seven and a half miles farther on we get a whiff of civilization at the turnoff of NM-537 leading north to Dulce (Figure 155). Here is the "bubble-type" Apache Nugget II Casino and the now-defunct Apache Flats Rest Area, which at one time

featured its famous "tepee potties." Then this was the only public rest stop along the entire 86-mile stretch between Cuba and Bloomfield and was often eagerly anticipated. In the mid-1990s the Jicarilla Tribal Council took note of the wave of Indian casino construction in New Mexico. In May 1996 the Jicarilla's Apache Nugget Casino opened for business in the capital of Dulce, 66 miles to the north, under the illusion that it would draw significant wealth to the reservation. However, Dulce is far from any airport and is located on a lightly-traveled highway. Not a good choice.

During the casino's first two years of operation it took in nearly $16 million in gross revenue but also acquired about $1.2 million of debt. The operation was plagued by inadequate accounting procedures, poor cost-control, nepotism, and a power struggle between rival tribal families (*Gambling Magazine* 1999). And then there was the difficult requirement that the tribe sign the New Mexico gaming compact and cough up 16% of slot-machine profits plus regulatory fees. In the fall of 1997 the Tribal Council decided it could not afford to sign the compact and in mid-1998 shut the casino down after only three years of operation (Jones 2003). The Tribal Council then considered exploiting the more heavily-traveled state route NM-44 and began construction of their Apache Nugget II casino at the intersection with NM-537 (Figure 155).

Prior to 1998 the New Mexico Highway Department had already spent about $1 million on the rest stop. Late that same year the Jicarilla Tribe cut a deal with the Highway Department: in exchange for the site the tribe would improve it and run it for ten years, and would also provide land for a Highway Department maintenance yard. The tribe immediately fenced off the area, demolished two of the four tepees, and began construction of the casino. In May 1999 the then-president of the tribe declared the deal null and void because—he claimed—that the tribal official who signed the agreement was not authorized to do so. This president resigned the next month and was later impeached (Jones 2003). For several years the Apache Nugget II and rest stop remained closed, also the three porta-potties that had provided the only service between Cuba and Bloomfield. Finally, in August, 2004, the Apache Nugget II was reopened for business, providing employment for 70 people. The collective "whiiiing" coming from the banks of 227 slot machines, perched in the midst of such wide open spaces, is no less than surreal. We'll revisit the Jicarilla Apache again in Chapter 23.

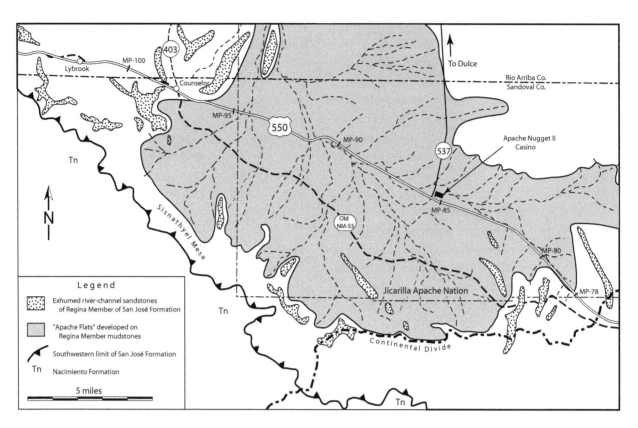

Figure 155. Apache Flats and original route of NM-44. (Modified from Baltz 1967; NMBG&MR 2003.)

Apache Flats

Most of the terrain from about MP-78 to MP-96 on US-550 just this side of Counselor consists of the mudstone flats and its associated alluvial debris (Figure 155). This stretch has been variably called "Apache Flats" (from the Jicarilla Apache Nation), "Tancosa Flats" (from the Tancosa Windmill located about 0.5 mile north of the casino), or the "Largo Plains" (from the upper Canyon Largo drainage). This stuff once had a very nasty reputation because when wet the old dirt roadbed converted to a sticky, tire-hugging, shoe-sucking goo.

The first road from Cuba to Bloomfield, then an "excellent dirt highway," was completed in 1929 (Ricketts 1930). It was designated as NM-55, and it only became NM-44 in the 1940s (Figure 155). The engineers chose an alignment well to the south of the modern one in order to gain some elevation and to cross the better-drained section of the flats. Even so, when wet it was a tough stretch to cross.

One notable experience in the flats was that of Jim Counselor, who with his wife Ann operated the trading post at what later became known as Counselor (Chapter 20). Ann vividly described the travail of Jim and a helper trying to cross the flats in a truck after the "Big Thaw" of the winter of 1937. The vehicle got bogged down up to the axles and even a team of mules couldn't budge it loose. Jim had to abandon it and slog the 14 miles on foot to the post. When he arrived his wife Ann described what walked through the door: "I was stirring up the live coals in the cook stove when the back door opened, and there stood Jim with an old gunnysack hung over one shoulder. He was absolutely the messiest sight I had ever seen in my life—from his big hat to his boots he was covered with brown adobe mud. It was even smeared in his three days' growth of beard. Great gobs of the sticky stuff still clung to his boots, making him look as if he was wearing a pair of snowshoes. His face was too dirty for me to read any expression there, but his tall figure dropped with dead-tired weariness" (Counselor and Counselor 1954). Today we whiz across Apache Flats without a second thought.

19

San Juan Basin Vegetation

Now we come to the subject of vegetation. The motorist may well ask, "Vegetation? What vegetation?" With the notable exception of the stretch from north of Cuba to the Continental Divide, the vegetation along here is certainly not impressive, but things don't have to be impressive to be interesting. The variable communities of vegetation along US-550 are cohesive enough to be mapped (Figure 156).

In effect, "reading" the characteristics of a vegetation zone is a proxy for measuring other environmental factors, primarily the moisture available to plant roots. This crucial factor is in turn governed by elevation, the underlying geology, and the legacy of previous disturbance, especially the grazing history. In short, the plant communities we encounter along our path reflect their natural environment and what had been done to it—however inadvertently—during the past 125 to 150 years.

First let's start with what exists today. As we've seen, Highway US-550 northwest out of Cuba makes a few ups and down on its gentle rise to the Continental Divide. Along the way it crosses a zone of piñon-juniper woodland with a sprinkle of ponderosa pine thrown in for tang. This is followed by a zone of mixed sagebrush/piñon-juniper woodland and by a zone of sagebrush plains. Piñon pine (*Pinus edulis*), by the way, is New Mexico's state tree. It often has a crooked trunk and reddish bark, grows very slowly, lives as long as 100 to 150 years, and reaches as high as 35 feet. It produces a tasty nut that is avidly gathered by both birds and people. It's usually found intermixed with juniper, both the one-seed (*Juniperus monosperma*) and Utah junipers (*Juniperus osteosperma*). The junipers are more drought resistant and usually extend to lower elevations than piñon (Elmore 1976). The smoke from burning juniper and piñon yields the unmistakable aroma that is the state's olfactory trademark.

If there is one distinctive shrub along this stretch, it's sagebrush, or more precisely, "big sagebrush" (*Artemesia tridentata*) (Figures 157A and

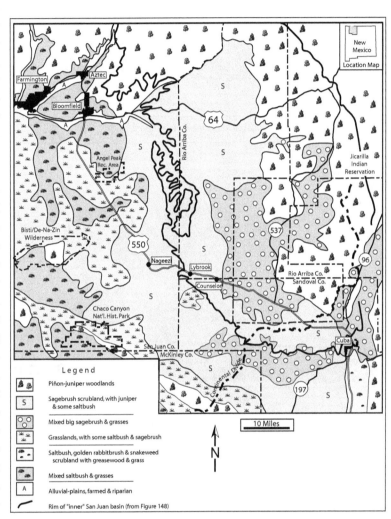

Figure 156. Simplified vegetation map of northwestern New Mexico. (Modified from Knight 1992.)

157B). The southern limit to the domain of the big sagebrush begins just south of Cuba (Figure 156). South of that point the dominant shrub is *Chamisa*, a.k.a. "golden rabbitbrush," which in the late summer turns large swaths of the countryside south of Cuba into a gorgeous shimmering gold. Big sagebrush, however, ushers in a world of gray. Big sagebrush gets the name "big" simply from its size, generally three feet but up to ten feet in thick soils. The generic name *Artemisia* is from Artemisia, wife of Mausolus, ancient ruler of Ceria in southwest Asia Minor. She was named after *Artemis*, the Greek virgin goddess of nature and of the hunt.

Detail of
Big Sage,
scale ~ 4" high

A. Typical close-up view of Big Sagebrush.

B. Typical field of Big Sagebrush.

Figure 157. Big sagebrush—trademark shrub of the San Juan basin. (Photos by author 2005.)

Sagebrush is not the same as "sage," which is a spice, nor is the same as "bigalow sagebrush" (*Artemesia bigelovii*), which is a different species entirely. The state of Nevada—appropriately nick-named the "sagebrush state"—has gone so far as to adopt the big sagebrush as its state plant because of its enormous range there. The species name *tridentata* means "three toothed," in reference to the three lobes on the tips of most leaves (see inset in Figure 157A). It maintains its leaves throughout the year. Because of the dense gray hairs on both sides of the leaves the plant has a silvery blue color. The species thrives in fine-grained, silty and sandy soils. It blooms in late summer and one mature plant may produce up to a million seeds. The leaves have a strong odor as a result of the aromatic, camphor-like compounds. The nutritive value for livestock is high, but its palatabil-ity and digestibility tend to be poor because of the high-volatile oil content. Many animals, however, will browse upon sagebrush when other food re-sources are scarce, especially in the winter.

Most highway travelers along US-550 how-ever are not browsers, but after a rainstorm they are treated to air richly perfumed by the unmistakable pungent aroma of sagebrush. The shrub is also a virulent source of hay fever. In pre-historic times big sagebrush was commonly used by many Native Americans. The wood was burned for fuel, and the leaves and seeds were eaten. The leaves were also used as a medicinal to treat coughs, colds, headaches, fevers, and just about everything else. The Indians would immerse themselves in the pungent smoke to alleviate the effects of an encounter with a skunk. Navajo weavers produced a yellow-gold substance for dying wool by boiling the leaves and flowers (Elmore 1976).

Pleistocene and earlier vegetation

As mentioned before, vegetation is controlled principally by the amount of moisture available to plant roots, but at a broader scale also by delivery systems of moisture-laden air masses. Major vegetation changes are regulated by climate changes, and the latter depend on world geography. In recent decades researchers have revealed the exquisite tango between global climate and geography during the past 50 Ma. Most relevant to our story, however, is climate change during the past 16 Ma. That pivot point, called the "middle Miocene climate transition," marks a profound shift from warm and humid to cool and dry conditions, and the beginning of more modern times. At 16 Ma, global oceanic circulation—the carrier of water masses—changed from generally east-to-west and parallel to the equator, to one involving pervasive meridional (south-to-north and north-to-south) movement. Meridional oceanic circulation de-livered deep, cold water from the north to the west coast of North America. This cold water welled up along the coast, creating cool, dry air masses that led to progressive cooling and drying out of the inboard Southwest (Chapin 2008). The climate change caused the demise of extensive forests and their replacement by mixed woodland-grassland country and increasingly by grassland savannas.

The changing vegetation facilitated an incredible diversification of large, hoofed herbivore mammals ("megafauna," of horses, camels, bison, etc.),which then evolved specialized teeth capable of chewing the abrasive grasses. The grasses themselves achieved amazing success in expanding their range by adapting a mode of continuous leaf growth (think of the weekly need to mow the lawn). This allowed them to recover somewhat from heavy grazing by the many large animals. The animals, ruminants ("cud-chewers"), had evolved in tandem to exploit the vast resource of the expanding grasslands. The ruminants are peculiar in that they evolved a stomach that allowed them to eat quickly and then retire to a place free from enemies where they could chew the food in safety. The stomach of ruminants is divided into four chambers. The newly-cropped grass passes into two of the chambers and is partly digested. Later, the food is regurgitated as a "cud" into the mouth, chewed again, and finally passed into the other two chambers.

By about 16 Ma the landscape of northwestern New Mexico had been eroded down from its former elevation of about 10,000 feet to something resembling the modern elevation (more in Chapter 31). The time 12 Ma introduced additional global oceanic changes as the Indonesian Seaway, which had previously been open to equatorial oceanic circulation, was closed down by the formation of the Indonesian archipelago, thus intensifying the components of global meridional circulation (Cather 2007; Chapin 2008). Beginning about that same time parts of western North America, including the Cascade Range of the Pacific Northwest and northern California, the Sierra Nevada of central California, and the Sierra Madre of northwestern Mexico, were uplifted to significant heights. The result was the creation of an enormous rain shadow to the east as these ranges wrung the moisture out of what humidity existed in the air masses coming off the Pacific Ocean (Dick-Peddie 1993).

A second major climate change, called the "terminal Miocene climate event," occurred at about 6 Ma, after the cooling and aridification of the Southwest had reached a maximum. By that time the mixed woodland/grasslands had thoroughly yielded to treeless, cool, arid steppes on the southern Great Plains, and to scrub deserts in the Southwest, including northwest New Mexico. This was the time of the mass extinction of many groups of the large land mammals. Also at 6 Ma the land mass now comprising the peninsula of Baja California separated from the west coast of Mexico, opening the Gulf of California (Chapin 2008). Thus was produced an open conduit for warm tropical air masses that could annually surge into a portion of the semi-arid Southwest, initiating the Mexican monsoon and vastly increasing the amount of erosion and the occurrence of stream piracy (Figure 15B). Last, but not all least, the Panama

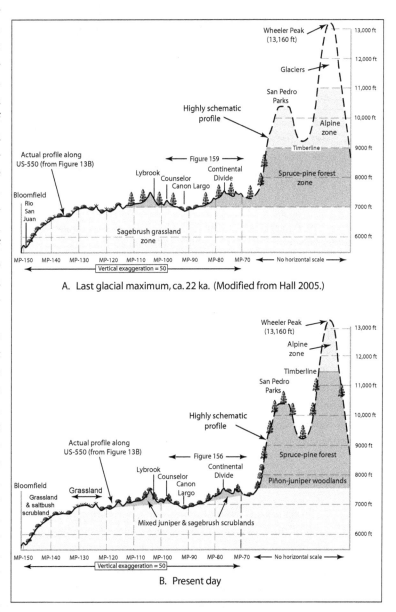

Figure 158. Schematic profiles of Pleistocene vs. present climate and vegetation zones of San Juan basin.

211

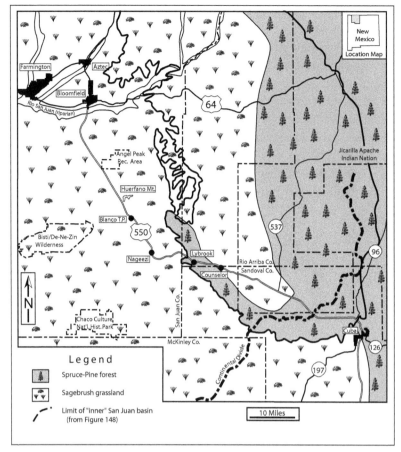

Figure 159. Primeval vegetation, ca. 22 ka. (Modern geographic features shown for reference; modified from Hall 2005.)

seaway was finally closed at 2.6 Ma after a gradual process extending over 10 Ma. This separated the Atlantic and Pacific oceans, greatly accentuated the global meridional oceanic circulation, jolted the global climate, and set up the modern relatively cold, dry world of the Pleistocene Epoch.

Major continental glaciation began in North America about 900-800 ka in the middle part of the Pleistocene Epoch and lasted until about 12 ka. At the peak of the final glacial period, about 22 ka, the vegetation of New Mexico had evolved to something quite different from that of today. Glaciers had formed on mountain peaks above 12,000 feet, including such places as Wheeler Peak (Figure 158A). The top of the spruce-pine forest zone was then compressed downwards to a 2,000-foot range from about 7,000 to 9,000 feet, in contrast to today's 4,000-foot range from about 7,500 to 11,500 feet (Figures 158B and 159). The vast reaches of piñon-juniper woodlands—so characteristic of large swaths of the state today—likely did not yet exist. The zone below an elevation of 7,000 feet was covered by a sea of big sagebrush (Hall 2005).

By about 8 ka, total rainfall in the Southwest had diminished to a point that forests began retreating northward and/or to higher elevations, leaving behind broad areas of grassland and only scattered woodlands where thick forests once prevailed. Winter precipitation was reduced or eliminated from much of the Southwest, and the summer, Mexican monsoon expanded (Figure 15B). As mentioned above, the monsoon fans north and northeastward from the Gulf of California and transports a load of moisture-laden air over a broad swath, including the southeastern half of Arizona and all of New Mexico. The pounding summer downpours in this swath stripped well-developed soils from their bedrock foundations and encouraged the spread of desert vegetation (Van Devender and Spaulding 1983). Deserts are newcomers to our world: they never existed before about 8 ka. In fact there have never been as many desert plant types as there are today. To put it more succinctly, modern desert-plant communities have just been born (Axelrod 1983).

"Primeval" vegetation

The scattered zones of woodlands, consisting of juniper, piñon pine, and minor ponderosa pine, were much more extensive in the San Juan basin some thousands of years ago than today. A great deal of timber was harvested by the Ancestral Puebloan people, especially during the major pueblo-construction episode in Chaco Canyon during the two centuries from about 950 to 1150 CE. It has been estimated that to construct the 13 large multistoried buildings and a host of smaller ones in the Canyon during these two centuries, the Ancestral Puebloans cut about 100,000 ponderosa pine trees (Betancourt and Van Devender 1981). It helps to appreciate the magnitude of this prodigious effort to remember that the work was carried out with stone axes and the stooped backs of men. And then there was the need for firewood. A computer model suggests that the Ancestral Puebloans would have used up all the timber in a 32,350-acre juniper-piñon woodland during this period, and that more than 200,000 trees were consumed. The juniper-piñon woodland has never fully recovered during the ensuing 850 years (Elias 1997).

Before the advent of stockmen to the area in the late 19th century, the vegetation was a stable assemblage of trees, shrubs, and grasses, ultimately controlled by the amount of moisture available. The last of the large grazing animals native to North America had long ago been hunted to extinction by early stone-age hunters. Only with the arrival of the Spanish settlers in the 16th century were large, domesticated animals—sheep, cattle, goats, and horses—reintroduced to North America. In the mid-19th century, when Hispanic and Anglo ranchers moved into the area, the grasslands and scattered woodlands were dominant (Figure 160). Over the past century and a half the woodlands have been greatly reduced in extent. Much timber was harvested for the construction of towns, mines, and railroads, as well as to provide fuel for homes and copper smelters.

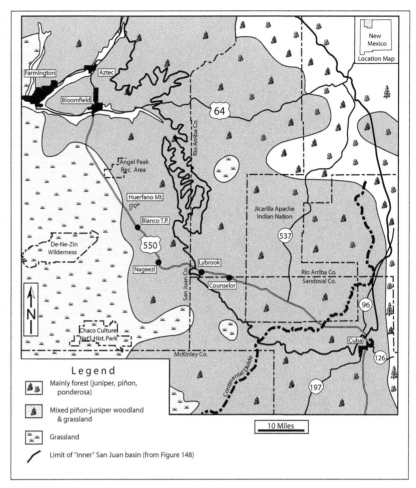

Figure 160. Vegetation, ca. 1880. (Modern geographic features shown for reference; modified from Gross and DickPeddie 1979.)

Like timber, grass is in effect money. Grazing animals, especially sheep and cattle, can convert grass to meat, wool, and hides, and those commodities can be traded or converted to money. These cud-chewing animals just loved the vast expanse of native grasses in New Mexico. An account from a certain Captain Johnston's 1848 journal gives us an idea of the once vast extent of grasslands: "The country from the Rio Grande to Tucson is covered with grama grass, on which animals, moderately worked, will fatten in winter . . . " (Leopold 1951). Another early traveler, Josiah Gregg, wrote that the high mesas of New Mexico were "clothed mostly with different species of a highly nutritious grass called grama, which . . . cures upon the ground and remains excellent hay—equal, if not superior, to that which is cut and stacked from our western prairies" (Moorland 1954). Even as late as the 1930s much of the range in parts of northwestern New Mexico was awash in grass. There was "no chico brush or sacaton bunch grass growing on alkali flats up here. Instead, there were great parks of black grama and wheat grasses" (Counselor and Counselor 1954).

Before the arrival of the cattle barons in the late 19th century the grasses of New Mexico were the forage of sheep—hundreds of thousands of them. Sheep had an advantage over cattle because they ate weeds and sagebrush when they had to, and watered on snow in the winter when they were far from water holes. Cattle couldn't do this. Also, a sheep only needed about 15 acres of New Mexico grazing land while a cow needed about 70 acres. The sheep were herded in scattered bands and did little lasting damage to the range (Kline 1978).

The sheep industry totally dominated the pre-Civil War agricultural economy of New Mexico. After the defeat of the Navajo in 1864 and the end of the Civil War in 1865 the way was open to the cattle industry. High beef prices during the war had depleted cattle in the North. After the war Anglo cattlemen, especially downstate Texans, found

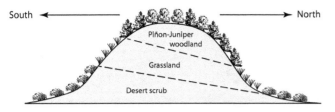

South ← → North

Piñon-Juniper woodland

Grassland

Desert scrub

A. Normal control of available moisture by altitude and exposure.

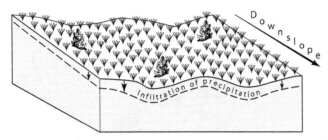

Downslope

Infiltration of precipitation

B. Undisturbed woodland savanna, with evenly-distributed infiltration zone and little runoff.

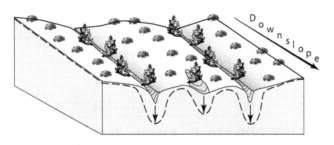

Downslope

C. Development of juniper-sagebrush scrub from overgrazed and eroded former savanna, with increased runoff, incised arroyos, and localized infiltration zone.

Figure 161. Origin of juniper-sagebrush scrubland. (Modified from DickPeddie 1993.)

themselves with five million head of cattle on their hands. The cattle drives to Kansas railheads went into high gear. At the same time many Texans brought their cattle to the ranges of New Mexico and prepared to stay. The arrival of the railroad in 1880 made it possible to ship cattle to eastern markets and enabled cattlemen to successfully compete with sheepmen. Another advantage of cattle was that they, unlike sheep, could be turned loose unattended. The cattle rancher needed only a log cabin or dugout for his headquarters (Beck 1962). The switch from sheep to cattle was on. In 1884 the territory had about a million cattle and 5.5 million sheep. Four years later there were 1.25 million cattle and 3.5 million sheep (Kline 1978). The rush of cattlemen, as well as homesteaders, into the territory after the war had created a concentration of cattle on the range like never before. The homesteaders were armed with barbed wire and competed with the cattlemen for water. The resulting high concentrations of animals and the close-herding put an enormous strain on the range (Kline 1978). The grasses were quickly consumed by the expanding herds of livestock.

The livestock completely altered the nature of the soil and the vegetation. Before disturbance, the distribution of the native vegetation was controlled by available moisture, which in turn was controlled mainly by a combination of elevation and exposure (Figure 161A). The grasslands occupied a zone midway between the piñon-juniper woodlands above and the desert scrublands below. In their zone the grasses had served as an effective water-management agent. Lateral flow of water was impeded by the grass and the water infiltrated directly into the root zone (Figure 161B). Overgrazing reduced the grass cover and exposed bare ground. Precipitation falling on this bald surface flowed laterally and concentrated into depressions, which then held more moisture than normal. Junipers were able to thrive in these over-watered depressions (Figure 161C). The reduced available moisture on the higher, stripped areas were now only capable of supporting desert shrubs. Erosion increased and arroyos were incised. The end result is a vegetation zone characterized by juniper in the lows and sagebrush or—in some places—golden rabbitbrush ("chamisa") with broom snakeweed on the highs. As the range of the native grasses was degraded, the juniper woodlands moved downslope and the big sagebrush moved upslope. In place of the grasslands we today have a huge swath of open juniper-sagebrush scrubland.

Big sagebrush is a very aggressive species. When on a roll, as today, it exploits its environment and crowds out the grasses, with which it once shared the range. Today the Bureau of Land Management (BLM) is attempting to expand the range of grasses by controlling the spread of sagebrush in certain areas. Herbicides and fire are the main controlling agents. Another, more low-tech controlling agent is a herd of goats! In fact, a business in Cuba, appropriately called "Western Weedeaters," has been successfully demonstrating the feasibility of setting free these critters onto the sagebrush. Unlike cattle, which are grazers, goats are browsers. They can be run into the same range and will not compete with the cattle. Although the goats need to be "trained," tended, and fenced, once set loose on a tract of range the sagebrush soon becomes history and the grass expands (*Cuba News* 2005d).

20

Counselor and the "Upper" Canyon Largo

Counselor is a tiny unincorporated community located at mile 97.7 on US-550 at the southern access point to Canyon Largo. It's the first settlement encountered after leaving Cuba, some 23 miles to the southeast. It consists of a trading post and house on the north side of the highway, and a small cluster of houses and buildings on the south side.

The trading post encloses a post office that serves more than 300 area families, and a private home is attached to the back. The place also serves as a bus stop. The name of the trading post is from Jim Counselor, an Irish-American sheepman and Indian trader in the first half of the 20th century. He established his trading post here in 1933. This chapter will deal with the Counselor area and the first 23 miles of the southern, upper headwaters of the Canyon Largo. The Largo is chock full of interesting things and stories, but I've set the range of discussion at an arbitrary 23 miles to keep this book focused on the US-550 corridor and its scope somewhat under control.

As we approach the Counselor post from any direction it seems about as remote as a place can get. The wind-swept expanses on all sides suggest loneliness in all its aspects — but people do live here and have so for many years. They are of course well-spread out, as the land dictates. The spot that would eventually become Counselor had been for some time a crossroads of sorts. The old state highway NM-55, built in 1929, once came up from Cuba from the southeast and forked at this place (Figure 11). The main NM-55 continued via the west arm of the fork and led to Bloomfield on the San Juan River. The north arm of the fork was the original NM-44, and it followed

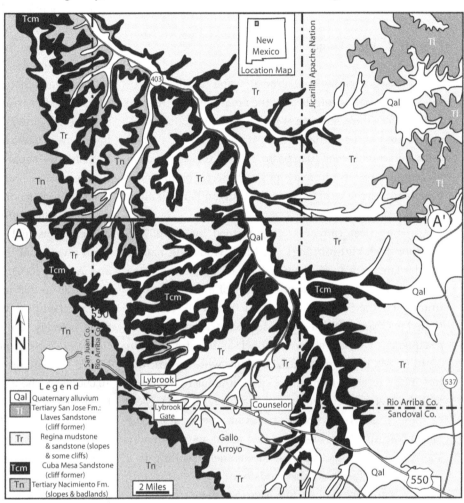

Legend

Qal	Quaternary alluvium
Tl	Tertiary San Jose Fm.: Llaves Sandstone (cliff former)
Tr	Regina mudstone & sandstone (slopes & some cliffs)
Tcm	Cuba Mesa Sandstone (cliff former)
Tn	Tertiary Nacimiento Fm. (slopes & badlands)

2 Miles

A. Surface geology. Modified from Manley et al. 1987.

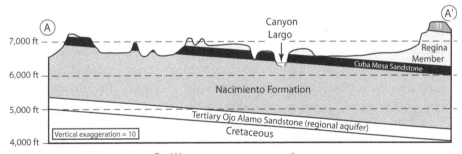

Vertical exaggeration = 10

B. West-to-east cross section.

Figure 162. Geology of the Counselor area and the "upper" Canyon Largo.

the path of least resistance down the 50-mile-long canyon appropriately called Canyon Largo (Spanish for "Long Canyon"), or sometimes simply "the Largo," to the San Juan River. In the 1940s the Largo route of NM-44 was decommissioned. The "new" NM-44 between Bernalillo and Bloomfield was designed as a straightened replacement for the old NM-55, which in turn then ceased to exist. The new NM-44 was redesignated US-550 in January 2000, and the old road down the Largo became today's NM-403.

"Upper" Canyon Largo

The geography and crucial water availability of this area are both controlled by the area's geology (Figure 162A). The Counselor area is located near the southwestern outcrop of the Tertiary-age strata that forms the inner, central part of the San Juan basin. Surface elevations drop more than 1,000 feet from a topographic crest on the inner basin's rim to the west down to the Canyon Largo bottom (Figure 162B).

The lower canyon wall is formed by the 250 to 300-foot-thick, cliff-forming Cuba Mesa Sandstone, named for the landform west of the Village of Cuba. The Cuba Mesa Sandstone is the lower member of a thick geologic unit called the San José Formation (Figure 4). As mentioned in Chapter 18, the Cuba Mesa Sandstone consists of a sequence of ancient stream-channel sands, formed about 55 to 51 Ma on a broad alluvial plain, with its headwaters in the recently-formed Rocky Mountains to the north and the Sierra Nacimiento to the east (Figure 151). Most of these streams—excepting the trunk stream along the Sierra Nacimiento front—wandered about and eventually stacked sand channel upon sand channel, resulting in a thick, seemingly-continuous sandstone.

Locally above the Cuba Mesa Sandstone of the inner canyon is an upper set of sandstone cliffs created by younger, stacked channel sandstones of the overlying Regina Member. Thus, in places the canyon walls have multiple components and steps, important to the early people who lived here (more below). Those canyon walls, especially in the heads of the canyons, form natural corrals for livestock, minimizing the need for fencing. This was just what livestock owners were looking for.

The Cuba Mesa Sandstone also acts as a massive water collector. Water from rainfall and especially snowmelt seep down through the porous sandstone until it encounters non-porous "tight" streaks, and then flows laterally downhill. Localized fractures and faults greatly increase the rate of lateral flow. Springs, called "spring seeps" or "drip springs," emerge at places where the sandstone is exposed in the canyon walls. Flow from these is generally in the range of one gallon per hour, but it is continuous all year (Counselor and Counselor 1954). At the tops of the mesas were vast parks of black grama and wheat grasses, ideal for grazing.

The Cuba Mesa Sandstone dips gently to the northeast and therefore plunges below the ground surface in that direction. The number of natural springs also decreases in that direction as the sandstone disappears. As mentioned earlier, where the Regina Member contains isolated bodies of stream-channel sandstones, which are thinner but in every other respect identical to the lower Cuba Mesa Sandstone below, there is created a "doublet" of cliffs—a lower one of Cuba Mesa Sandstone and an upper of Regina sandstones—separated by an area of flat ground, a ledge, that became very attractive to early inhabitants looking for defensible sites not too high above the canyon floor.

The flat-bottomed flood plain is geologically a fairly modern construct. Oil and gas wells drilled in the flood plain reveal that the sand is 100 to 200 feet thick (Dugan and Arnold 2002). Evidently an earlier river cut a valley some 100 to 200 feet into bedrock, probably during the Pleistocene Epoch when there was more precipitation than there is today. Canyon walls in the upper reaches of the Largo are now generally about 350 feet high. Therefore, before the canyon was partly backfilled during the past 10,000 years the canyon walls must have been about 500 feet high.

The Canyon Largo was early on recognized to be a natural transportation corridor (Figure 5). In 1859 a U.S. Army lieutenant William H. Bell explored the Largo to investigate the feasibility of wagon road to military posts and settlements in the San Juan River Valley. Bell found the route well-traveled even then. In 1860 gold was discovered in the San Juan Mountains of southwestern Colorado. Access to the gold fields from New Mexico was both via Abiquiú and Tierra Amarilla, and Abiquiú and the Canyon Largo. The going was rough and the increased traffic demanded better roads. In 1876 the Santa Fe, Abiquiú, and Cañon Largo Toll Road and Turnpike Co. was incorporated to construct a road down the Largo to the San Juan River (Torrez 1988). The flat-bottomed canyon floor provided the easiest and most direct access from the Rio Grande Valley to the San Juan River Valley well into the 1920s. It also served as the conduit for early ranchers leading livestock into the area. It was therefore no accident that the original course of state highway NM-44 was routed down the Largo from the high ground near the present Counselor Trading Post. Once through the canyon, NM-44 accessed the main road to the town of Aztec and the railhead of the narrow gauge Denver and Rio Grande Railroad.

The Largo route though was treacherous, sandy, and subject to violent summer flash floods. At the time the state highway department spent little time or money on maintenance because of the utter futility of controlling the shifting sands. There wasn't a single bridged side canyon or arroyo the entire distance. In places the road inched its way around a scratched-out, one-way roadbed, called a "dugout," around a protruding mesa foot. Much of the day-to-day maintenance was done by the local ranchers who depended on the road. When it rained the way became nearly impassible, and in heavy snow it was unthinkable. In such conditions, to somehow get from Aztec to a ranch in the upper Largo required a ride on the Denver and Rio Grande narrow-gauge railroad through Colorado and down to Santa Fe, a bus ride south to Albuquerque, and a hitched ride on a feed truck or whatever up to Cuba and beyond.

The land of the Largo is harsh and could (and still can) only support a handful of families scattered thinly up and down the canyon. The area was not suitable for irrigated farming but was well suited for grazing cattle and sheep. The homesteaders all had similar essential needs in common including water, access to supplies, access to a doctor, and, for some, access to a school.

This remote, silent, and today virtually abandoned country has been given a voice by an out-of-print book written by Jim Counselor and his wife Ann (Counselor and Counselor 1954). entitled *Wild, Wooly and Wonderful* (first mentioned in Chapter 18). The autobiographical book is an odd piece of work in that it only covers their years in the Largo from 1930 to 1942. Amazingly, in its 392 pages of chronological narrative there is only one date cited, the year 1935, which appears on page 347! To understand the book's chronology one must plod backwards and forwards from this year and reconstruct the narrative based on the known dates of cited events. Their book (exceptions cited) is the main basis of the story that follows.

Jim and Ann Counselor

James Eugene Counselor was born on June 2, 1892, in the tiny town of Guadalupe (U.S. Army Registration record) between present-day Manassa and Conejos, Colorado. His father, Eugene S., was born in New York. His mother, Jadill, was from Michigan. For an unknown reason, in 1900 young James was living with his paternal grandmother in Michigan (1900 U.S. Census Record). In 1917 James called himself an independent cattleman and lived in his father's household (U.S. Army Registration record). That same year he enlisted in the army, served two years during WWI and fought in France's Argonne Woods.

Upon his discharge in 1919 he returned to his parents' home in Colorado. Soon after that, for uncertain reasons, he settled in the vast expanse of the Canyon Largo and established a sheep ranch in Rincón Largo, a side canyon about 12 miles north of the present-day Counselor Trading Post, and a couple of miles off the main wagon and stage route that connected population centers in the Rio Grande Valley to those to the northwest in the San Juan River Valley. In 1923 he married Beatrice, a Kansas girl nine years his junior (1930 U.S. Census Record). Beatrice must have either died or she divorced Jim early in 1930 because that summer Jim met his future wife, Ann. Ann was born near Chicago on April 1, 1897. She was 33 years old and, in her own words, "would never win any prizes in a beauty contest." With blunt honesty she added that she had a "prominent nose." Ann was visiting an older female friend, a Mrs. Hall, who ran a boarding house in Albuquerque. While waiting for Mrs. Hall to meet her at the train station Ann saw a "tall, broad-shouldered young man," who proceeded to give her what she described as "the once-over." For her it was love at first sight.

In the nick of time Mrs. Hall arrived and spirited a bedazzled Ann off to her home in the city's "eastern heights." The next day Mrs. Hall received a phone call from the "young man," Jim Counselor, asking about her house guest. He came to call. It seems Jim was in town for a few days to attend a celebration called the First American Fiesta, and he asked Ann to join him. At the end of their get-together he astounded her by announcing, "Get yourself a good night sleep, Ann, we're leaving for the ranch at sunup!" He had invited her and Mrs. Hall to drive with him to his sheep ranch in Canyon Largo, over 100 miles away. The guy certainly knew what he wanted!

The next day they sped over the cement road in his car to Bernalillo, turned northwest on a "rough, bumpy dirt road" and crossed the Rio Grande on the "shaky old wooden bridge." The plodded past the ruins of Tiguex (present Coronado State Monument), passed the Santa Ana and Zia Pueblos, crossed the Rio Salado at the Village of San Ysidro, and there refilled their waterbag at the only store (the Seligman store?). They proceeded up the Rio Salado anticline, passed the "white wall" (*Cuchilla Blanca*, Figure 78) of Todilto Gypsum forming the anticline's northern exit gate, ogled the irresistible Cabezón Peak to the west, and arrived at the area's metropolis of Cuba late in the day. They drove up the town's "one staggering street, through flocks of scrawny chickens and yelping dogs and a pack train of burros." After supper at John Young's Hotel they continued on to the ranch on the Largo. It had been one very long, hard day.

The ranch consisted of a "comfortable four-room adobe ranch house, with its beamed ceiling and Navajo-rug-covered floors and corner fireplace," and a "bunkhouse, garage, blacksmith shop, granary, chicken house, and horse corral." The ranch house was supplied with water piped from a drip spring on the mesa behind the house. Such springs were the heart of the ranching economy. The quality was so pure that it was safe for drinking and could even be used in automobile storage batteries. The ranchers made due with the low flow rate by constructing gathering systems and storage tanks.

For the next several days Jim regaled her with blue skies, beautiful mesa country, horses, livestock, and country meals. She was instantly taken by the desert, and by Jim Counselor. At the end of the stay he proposed and she accepted. They were married in Durango, Colorado in December 1930. The first 24 hours of their married life, according to Ann, set the pattern for the rest of their years together: "lots of laughter, some hard work spiced with a dash of fear now and then, and seasoned with just enough trouble and hardship to make the years of our married life something worthwhile and good." They returned home to the ranch where Ann learned to be a sheepwoman and a sheepman's wife.

The appearance of the Canyon Largo as a dry, dusty place belies the fact that it is a natural water funnel. The intense, summer monsoonal rainstorms can suddenly put an end to the tranquility. The Largo drainage basin has a huge extent and considerable relief. The rainwater accumulates faster than the land can absorb it and instead shoots down the side canyons until the volume is concentrated by the main canyon into a very dangerous deluge. The Counselors tell a story about viewing a storm in progress off to the east on the Jicarilla Reservation and driving out to the

main canyon to "watch the wash run." They arrived just in time to see a small black car start to cross the river bed, in which the water at the time was about knee deep. As the driver neared the midpoint of the wash they heard a roar somewhere in the distance. The car's motor sputtered and died. Jim plunged into the shallow water and dragged the driver out of the car and to the bank. Just then, Ann writes: "Rushing down upon the two men and the car in the middle of the stream, was a wall of water that seemed to fill the entire channel of the wash." And then this:

"With a roar, the wall of water hit the car, turning it over and over, as it rushed down the wash. The deep, roaring, tearing sound of rushing water was all around us now. The channel of the wash was bank-full of dirty, foaming, whirling water, with waves big enough to swamp a rowboat. Trees, brush, logs, and a dead calf went rushing by. Big pieces of the adobe banks were undermined and fell with loud crashes into the flood. Like a savage, growling beast, it devoured everything in its path."

When wet, the sand of the Largo valley floor eats vehicles! If a vehicle becomes stuck the sands are washed out from under by the current it until the vehicle gradually sinks, sometimes out of sight. Every Largoan has his/her own favorite story. This recent account captures the flavor quite well:

"These people tried to cross a streambed with their new four-wheel drive truck a few hours before a storm. It didn't look like much water, but the streambed was washed out under what looked like a little puddle, so they got stuck in the arroyo bottom. It started to rain. ... the flash flood (the only way our streams flow) sounded like a freight train coming down the canyon. They got out and huddled under a rabbitbrush as the stream cut away the ground under their truck. Their nice new four-wheel drive truck rolled over and started sinking. Before the night was through, the only thing sticking out of the sand was the bottom of the tires" (Irick and Irick undated).

State highway NM-55 was the main road between Cuba and Bloomfield that accessed the upper Canyon Largo. The road was built of dirt and gravel by the New Mexico Highway Department in 1929 (Ricketts 1930). Heavy supplies had to be hauled by wagon, usually from stores in Cuba or from as far away as Albuquerque, or, less desirably, from Aztec. Typical necessities included coffee, flour, sugar, dried fruit, lard, canned milk, and tobacco. Another need was kerosene for illumination. The kerosene was the manufactured type made from coal and was in fact called coal oil and shipped in drums (Waybourn and Horn 1999). The nearest doctor was located far away in Aztec.

During the "Big Snow" of the winter of 1931–1932 the Counselors and many other ranchers lost a great deal of livestock. Travel down the Canyon Largo at that time was out of the question. The canyon was closed. After tending to their livestock as best they could they retreated to Young's Hotel in Cuba for the winter.

Upon return in the spring of 1932 they counted their losses, and they were large. Jim supposed that they would have to "starve it out" for two or three years to build up the herd. Only then would they restore an acceptable quality of life. As 1932 progressed the Great Depression deepened. Compounding their losses of sheep in the Big Snow were the low prices for lambs and wool. They worried terribly about being able to keep their ranch. Then Jim learned that a trading post nearby was for sale. For years the area's mail drop was at Julio Montoya's goat ranch, about a half mile south of the present post where the old state highway crossed the Gallo Arroyo (Figure 162A). In about 1932 a character named Peg Leg Willie (Billie Rice) had opened a dirt-floored store and bootleg joint on the goat ranch, and the next year decided to sell out. Jim decided to buy out old Peg Leg and build a store just up the road from it. He and Ann drove their sheep herd to the Denver and Rio Grande narrow gauge railhead at Chama and sold it. They sold their ranch house and land to the U.S. government, and then proceeded to build the house and store at the present site of Counselor. Their post opened for business in December 1933 (Figure 163A).

The Counselors' book tells an interesting story about the post's first piece of business. The grand-opening party was attended by people who came from miles around. Around midnight the guests heard a crash from outside the kitchen, followed by a string of muffled oaths in Spanish. Rays of flashlights lit up the face of little Mariano Otero (a neighboring rancher) in the bottom of the dry cistern. A number of hands reached down and grabbed Mariano by the overalls to haul him up when everyone heard a rip, followed by him yelling to stop. A pair of new overalls thus became the first sale.

Jim and Ann dealt extensively with the Navajo, and became their trusted friends. He soon began to be called *Bilagáanasnééz* by the Navajo, for "Tall White Man," and Ann became known as *Sima Yazzie*,

or "Little Mother." In the mid-1930s things rapidly turned sour for the Navajo. There were rumors that the federal government would soon require them to reduce their herds. At this time an imposed livestock reduction was considered necessary to rescue the badly abused public domain. This did not take into account the inconvenient fact that a sufficient number of livestock was life itself to the Navajo. At the store one day, while a group of Indians were discussing the potential impact of such a drastic policy, their spokesman asked Jim point-blank, whose side would he take? With no hesitation he declared that he would "stand by the side of my Navajo brothers." The Navajo never forgot. He was now one of them.

Jim and Ann had no children. Her one pregnancy ended tragically in early 1934. After struggling to free a mud-bound automobile on the road home from the doctor in Aztec to save the life of a Navajo girl, she began feeling poorly. A couple of days later she felt worse. After a frantic 75-mile drive back to the doctor in Aztec her only child—a son—was born two months before his time, dead.

A. North view of original Counselor Trading Post in the 1930s. (Photo courtesy of Aragon 2002.)

B. Northeast view of Counselor Trading Post. (Photo by author 2005.)

Figure 163. Counselor Trading Post.

In 1935 the Indian Department indicated that stock reduction would in the near future become official policy for the Indians of the Eastern Navajo jurisdiction. The Counselors helped the Indians prepare a petition to Washington protesting the measure, to no avail, but his efforts added to the growing respect by the Navajo. In about 1936 economic conditions began to improve for a time. Several rainy seasons had resulted in good range feed and abundant harvests, the market for lambs and wool increased, and livestock herds grew. Trading business increased as well and the Counselors began to outgrow their store. They constructed a new building—a combination of store, warehouse, rat-proof flour room, and small office.

Later that year, though, word came that the stock reduction was now official policy and that the Navajo had to sell 50% of their goats to the government for a pittance, reducing the animal numbers often below that required for subsistence. Every Navajo family tended a herd of goats. Goats were both their "meat markets and milkmen." This food source, when combined with a little harvest of corn and beans and with cash earned from rug weaving and rental of tiny allotments to non-Indian sheepherders for winter pasture, allowed the Navajo to eke out a living. The stock reduction threatened to utterly destabilize the Navajo economy. Despite the fact that the Bureau of Indian Affairs later admitted that the stock reduction in the Eastern Navajo district had been a terrible mistake, the damage had been done. Instead of pelts, the Indians began to sell more and more jewelry, saddles, bridles, and wagons to buy food. Wage work for the Dulce Indian Agency at the Jicarilla Apache reservation in the Emergency Conservation Work (ECW, later to be known as the Civilian Conservation Corps, or the CCC) became an important ingredient in the local economy.

Conditions limped along until Pearl Harbor on December 7, 1941. Some 3,600 Navajos served in the military during WWII. One of those was a Navajo boy named Leo Sam. Ann transferred some of her love for her dead son to Leo. In 1942 she received a telegram saying that Sergeant Leo Sam had been killed in action in the South Pacific. This was the final straw. At this point *Wild, Wooly and Wonderful* abruptly ends. In about 1944 she and Jim sold the store and moved to Albuquerque. She had had enough of the desert.

Fully 305 pages of the book's 392 pages describe the years 1930–1933 in engaging prose. During this time Ann had written detailed weekly letters to her older sister Matilda in Chicago, and also kept a type-written dairy. The book's narrative is in Ann's voice, and she likely referred to these written accounts when much later she penned the book, with Jim's concurrence. From the year 1934 onwards the narrative speeds up. For example, the year 1934 was covered by 36 pages, 1935 by nine, 1936 by 27, 1937 by ten, and the years 1938–1942 by five. As their "romantic" period ended and their lives became more concerned with economics and politics involving the Navajo, her attention to her letters and diary evidently flagged.

In Albuquerque Jim established the first freight and bus service between Albuquerque and Farmington, known as the San Juan Basin Lines, Inc. He operated it between from 1947 until 1959 when he was bought out by Trailways. Jim Counselor died on August 24, 1966 in Albuquerque at the age of 74, and his ashes were interred in Albuquerque (*Albuquerque Journal* 1966). Ann continued her work on behalf of the Navajo. Since tuberculosis was a major health problem for the Navajo, many Indian children were taken from their homes and sent to a TB sanatorium in Albuquerque. Ann could speak Navajo and spent many days with the children. She dedicated more than 20 years of volunteer work for the American Lung Association. Ann died on December 30, 1981 in Albuquerque at the age of 84, and her ashes were interred in her home town near Chicago (*Albuquerque Journal* 1982a, 1982b).

Leonard Taft takes over

Leonard Taft, a wealthy pharmacist from Farmington, bought the Counselor Trading Post in the late 1930s (Linthicum 2003) or early 1940s (suggested by Counselor and Counselor 1954). Taft also owned trading posts at Bisti, south of Farmington, and at Blanco, west of Counselor. Harold McDonald had come to the area, probably in the late 1930s, and Taft soon gave Harold a job working behind the counter. One day McDonald met a young woman at the post who was from a local ranching family, and they were married soon afterwards. In the late 1940s Taft took on Harold McDonald and his new wife Jane ("Turk") as partners, and sold the post to them in 1949.

The present metal-frame store building was built in 1965 when the earlier store was razed (Kelley and Francis 2006). In 1966 the McDonalds bought out Taft's interest in both the Blanco and Counselor enterprises (Donovan 2003). Harold doubled as a railroad agent for the Union Pacific Railroad and would find employment for the area's Navajo men (Aragón 2002).

Harold and Turk were well respected by the Navajos. The couple was known for hosting a huge Christmas party on the Saturday before Christmas, and in some years over 1,200 attended (Moore 1999). The Navajo name for the place is *Bilagáanasnééz*, which means "Tall White Man." Harold McDonald was six feet tall (Linthicum 2003). It's interesting to note that this is the same name bestowed earlier on Jim Counselor.

On September 30, 1988, as Harold was closing up the Blanco Trading Post, he was murdered during an armed holdup. The 16-year-old thug, a local from neighboring Nageezi, pled guilty in the belief that he would only receive a five-year sentence and be out by age 21. He was sorely mistaken. After a change of venue from San Juan County to Gallup he was the first juvenile in the state prosecuted as an adult and in 1990 was sentenced to life in prison. Harold's daughter, Kelly Aragón, is convinced that there were others involved in the crime.

After Harold's death, Turk sold the Blanco post and concentrated her efforts in Counselor. The mission church across the road from the post moved out and Turk soon tired of running the business. In 2002 she decided to put the post, as well as the mission and school buildings, up for sale. Turk died in October 2002 at age 72 (Linthicum 2003).

The McDonald siblings consist of four brothers and a sister, Kelly McDonald-Aragón. Kelly was raised in the Counselor area. She worked for the store all her life and had run it with her mother for several years. After Turk's death a disagreement arose between Kelly and the brothers about management of the store and post office. In the course of the dustup, Kelly moved the post office out of the store into another building. In September of 2003 the brothers put the entire town up for sale—asking price $2 million. The property consists of 320 acres of land, one trading post, one Christian mission school, five houses, and interest-free accounts of some 300 Navajos who live in the area (Linthicum 2003). The trading post itself is the core of the community. The building behind it was Turk's house, and the building to the east of it was Kelly's house in the old blacksmith shop (Figure 163B). In April 2007 the Navajo Nation purchased the property for a tidy sum of $1.25 million (Linthicum 2007).

21

Side Journey down the Canyon Largo

The access road to the Canyon Largo is the dirt State Route NM-403, which begins its northern course just to the east of Counselor Trading Post. Today the canyon and the road pass through the heart of the huge Blanco Gas Field and a number of smaller oil fields. The area and its many side canyons are laced with a network of roads leading to individual oil and gas wells. These roads must be kept open to field service trucks, so getting around the canyon is not that difficult.

With very few exceptions, the area is uninhabited, but it is important to remember that it was not always so. During the early decades of the 20th century, especially following the 1916 passage of the expanded Homestead Act that gave stockraisers access to 640-acre homesteads (in contrast to the 160 acres allowed under the original Homestead Act of 1862), the canyon was a magnet for settlers. These decades also coincided with a period of abnormally high precipitation, a situation that was destined to change during the 1930s. But at the time the canyon must have appeared like a very good place to live.

Today this silent region has a fascinating historical ambience. In dry weather a leisurely drive down the entire length of the Largo to the San Juan River—with short side trips—can be a pleasant day's outing. The text below only deals with the upper 23 miles and flags some of the more interesting historical sites from south to north (Figure 164).

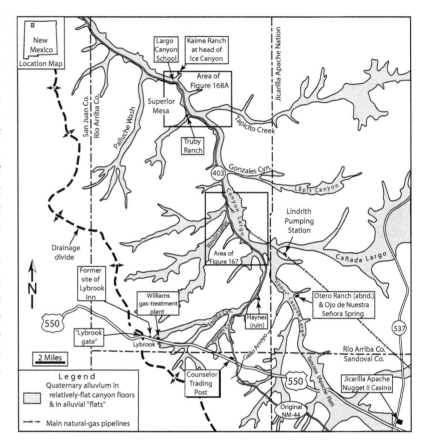

Figure 164. General geography of the Counselor area and the "upper Canyon Largo.

Haynes

The ruined old Haynes trading post and stage station is located just east, and barely visibly, off the Largo Road (NM-403) in a little *rincón* (Spanish for "corner" or "nook") in the sandstone cliffs, 5.5 miles north of Counselor (Figure 164). The ruins consist of massive masonry walls enclosing a 75 x 20-foot block of four large rooms. The huge roof lintels have collapsed down into the rooms, and the adobe portion in front has long since returned to the elements (Figures 165A and 165B).

John Roger "Doc" Haynes was born in 1871 in or near Owensboro in western Kentucky. His father was a physician, and the son chose to follow his father's example. John arrived in New Mexico sometime before 1900, because in that year we find him as a young, unmarried physician living at an Albuquerque boarding house (1900 U.S.Census Record). Sometime between 1900 and 1910 he arrived in the Largo to mix medicine with trading (McNitt 1962). His married life seems to have been a bit messy. He married a Hispanic girl from Albuquerque named Juanita Armijo. By 1910 they were living in the Largo country with a young daughter, Juanita Helen (b. ca. 1905), a son, John Roger Haynes,

Jr. (b. 1909), and John Sr.'s younger brother Samuel H. Haynes from Kentucky. Sometime during the next decade Juanita moved with her son, John Jr. (her teenaged daughter may have remained with her father) to live with her uncle and her siblings in Albuquerque (1920 U.S. Census Record). Juanita died in 1930, and John Jr. the next year.

One source (Kelley and Francis 2006) states that Haynes took over a store in the Largo owned by Mariano Otero, Sr., who died in 1904 (see Otero Ranch below), and that he operated a post office out of it from 1908. In his autobiography, Albuquerque business leader Roy Stamm (Stamm 1999) mentions that in 1933 he ran into an old acquaintance, a "Dr. Haynes," who then was a doctor living in Albuquerque, and who some years before "had thought it expedient to go to the country where he opened a trading post and post office, Haynes, near the Navajo Reservation and practiced his medical profession on the side." The Largo road back then was a well-traveled stage road connecting Durango with Santa Fe, and only later became NM-44. Haynes kept his store well stocked with wares hauled in from Farmington (Kelley 1977).

There is considerable confusion about the location and dates of operation of Haynes due to the fact that there were probably two Haynes Trading Posts. Doc Haynes' younger brother Samuel purportedly was the first postmaster from 1908 to 1929 at a post office on the old dirt road (NM-55) that ran from Cuba to Bloomfield (Pearce 1965). The road to Chaco Canyon branched off to the south at that point. A New Mexico transportation map (Federal Works Agency 1946) shows a "Haynes" located at that place. Somewhere in this chronology this post became known as the Otis Trading Post. More recent maps (USGS Blanco Trading Post 7.5-minute quadrangle, 1966; BLM Chaco Canyon quad, scale 1:100,000, 1997) show a Tsah Tah Trading Post at this site. Another source cites a Samuel H. Haynes as the postmaster of a post office somewhere in Canyon Largo, also from 1908 to 1929 (Julyan,1996). Perhaps Samuel worked with his brother for a time at the post on the Largo road before moving to his own post to the west. Sifting through all these conflicting data I tentatively conclude that our John Roger Haynes indeed ran the facility on the Largo road from 1908 to 1929, and that Samuel ran the one to the west during the same years.

Haynes died in Albuquerque on December 28, 1934 (*Albuquerque Journal* 1934). His daughter (Juanita Helen?) transported his body to Los Ojos, in Rio Arriba County for burial. A visit to the *Cemetario Católica* at Los Ojos reveals a derelict plot with three headstones: one for John Roger Haynes (1871–1934), his wife Juanita Armijo Haynes (1882–1930), and his son, John Roger Haynes, Jr. (1909–1931). The loss of a wife and a son, even though they were evidently estranged, within a year of each other probably exacted a toll on Doc's health. We don't know what happened to daughter Juanita Helen. Nor do we know what ties Haynes had with Los Ojos.

Jim Counselor had worked for Doc Haynes for a time in 1919 after his discharge from the military. Counselor wrote, "Doc treated Mexicans, Indians, and a few whites for whatever ailed 'em. He had ever' damn thing—sheep, mules, cattle, Indian ponies—but he wasn't a rancher, he was a trader" (McNitt 1962). The Counselors, in their book

A. Northwest view of remains of Haynes Trading Post (rock-walled back rooms of original structure). (Photo by author 2001.)

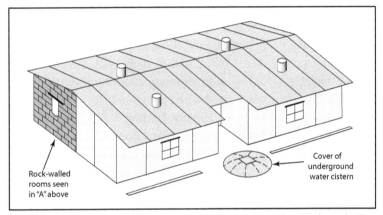

B. Southwest view of "reconstructed" Haynes Trading Post. (Drawing by author.)

Figure 165. Haynes Trading Post.

(Counselor and Counselor 1954), describe the place as they saw it in 1933:

"In the old days, this famous trading post had been an elaborate establishment, center of the commercial life in this part of the desert. The huge main fortresslike building, whose adobe walls were twelve inches thick, had once housed store, post office, barber shop, and restaurant. The kitchen was immense, and the living room about the size of a hotel lobby. At the back of the building were four huge bedrooms. Every room except the kitchen had been heated with fireplaces that used to keep one hired man busy all winter supplying them with wood. Many other buildings, corrals, and sheds, scattered around in shabby disrepair, now made the place look like a ghost town."

Doc Haynes was a big man, 6 feet 4 inches, and weighing about 300 pounds, and he must have struck an imposing profile. The Counselors, in their *Wild, Wooly and Wonderful* (1954) tell a great story, probably dating from about when Jim worked with Haynes in 1919, about a gasoline sale at the "Casa Haynes Trading Post." Gasoline had to be hauled in five-gallon cans from Albuquerque by freight wagon, and the trip sometimes took two weeks. Haynes was the only gas stop between Cuba and the San Juan River, and Doc was not about to give the stuff away.

"A slick-looking dude from New York, dressed fit to kill in white linen knickerbockers, a pair of two-toned long baseball socks, and wearing a cap, drove up in a big black automobile and ordered Doc in a kinda 'my good man' tone of voice, to fill up his car with gas.

"Doc looked him over good-naturedly, wondering, I reckon, how he ever happened to get sidetracked way out here in the desert. He even started to tell him what the price of gas was, but the fellow bawled him out for not hurrying up the servicing of his car. So Doc kept his advice to himself and uncorked three five-gallon cans of gas, which just about filled up the gas tank of the car. When he handed the dude five silver dollars, the change from the $20 dollar bill the guy gave him, the New York kid exploded. Doc listened to him rant and rave for a while, then he went back into the store and sat down, not saying a word. But I knew he was pretty mad, because his face was getting awfully red.

"The dude followed him into the store and slammed the five dollars down on the counter and started in orating again. He talked like a vaudeville Englishman and I really enjoyed listening to him. But doc wasn't listening, he was whistling through his teeth like he did when he was studying over a proposition. Finally he picked up the silver and tossed the $20 bill on the counter. Then he picked up a rubber hose and went out and siphoned his 15 gallons a gas out of the guy's car! . . . The dude cooled off, paid for the gas, and left." He paid $20 (about $200 today) instead of $15 for the gas because "the dude made the mistake of squawking too loud and too long."

An arroyo with vertical walls 20 feet high has eroded its way eastward and past the north side of the station. It is quite possible that the unstoppable headward advance of the arroyo so degraded the access to the Haynes Station that by about 1929 Doc was encouraged to abandon the post and move to Albuquerque.

Haynes is located on State Land (Sec. 36-T24N R6W) and it seems odd that Doc Haynes would have chosen this site to begin with. In 1898 the U.S. Congress, via the Ferguson Act, had granted Sections 16 and 36 in each 36-square-mile township to New Mexico Territory for the benefits of education. If Doc Haynes had settled in the area during the very early 1900s he must have been aware that he was moving onto what would become state land . . . or maybe he wasn't because New Mexico didn't achieve statehood until 1912. Could it be that the inconvenient fact of being located on state land was an additional inducement to pull up stakes?

Otero Ranch

A little more than 8.5 miles down the canyon from Counselor and past the Haynes ruins we encounter a junction. The main Largo road (NM-403) goes to the left (northwest) on its way down to the San Juan River Valley. The road to the right angles back to the old Otero Ranch. This ranch had its headquarters at the site of an old, and quite reliable spring called the *Ojo del Nuestra Señora* (Figures 164 and 166).

The story of the ranch is an interesting tale indeed. Mariano S. Otero, Sr. (1844–1904) was one of New Mexico's leading stockman, a banker, the one-time owner of the huge Baca No. 1 Location in the Jémez Mountains, and a delegate to the Territorial Legislature. In 1876 *Don* Mariano's men were herding 50,000 sheep in the hills near Nacimiento (later renamed Cuba). The stockmen got into a gambling binge with a group of Navajos. After the stockmen had lost everything they had, the Navajos suggested that they may take their sheep as well. The very frightened herders abandoned the herd and walked all the way back to *Don* Mariano's home in Bernalillo (White and White 1988).

A thoroughly exasperated Mariano organized a heavily-armed posse and stormed back to the scene of the crime. With difficulty his men gathered up most of the lost sheep that were scattered about the area and took them 20 miles west to the Ojo del Nuestra Señora spring, where they found a guy running a herd of goats. Mariano made an arrangement with the goatherder (perhaps "an offer he couldn't refuse") and took over the spring. After watering the sheep for 30 days his men began the chore of constructing the Ojo de Nuestra Señora Ranch (White and White 1988). The ranch later became known as the Otero Ranch, and the spring as the Otero Ranch Spring (Figure 166).

After *Don* Mariano's death in a carriage accident in 1904, the ranch was taken over by his son, Mariano S. Otero, Jr., who ran it as a cattle operation. Mariano S. Otero, Jr. was Jim Counselor's neighbor. Jim was a sheepman and Mariano was a cattleman, and despite the myth of incompatibility they became close friends. Mariano Jr. was born in Bernalillo in 1879. He was a short man and a dozen years older than Jim Counselor. He was educated at the Lexington Military Academy and later (1892–1894) at Notre Dame. Like his father, he ran several thousand head of cattle in the upper Largo and controlled a large segment of the range west of the Continental Divide (Hernández

2000b). In August 1940, while climbing on his horse at his ranch, he collapsed. He was rushed to Farmington and died there two weeks later (*Albuquerque Journal* 1940d; *Albuquerque Tribune* 1940). His wake was held in the Counselor blacksmith shop (Aragon 2002). The Otero Ranch house is today abandoned, falling down, and located just inside the Jicarilla Apache Reservation (Figure 166).

Lápis Canyon

The Counselors' book (Counselor and Counselor 1954) provides another interesting piece of Canyon Largo history that will not be found anywhere else, and that is the story of *Lápis Canyon* (Figure 164). *Lápiz* is the Spanish word for "pencil," and in their book the Counselors refer to a "Pencil Spring." However, no spring of this name appears on the USGS topographic map (7.5 minute González Mesa Quad 1963). The map does show a Gonzáles Spring located at the mouth of Lápis Canyon, and I suspect it is the same as the Counselors' Pencil Spring. According to their account, a white trapper "years ago" had his camp at this spring. His name was Dunlap, but the closest the Indians could get to the pronunciation was *Don* Lápiz, or Mr. Pencil!

Canyon Largo families

As mentioned above, today the upper canyon is virtually uninhabited. This is a far cry from the halcyon days of the 1920s and 1930s. After several decades of adequate rains and abundant snowpack, by the late 1920s drought had set in. The once lush grasses on the mesa tops were taken over by sagebrush. The last wool-growing seasons were in the early 1930s, and the last cattle drive to Durango was in about 1940 (Young 2004). Most of the ranchers bailed out during the 1930s, pounded by the combination of drought and the Great Depression. Three of the departing families—Tafoya/Martínez, Apodaca, and Martín—left their ghostly traces. Two other families—the Trubys and the Kaimes—hung on for a while, but only the Trubys have a presence today.

Tafoya/Martínez

This family's homestead is located on BLM land about 0.5 mile west up Tafoya Canyon, and is today protected as a historical site.(No. 1 in Figure 167). In 1904 Margarita Maestes Martínez, a widow

Figure 166. Southeast view of the Otero ranch-house ruin. (Photo by author 2004.)

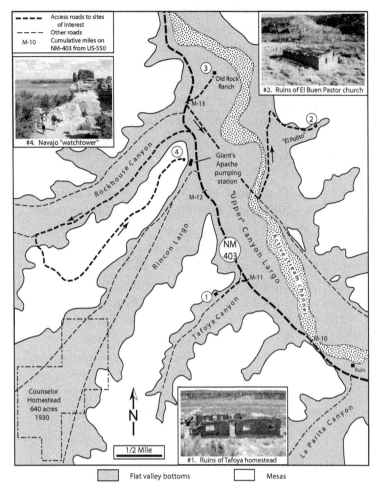

Figure 167. Main sites of interest in "upper" Canyon Largo.

living with her family in Corrales, moved to the upper Largo and filed on a 160-acre homestead. Her son built a one-room house near a spring in a side canyon, then called *Rincón del Burro* (Spanish for "Corner of the Donkey"), but today known as Tafoya Canyon. In 1914 Margarita's grandson, Mike, was born at the homestead. After his father's death during the influenza epidemic of 1918, Mike and a brother were sent to live at the St. Anthony's Orphanage in Albuquerque. After Margarita's death 1919, Mike's aunt and uncle, Isabel and Luís Tafoya, took over the homestead and in 1923 received a patent on the plot in their name. Mike then returned to his family in Durango but spent his boyhood summers in the canyon. He eventually moved to Aztec (Young 2004).

Today the roof is gone but the sturdy walls have held up. A walkover of the site reveals something of the construction methods of the time. Most of the homes built in the canyon were of sandstone blocks obtained locally. The wooden doors and windows were made of milled lumber that was hauled in by wagon. The walls were finished with gypsum plaster and whitewash. The dirt floors were hardened by generous application of gypsum plaster, or—somewhat less desirably—by a slurry mixture of dried cow dung. The sloped roofs were supported by *vigas* (main wooden beams), usually of Douglas fir or Ponderosa pine, and *latillas* (minor beams) of cottonwood. Characteristically, the Tafoya house incorporated a huge sandstone boulder into one of its walls (Young 2004).

Apodaca

In the late 19th century Martín Apodaca homesteaded in a *rincón* on the east side of Canyon Largo, called *El Polito* (No. 2 in Figure 167), and raised sheep. The name of the place either refers to a "little pole" or perhaps a misspelled *El Pollito*, "little chick." An excellent spring was located at the head of the rincón. Apodaca donated a little parcel of land for a church, *El Buen Pastor* ("the Good Shepherd"). The priest from Blanco, on the San Juan River to the north, came down and conducted services every two months (Young 2004). Like the Tafoya house, the church is a protected site on BLM land. Today one can drive most of the way to the site and walk a short distance to the remains of the building.

Martín

In about 1882, Nestor Martín built his Old Rock Ranch (No. 3 in Figure 167). He was a sheepherder, and later sold the property to strongman Mariano Otero, Sr., whose cowboys were then running cattle from Gallup and the Mt. Taylor regions to his ranch in the Largo. Only a portion of one wall still stands, but the lonely place is worth the stop.

Truby

The one family name, after Counselor, that is intimately associated with the upper Canyon Largo is Truby. The Truby Ranch is located at mile 20.6 on the west side of the canyon (Figures 164 and 168A). Today it is the only active

226

ranch in the upper part of the Largo and it runs about 50 head of cattle (Henry 2004). How the Trubys came to the Largo is an interesting tale. The family was originally from Texas, where they had raised cattle. In 1899 they arrived in southwestern Colorado. That was the year that a former slice of the former Ute Indian Reservation, the "Ute Strip," was opened for settlement. The Trubys homesteaded near today's Bondad, Colorado (about 20 miles north Aztec) and ran cattle in the rangeland in and around the Animas River Valley. They soon began butting heads with another ranching clan, the Cox family.

The Coxes had also come from Texas, but somewhat earlier. In the late 1870s they had settled in Cedar Hill (then called "Cox's Crossing") along the northern New Mexico portion of the Animas River Valley. In 1904 they switched from cattle to sheep, and perhaps this "unconscionable" act launched the feud between the two families. More likely there was bad blood between the two for some reason or another even before this event. Over the next several years the strife between the families festered. It came to a head in 1911 at a chance encounter. Escalating words led to gunfire and when it was over Ike Cox had shot Bill Truby, 27, dead. Later that year Cox was tried in Durango and acquitted. But that was not the end of it. The next year, 1912, the two factions ran into each other again and shots were fired. And again, another Truby boy, Sam, 22, was dead. A second trial was held in 1913 in Durango and a second "not-guilty" verdict was granted (Craig 2002).

At this point the matriarch of the Truby family, Elizabeth, announced that she had had enough. In early 1914 she and her remaining three sons moved up into the Canyon Largo and filed on three homesteads (Craig 2002). Life in the canyon was good in those first days. The rains were adequate and the cattle business was healthy. Potable water could be had by hand-digging shallow wells in the river alluvium. However, if medical attention was needed, either a frantic race to Aztec or a prayer was called for. Not until 1928 was there a motor vehicle in the canyon. About 1.5 miles north of the Truby ranch was a one-room schoolhouse. In those days eight students or more were required to open a school, and the teacher had to stay with one of the local families (Henry 2004). The school was shut down in the early 1930s.

The Truby ranch is located at the mouth of Cíbola Draw (Figure 168A). One source refers to this draw as "Trubey Canyon" [sic], and notes that the brothers Henry and John ran a small trading post here (McNitt 1962). Supplies for the ranch and the store were hauled in from Aztec, a two-day trip each way. The store was shut down sometime before 1931 because it was too difficult to get those supplies (Henry 2004). The youngest surviving Truby son, Harold, and his wife, Beth, owned and lived on the site. Harold was killed in an accident in 1977 and Beth carried on until she suffered a stroke in 2003. The ranch is still active but is being run from a distance by Beth's three daughters. The ranch house is now semi-deserted (Kaime 2004).

Kaime

In the 1880s three Kaime (pronounced "Came") brothers moved west. One, Edwin, settled in Denver and the others in California. Edwin's son, Clifford Kaime, was born in St. Louis, Missouri in 1887 but grew up in Denver.

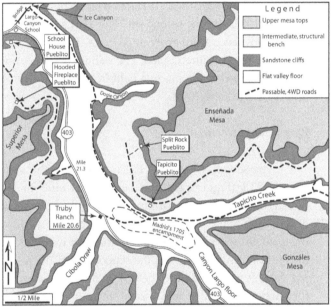

A. Truby Ranch area and site of 1705 Spanish/Navajo battle (location in Figure 164).

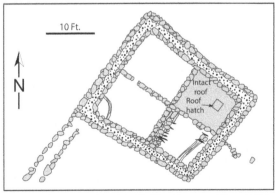

B. Plan of Tapicito Pueblito. (Modified from Towner and Dean 1992.)

Figure 168. Main sites of interest in "middle" Canyon Largo.

After high school he rode a horse down into New Mexico and served in the calvary during the time of Pancho Villa. Later he bought a ranch in Hatch, near Las Cruces, New Mexico. There he met Gladys Hersey, the daughter of an army (also calvary) man. In 1916 they loaded their belongings on a railroad boxcar and moved to Albuquerque, where they looked for a ranch to buy. They found one in the Cabezón area and stayed there about three years. Then they moved to the Largo and purchased a total of about 1,240 acres. They built a ranch house at the head of Ice Canyon, near a spring, about two miles up from the Canyon Largo (Kaime 2004; Figure 164).

Clifford and Gladys had two daughters and two sons. In 1948 Clifford deeded the ranch to the sons, Edwin and George, and the two divided up the mesa above the ranch house between them. Clifford and Gladys moved to Durango. Clifford committed suicide at age 82 in 1971, and Gladys died in 1989 at age 91. The first ranch house burned down in 1957, and the second on New Year's Day in 1960 while the Kaimes were away in Arizona. Today the existing ranch house is unoccupied. The Kaimes sold the ranch in 1992 and today the place is held in the name or Robb Enterprises (Kaime 2004).

Largo Canyon school

The school, located on the west side of the valley at mile 23.1 (Figures 164 and 168A), was built in 1954 during the height of the drilling and development boom of the huge Blanco Gas Field. It provided for the kids of the gas-field workers. In 1983 a gymnasium was added to the eight-room building but the school closed the following year. It was deeded to the school district and put up for auction. Over the next several years it went through a number of hands until 1997, when a couple, John and Patricia Barlow-Irick, purchased it. The Iricks are pursuing their dreams. They were married in 1993. John is a metal-sculpture artist and a Certified Public Accountant. He aspires to build a welding workshop and artist's paradise. Pat is a self-described free spirit and is working on a Ph.D. in plant taxonomy. Her dream is to write, grow fantastic gardens, and ride mules through the canyon. He wanted water, she wanted desert, neither wanted neighbors or deadlines. Over the following years the Iricks transformed the old school into a retreat with workshops, living quarters, and guest accommodations, and renamed it the Largo Canyon Station, a.k.a. Largo Canyon School. They have made the fixed-up old school into an eco-friendly resort facility for artists, writers, and small groups (Irick 1997).

The Iricks also for five years had operated a burrito place called the Navajo City Roadhouse, located at MP-87 on US-64 near Blanco at the mouth of Canyon Largo.

The *Dinétah*

This is part of the Navajo ancestral homeland, the *Dinétah* (more in Chapter 24). The Athabascan-speaking Navajo moved into the headwaters of the San Juan River probably about 1500 CE and their settlements centered on Gobernador Canyon and Canyon Largo. The Dinétah is where the Navajo first encountered the pueblo Indians and where the Spanish first encountered the Navajo. The Spanish were driven out of New Mexico after the Pueblo Revolt of 1680. When they returned in 1692 to stay, many pueblo Indians fled west to the Hopi country in Arizona and perhaps a few hundred, particularly from Jémez Pueblo, took refuge to the northwest in the Dinétah (Marshall and Hogan 1991). The pueblo refugees lived with the Navajo for a decade or two, and probably intermarried. A growing fear of the Navajo's traditional enemies to the north, the Utes, led the Navajo to construct their homesites in defensive locations on mesa tops. During this period, the so-called Gobernador phase of Navajo history, many of these homesites, or "pueblitos," were constructed. By about 1750 the people abandoned the Dinétah, probably due to increased pressure from the Utes, and moved to the west and southwest. Today most of these pueblitos are located on BLM land and many have been studied and excavated (Linford 2000).

The Pueblitos

There are four pueblitos in the area near the Truby Ranch (Figure 168A). All are easily accessible via four-wheel drive, high-clearance vehicles. Two of these—Tapicito and Split Rock—are located on the east side of the canyon opposite the Truby Ranch. The other two—Hooded Fireplace and Largo School—are located on the west side of the canyon via a steep road about 0.6 miles north of the Truby Ranch.

The Cuba Mesa Sandstone forms the lower cliff and the inner wall of this part of the canyon. A sandstone body of the Regina Member, here stacked on top of the Cuba Mesa Sandstone, gives the area an interesting double-cliff character. The four pueblitos are all located on top of the Cuba Mesa and below the upper, Regina cliff, and are set back in a way such that they are, importantly, not visible from the canyon floor.

In addition to the four pueblitos there is an intriguing "watchtower" overlooking the Largo Road near mile 12.5. We'll talk about the watchtower first because it's the first one encountered on the drive north down the Largo.

"Watchtower" ruin

This ruin—not a pueblito—is located on top of the Cuba Mesa Sandstone cliff on the west side of the canyon, and hovers over the Giant Apache pumping station and tank battery at mile 12.3 (No. 4 in Figure 167). This structure is accessed by a side road, 0.3 mile north of the pumping station. The road goes southwest for two miles down the appropriately-named Rockhouse Canyon, then turns south, climbs to the top of the cliff, and doubles back to the northeast. The road ends at barricade, but a foot trail leads a short distance to the ruin at the point of the cliff. The watchtower is perched atop a pinnacle of rock and offers a superb view of the upper Largo.

Tapicito Pueblito and the battle of 1705

Directly across the Largo from the Truby Ranch is the mouth of Tapicito Creek. High above the valley floor on top of the first ledge and out of site is the Tapicito Pueblito (Figure 168A). Three centuries ago this was the site of a sharp and nasty battle. In August 1705 a large punitive military expedition, led by Roque Madrid and authorized by the Spanish governor, arrived here. Madrid's recently-found journal has been translated from the original Spanish (Hendricks and Wilson 1996) and it tells this tale. His assigned mission was to punish the Navajo deep within their homeland in retaliation for raids on the Rio Grande settlements and to discourage such future behavior. His muster included about 400 men including soldiers, citizen militia, and pueblo Indian auxiliaries, and about 700 horses. The shear size of the expedition limited the line of march to wide paths with promising sources of water.

The expedition left the plaza of San Juan Pueblo (just north of present-day Española) on July 31, 1705, and in 20 days made a circuit of about 312 miles. The column approached the Dinétah from the north. After two short battles with the Navajo, Madrid marched down Albert Canyon (a north fork of Tapicito Creek) to its junction with the Tapicito and west to the Tapicito/Largo confluence. There they camped for three days. They called the camp *Nuestra Señora de Guadalupe* (Hendricks and Wilson 1996).

At the time Tapicito Canyon was presumably a wide bottomland with abundant water, pastureland, and Navajo corn fields. On the north side of the confluence was the high Enseñada Mesa that loomed some 500 feet above the canyon floor. On a topographic bench, below the mesa top and about 250 feet above the valley floor on the top of the Cuba Mesa Sandstone was a Navajo pueblito. The pueblito itself was not visible from the canyon floor. About 30 Indians shouted down to Madrid from the bench, asking for peace. Madrid kept the Navajo occupied and sent most of his force down-canyon to feign a retreat while he ordered 100 of his pueblo auxiliaries to circle around, find a way to the top, and trap the Navajo on the bench. Entrap them they did. Afterwards Madrid wrote cold-bloodedly:

"Two who attacked in this direction died at our hands and the rest above, by lance and bullet wounds. Of the more than 30 that were there, no more than five escaped, not counting two who in a great fury threw themselves over the edge, or the many more deaths that the Indian allies will have carried out among the women and *chusmas* [non-combatant women, children, and/or elderly] over the distance they covered on top of the mesa. This is their [the pueblo allies'] custom, and no matter how I reproach them, they neither take heed nor pay attention unless Spaniards are present. When this battle was over I retired to lay waste to corn fields" (Hendricks and Wilson 1996).

Today the ruins atop the bench, the Tapicito Pueblito (officially known as site LA2298), is one of the best-known pueblitos in old Navajo country (Figures 168A and 168B). The structure was first reported in 1941 (Powers and Johnson 1987). Tree-ring dating suggests a construction date of 1694 for the main structure. Madrid makes no mention of it, but it must be remembered that the pueblito is not visible from the canyon floor. Some authors therefore assume that the structure did not exist in 1705 and suggest that it was constructed later using recycled vigas (Hendricks and Wilson 1996). However, based on three sets of tree-ring samples collected over 50 years, others conclude that the structure was indeed constructed in the 1690s and that it makes up the oldest accurately-dated Navajo pueblito (Towner and Dean 1992). The site was likely chosen for habitation due to a secure nearby source of water. Water running off the cliff face of the "slickrock" Cuba Mesa Sandstone has eroded several *tinajas* (Spanish for "large earthen jars"), or tank-like depressions worn into the sandstone. Two of these *tinajas* are quite deep and contained hundreds of gallons of water even during the dry season (Marshall and Hogan 1991).

The Tapicito Pueblito is different from other Dinétah pueblitos. Architectural details, such as rubble-core/veneer-wall construction, and a high concentration of non-local ceramics, including some from Zia, Cochití, and Pecos Pueblos, make this site unique (Towner and Dean 1992). The small Tapicito Pueblito is the only known pueblo refugee community in the Dinétah, which makes the ferocity of Madrid's pueblo allies against what may have been their own kin all the more ironic.

After this battle Madrid led his force south. He stopped at a well-known spring called *Ojo del Nuestra Señora* (near what would become the Otero Ranch) before heading southeast toward the Jémez River Valley and home. This spring was also cited in 1745 by three Franciscan friars on a mission to the Navajos in their Canyon Largo mesa-top settlements (Reeve 1959). In 1875 a surveyor with the U.S. Army's Wheeler Survey described the spring in this way: "a fine bubbling spring situated in a drain running into Cañon Largo. The ground round about is a marshy; large bowlders [sic] are scattered over the ground, low mesas of sandstone enclose the drain, and numerous trails concentrate at this one volcanic spring in the desert" (Hendricks and Wilson 1996). The surveyor was however mistaken about the spring's volcanic nature.

A. East view from trail below (pueblito on top of huge block in upper center).

B. West view from above. Logs are about six feet tall; parking area in distance.

Figure 169. Split Rock Pueblito. (Photos by author 2004.)

Split Rock Pueblito

Split Rock Pueblito is located about 0.5 mile north of Tapicito (Figure 168A). A small BLM sign is located on the east side of the road and a parking area is on the west side. A trail beyond the sign leads east about 0.2 miles to the base of a cliff. It's difficult to spot the ruin until you're almost right at it (Figure 169A). The barely recognizable, four-room ruin is on the top of the split boulder (Figure 169B). A vertical fissure propagates from the main sandstone bed below, upwards through a small erosional remnant of a sandstone bed above, i.e., the "boulder" — hence the name Split Rock. This pueblito is unusual in that it has no nearby *tinaja* or other surface water supply. Its inhabitants likely got their water from the *tinaja* at Tapicito to the south (Marshall and Hogan 1991). The BLM stabilized the ruin in 1975, and installed an explanatory sign at the cliff's base. The ruin is assessable via a tricky scramble up a rubbly recess in the cliff a short distance to the north.

Hooded Fireplace Pueblito

Hooded Fireplace is located on the west side of the canyon. Access is via a steep dirt road branching west from the Largo road at mile 21.3 (Figure 168A). The road ascends the bench and then veers off to the north. After about two miles a bleached BLM sign and a trail sign-in box are on the left. The ruin, not visible from the parking area, is reached via a very short trail. The six-room pueblito is located on a north-facing point of a bench on the west side of Superior Mesa. Several *tinijas* in the slickrock Cuba Mesa Sandstone north of the site were the likely source of water for its inhabitants.

School House Pueblito

The School House Pueblito is located about 0.6 mile north of Hooded Fireplace (Figure 168A). A small parking area is located on the east side of the road and an old road (now blocked off) leads about 0.25 mile to the cliff rim. A hike for 100 yards or so to the north along the cliff rim leads to the ruin. The two or three-room ruin is located on a small promontory of the Cuba Mesa Sandstone cliff overlooking the main Canyon Largo, with Ice Canyon to the east and the old Largo Canyon School directly below. A group of *tinajas* west of the site in the Cuba Mesa Sandstone provided the water supply.

As US-550 proceeds west from Counselor the roadbed traverses the up-and-down topography developed atop the channel sandstones and intervening flood plain mudstones of the Regina Member. The sandstone rim up ahead at Lybrook is called Crow Mesa (Figure 170). It forms the boundary between two major viewsheds. Behind us, east of the crest of the highway at MP-104, we are bathed by the very extensive viewshed of the Sierra Nacimiento. Beyond this point we fall into (intermittently at first) the viewshed of Huerfano Mountain. The crest also marks the divide between the Canyon Largo drainage basin to the east and the Blanco Canyon drainage basin to the west. The aspect of our drive changes so significantly here that I think it is appropriate to dub this place the "Lybrook Gate."

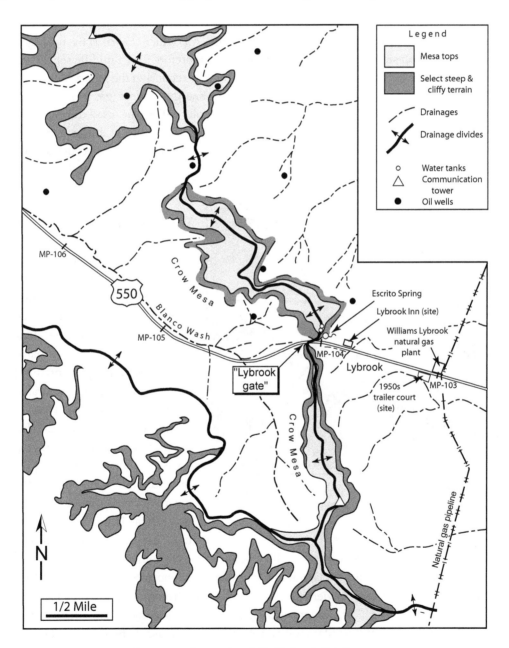

Figure 170. Geography of the "Lybrook Gate" area.

The competing headward reaches of the two drainage systems (Canyon Largo vs. Blanco Canyon) have almost breached the divide and have lowered it such as to produce a natural passageway. Add a reliable spring and we have a perfect place for a settlement, and for the path of a highway. The spring is named *Escrito* (Spanish past-participle verb for "Written") Spring The Spanish translation just doesn't seem right. The name is more likely a futile Spanish attempt to spell the formidable Navajo word, *Tódóó Hódik' áadi*, meaning "Slanted Water," or "Spring from a High Place" (Linford 2000).

Today the spring is not used. Rather, the local water supply is provided by a 1,500-foot-deep well. The water is pumped up and collected in the tanks perched on the mesa-top at the Lybrook Gate on the west edge of the tiny village. This same well supplies the Counselor Trading Post via a six-mile-long pipeline. Along this distance the water falls some 400 feet in elevation, resulting in some 70 pounds per square inch of pressure at Counselor, "enough almost to blow you out of the tub" (Aragon 2002).

The "town" of Lybrook

Lybrook is a tiny, loosely bound, unincorporated community. An outlier of sorts is the new Lybrook Elementary School at mile 99.5 on the south side of the highway. This is a K-8 (therefore also a middle school) facility with about 88 Navajo students. It is one of only three, very-spread-out elementary schools in the Jémez Mountains School District. The community itself "begins" (kind of) at MP-103, at the William's Energy Services (gas processing) Plant on the north side of the road. Also on the south side is the derelict remains of an old trailer park used during the 1950s to house oil and gas field workers and their families—a gaunt reminder of the hustle and bustle this area experienced during the gas boom. One wonders about the lives and stories of the people who lived here five decades ago who are now long departed (Figure 171).

Figure 171. North high-oblique aerial view of Williams natural gas plant and abandoned 1950s trailer park. (Author by author 2001.)

Williams Gas Processing Plant

The tangle of pipes, valves, and tanks at the gas plant belies a highly organized, engineered system. The Williams gas processing plant is a sort of "tip of the iceberg"—and essentially the only tip we'll see along US-550—of the enormous San Juan basin natural gas industry. The primary purpose of this plant is to produce pipeline-quality natural gas from what it receives from the huge Blanco Gas Field to the north (more in Chapter 32). The plant processes, or "cleans," the natural gas delivered to it via a vast network of gathering lines that connects the individual gas wells to the plant. Processing removes water, contaminants, light-weight natural-gas liquids (NGLs), and heavier gasses to produce a "dry" methane natural gas suitable for burning. Without processing, the gas from the wells would corrode or foul the major pipelines. The treated pure gas exits the plant via an output or "tailgate" lateral line to the main Public Service Company of New Mexico (PNM) pipeline south of US-550. The NGLs and heavier gases that are stripped from the gas are piped as a natural-gas liquids blend, mainly ethane (C_2H_6), propane (C_3H_8), butane (C_4H_{10}), and natural gasoline $C_{4+}H_{10+}$) via a dedicated 12-inch Williams pipeline. This unimposing plant thus plays a vital role in providing usable fuels to the state's major population centers in the Rio Grande corridor.

Lybrook Inn

Lybrook's item of greatest interest is a historical site located a little less than a mile beyond the gas plant and just east the Lybrook Gate at MP-104. Look on the north side of the highway at mile 103.8, near the foot of the cliff above which are perched the two water tanks. The words, "Lybrook Inn," are sketched into the cliff face (Figure 172). East of the cliff, with a little patience one can make out the subtle remains of the grand old Lybrook Inn. The site today consists of only fragments of a foundation and a loose pile of sandstone blocks (Figure 173A).

The name Lybrook comes from William A. Lybrook. His aunt was a sister of Richard Joshua ("R.J.") Reynolds (1850–1918), from North Carolina, the founder of the Reynolds Tobacco Company. In 1913 Reynolds introduced Americans to the first modern cigarette, Camel. He soon became the cigarette king. Before the U.S. entered World War I in 1917 Americans smoked about 25 billion cigarettes; after the war that number had shot up to 53 billion, and about 40% of that number was Camels. The increase was attributed to war tensions and the emancipation of women (Dykeman and Stokely 1968). Needless to say, uncle R.J. was a wealthy man.

William ("Will") A. Lybrook was born in about 1873, one of eight children. As a young man he left home and enlisted in the Navy during the Spanish-American War. After the war he earned a degree in mechanical engineering from Emory University in Atlanta. His overbearing uncle R.J. offered him a position with the company, but Will turned him down. Instead he left to work on the Panama Canal and then wound up in Oklahoma, where he and his new wife, Lottie, bought a ranch and where he worked in the oil fields as an engineer. For some unexplained reason they drove their livestock cross country to southwestern Colorado and settled along the La Plata River. In about 1918 they moved to Farmington when the San Juan basin's first gas boom attracted skilled oil-field people. There for a while he operated a boarding house, but it soon burned down (Atteberry 1990).

Figure 172. Remnants of Lybrook Inn sign in cliff behind site. (Photo by author 2000.)

A. Northwest, high-oblique aerial view of former site of inn. (Photo by author 2001.)

B. Northwest view of the inn during its heyday, ca. 1960. (Drawing by author from photo by Atteberry 1990.)

Figure 173. The Lybrook Inn.

By about this time he must have heard about the good ranch country that was located to the east, on the far, east side of Crow Mesa. He made his move and in the early 1920s homesteaded on 640 acres at the present site of Lybrook. Later he purchased homesteads filed on by his father-in-law and brother-in-law (a common practice). He also bought the homesteads of others willing to sell and soon had accumulated some 9,000 acres in San Juan, Rio Arriba, and Sandoval Counties (Atteberry 1990).

Its location in the southwestern corner of Rio Arriba County guaranteed that this place would always have certain intrinsic problems. Supplies and medical attention could be gotten in Aztec, about a day's trek over the tricky Farmington road or even longer down the sandy Largo road. Lumber could be gotten from Cuba, most of a day's trip if Apache Flats were hopefully dry, and of course Bernalillo and Albuquerque were only about another day's trip over a relatively good road. But legal issues, such as the paying of property taxes, would have to be dealt with at the county seat at Tierra Amarilla, perhaps two days away over 120 miles of bad roads.

Will's first order of business in his grassy kingdom was to build a log cabin just north of where they later built their big house. Afterwards the cabin served as a storeroom (Kelley and Francis 2006). Next he set about building a big two-story house (Figure 173B). Large ceiling beams—*vigas*—were hauled by wagon over the primitive road from Cuba, and sandstone blocks were quarried from the nearby sandstone cliffs. After three years the house was completed. Will called it the "big house." And big it was. The family quarters occupied the west side. They consisted of six bedrooms, two-and-a-half baths, kitchen, living room, and dining room. The east side was for business. The main floor housed a trading post, and five bedrooms and a bathroom for hired hands occupied the second floor. The huge attic was for storage, and the basement had a steam boiler to heat the place when guests were on hand. When the family was at home, heating was provided by burning wood, and light was supplied by kerosene lamps. Water came from the nearby Escrito Spring. A tunnel had to be dug through the sandstone to access the spring via an outlet, down which the water flowed to the house by gravity (Atteberry 1990). Jim Counselor, the sheepman from Canyon Largo, and two of the Young boys from Cuba worked on the project (Hernández 2000b). The family, now consisting of Will, Lottie, and three sons, settled in.

Lybrook tried to persuade two of his older sisters to move from North Carolina and live with him and his wife. He even built a pair of rooms in the house in anticipation. They did come to visit but declined the offer to remain in sagebrush country. But Will kept himself constantly busy. He built a pigeon house out back, brought foxhounds for sport and hogs for meat from back east. Then in 1925 Lottie died. Will sent his sons to boarding schools in the Farmington area, but—at the insistence of his sisters—eventually sent them to schools back east. He was soon a lonely man (Atteberry 1990).

Fortunately his brother Sam, who had come to visit in 1923, had stayed on. Sam homesteaded at the mouth of Rincon Largo, just down the canyon from Jim and Ann Counselors' place, but apparently didn't prove up the land. Will Lybrook died of pneumonia in 1935 and was buried in Albuquerque. Sam moved into the big house and sold the trading post's accounts to Jim Counselor, who in 1933 had opened his post six miles down the road, and ran livestock on the Lybrook Ranch until Sam's death in 1939. The guardians for Will's three sons leased the house to the Bureau of Indian Affairs, which used the building as a warehouse (Atteberry 1990).

The Lybrook boys served overseas in World War II, and when they returned home they found the place gone-to-seed. The local Navajos—according to their belief—considered it a "dead man's house." They never expected it to be inhabited again and treated it accordingly. Son Robert was saddled with the chore of paying the back taxes, but the property was spread over three counties and the taxes had to be paid in the county seats of each: Aztec, Tierra Amarilla, and Bernalillo. When he arrived in Bernalillo he learned that a certain Dr. Haygood had paid the taxes while he and his brothers were away. The doctor believed that the house was located in Sandoval County, and demanded title to the house. But he was wrong: the house was located in Rio Arriba County. The brothers took the issue to court and won, but they had to repay the doctor, with interest. Ironically, the Lybrooks sold the house a few years later to a man, who in turn sold it to the doctor. Haygood got his house (Atteberry 1990)!

Only then did the Lybrook house become known as the Lybrook Inn. In the early 1950s Haygood acquired a liquor license, converted the trading post to a bar and restaurant, and rented the rooms to gas-field workers. This was the beginning of the second, really big boom in the gas patch. By the middle 1950s Haygood began to hear the siren song of possible riches in the oil and gas business. He traded the inn and 320 acres for a drilling rig in California. To his dismay the rig turned out to be a pile of junk, but—the court decided—it may have been a heap but it was a drilling rig. The happy new owner of the inn, a man named John Kennedy, ran the establishment until 1966 (Atteberry 1990).

In October of that year a fire broke out. Fire crews from Southern Union Gas Co. rushed to the site but—inexplicably—Kennedy waved them off, despite a huge collection of Navajo rugs on the second floor going up in smoke. Something truly wasn't right here. The Lybrook Inn burned to the ground. Since then the property has changed hands several times. An old abandoned saloon today occupies the site (Atteberry 1990).

There is an interesting postscript to the Lybrook story. It is difficult to explain how Will could summon the resources to build and maintain such a huge house. Some believe that he—and perhaps his brother Sam as well, who was quite a drinker, and even his sons—were "remittance men," i.e., scions of a wealthy family paid to stay out of the way and out of mischief (Hernández 2000b). Regardless, there is little left of him today except his family name.

Escrito Trading Post

The present trading post is located at mile 103.6 on the south side of US-550. In 1946, O. Chapman gave 150 horses to brother-in-law Sam Lybrook (Chapman's sister was Sam's wife) as payment for a plot of land at Escrito Spring, west of the Lybrook Inn, and built a store, café, and dwelling the next year. The facility was located on the north side of the highway, near the base of the cliff below Escrito Spring. The original store consisted of a single structure with a dirt roof covering the store, storage room, kitchen, two bedrooms, and a hogan-shaped café. Water was hauled in from Escrito Spring, a Navajo sacred place (presumably without Navajo consent). From 1951–1954 he leased the store to Jim Mauzy (Kelley and Francis 2006).

Widening of NM-44 in 1954 led to razing of the store, so Chapman built another one on the south side of the highway. In 1955 Mauzy bought some deeded land on the south side of NM-44 and built a store just east of Chapman's. Part of the Chapman store, built of ammunition cases, served as a school, church, and storeroom. In 1966 Chapman's store burned down (two small boys died in the fire) and it was replaced by the current building (Kelley and Francis 2006).

Jim Mauzy was a local guy whose parents had homesteaded around Lindrith. In 1967 he moved to Bloomfield so his children could go to school, and he started a dry goods store there. He sold his Lybrook store to F.T. Akins, who in 1969 sold it to Roland "Buddy" Spicer. In 1974, Chapman's son Al took over the Chapman store and in 1977–1978 Al bought the Mauzy store from Spicer (Kelley and Francis 2006).

Indian Country I—the Jicarilla Apache Nation (*Tinde*)

For the lonely 17-mile stretch (MP-78 to MP-95; Figure 155) between Cuba and Counselor, US-550 traverses the completely uninhabited southern edge of the Jicarilla Apache reservation (Figure 22). The term *Jicarilla* is Spanish for "Little Basket" or "Little Basketmaker." However, the Jicarilla people call themselves the *Tinde*. In 2001 the Jicarilla Apache tribe officially changed its name to the Jicarilla Apache Nation.

The semi-desert part of the reservation contrasts sharply to the northern, more mountainous and forested part. Virtually all the tribe's people live to the north near the vicinity of the tribal capital town of Dulce, 66 miles to the north of US-550. It has also been in the north where the tribal administration has been acquiring significant new additions of land. Today this huge reservation astride the continental divide consists of about 875,000 acres.

Most of the level ground east of Counselor, sometimes called Apache Flats, is underlain by mudstones of the Regina Member of the San José Formation. Looking across these empty flats it's easy to write off the land as being worth little or nothing. However, not far off to the north of US-550 this chunk of open terrain morphs to the Jicarilla tribe's virtual money machine in the form of hundreds of oil and natural gas wells.

Most of the historical account below is taken from Tiller's *The Jicarilla Apache Tribe, a History, 1946–1970* (1983), and *Time-Life Books* (1993, 1995). So let's set the historical stage.

Pre-Spanish period

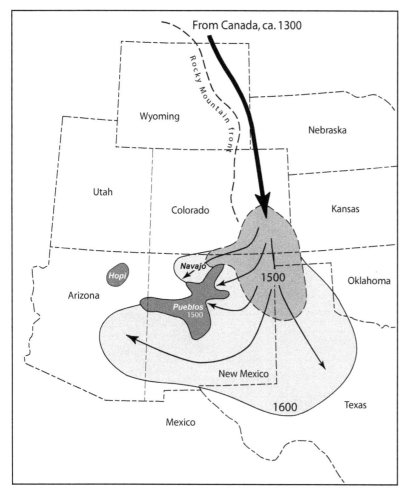

The southern plains of the United States are that great swath of the high plains that abuts the Arkansas River on the north and sweeps across southeastern Colorado, eastern New Mexico, southwestern Kansas, the panhandles of Oklahoma and Texas, and extends into central Texas. This is the vast short-grass country that once supported huge herds of grass-chomping buffalo. Beginning about 1300 CE, groups of Athabascan-speaking Apache people migrated south from Canada, following the migrating buffalo herds along the eastern flanks of the Rocky Mountains (Figure 174). By about 1500 CE they had arrived in what would become New Mexico and had fanned out as seven distinct groups: Western Apaches of eastern Arizona, Chiricahuas in southeastern Arizona, Mescaleros in southeastern New Mexico, Kiowa Apaches, Lipans, Navajos, and Jicarilla (Figure 175).

Figure 174. Migration of southern Athapascan people, ca. 1300–1600. (Modified from *Time-Life*, 1993, 1995.)

Spanish period, 1541–1821

Spanish contact with the Apache began in 1541 when Coronado and his men ventured onto the eastern plains. By 1600 the Spanish referred to these people generally as *Querechos*, and later as *Apaches* from the Zuni name for them, *Apacu*, meaning "Stranger" or "Enemy." Prior to the Spanish reconquest in 1692 the Jicarilla numbered about 10,000 (in contrast to 330 in 1897, and to about 2,500 presently). Before their acquisition of the horse, the Apache planted crops on a limited basis and gathered wild foods. This led to a somewhat settled village life. Most of the year, from fall through spring, they hunted buffalo on foot with arrows and lances, and moved with the herds. By about 1700 CE the Jicarilla were considered a tribe distinct from the others. In that year the Spanish governor had ordered that the head of an executed criminal be displayed on a post in Taos as a warning to the "apaches of la Xicarilla" (Tiller 1983).

During the late 17th century and the early 18th century three seminal events occurred that would greatly influence the Jicarilla's history. First, the Apaches' acquired the Spanish pony from the Spanish-controlled pueblos in the mid-17th century. Before that time they used dogs as beasts of burden. The dogs dragged loads on primitive vehicles consisting of two poles with a small mounted platform, but they were not very effective and their range was

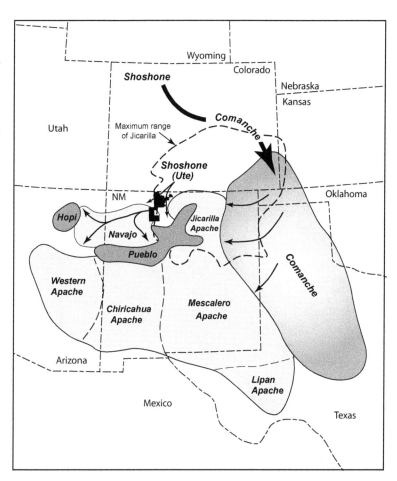

Figure 175. Indian migration, 1680–mid-1700s. (Modified from *Time-Life*, 1993, 1995.)

quite limited. Furthermore, dogs competed with the people for hard-earned meat. On the other hand horses ate grass, and horses could carry a man on its back. The horse, which the Indians compared to their dogs by calling them Mystery Dogs or Big Dogs, thoroughly revolutionized the life of the Apaches. Mobility allowed the people to efficiently move with the herds and to kill larger number of animals. More meat translated to reduced reliance on agriculture, a more nomadic lifestyle, and increased population. After the Pueblo Revolt of 1680 the Apaches obtained hundreds, if not thousands of horses left behind as the Spanish fled south. As mounted warriors the Apaches became a formidable adversary and controlled the southern plains, the territory the Spanish referred to as *Apacheria*.

Second, by the early 18th century other Plains tribes began to acquire horses as well. A group of Shoshone people migrated onto the plains from the Rocky Mountain valleys as competitors to hunt the huge bison herds. These people would eventually be called the *Comanche* (Figure 175). Superb horsemen, they learned how to shoot their arrows from the back of a horse, much like the mounted nomadic steppe warriors who terrorized the civilized world of Europe and west Asia a thousand years ago. In contrast to the Comanche, the Apache people dismounted before shooting and were therefore not as effective as warriors or hunters. Expansion of the Comanche-dominated high plains put pressure on the Apache to the west. By 1718–1719 the Jicarilla were in full retreat from their traditional hunting grounds. A number of bands merged to form the modern Jicarilla tribe. The variable nature of the terrain they occupied forced a social subdivision of the Jicarilla into two entities: the plains people, the *Llaneros*, and the mountain-valley people, the *Olleros*. The former occupied the region of the present Mora, San Miguel, and Colfax Counties, especially the Cimarron Valley. The latter occupied the upper Rio Grande Valley and its tributaries, i.e., the Abiquiu and Tierra Amarilla area (Figure 176). The two groups however shared culture, language, and religion.

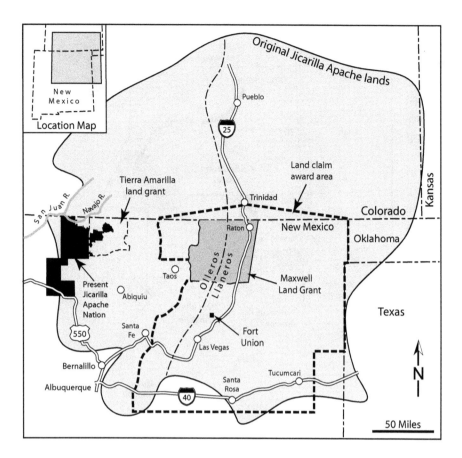

Figure 176. The Jicarilla people, ca. 1700–present.
(Modern geographic features shown for reference; modified from Nordhause 1995.)

Third, the plains tribes acquired firearms and ammunition from the French, who were exploring and trading in the upper Mississippi Valley, and some fell into the hands of the Comanche. The Jicarilla were forbidden by the Spanish to acquire firearms, and therefore the Jicarilla became sandwiched between the Spanish to the south and their arch-enemies the Comanche to the east. Thus began the Spanish practice of listing the Jicarilla as "friendly Indians," and using them as auxiliaries and pawns.

All during the mid-18th century the Spanish struggled with the Comanches. Finally, in 1786 the Comanches sued for peace, and thus began a 35-year period of relative tranquility. The Jicarilla remained on good terms with the Spanish due to their long mutual association. Many Spanish words entered the Jicarilla vocabulary, pueblo customs entered the Jicarilla culture, and—however slightly—Christianity made some inroads in Jicarilla life.

Mexican period, 1821–1846

During the Mexican period Jicarilla/Mexican relations deteriorated and Jicarilla raids on settlements increased. The Mexican government had granted land carved out of Jicarilla territory to Americans in an attempt to promote peace and economic prosperity along the northern frontier. In 1841 the 1.7 million-acre Beaubien and Miranda Grant was issued in northeastern New Mexico, and in 1845 the grant was purchased by Lucien Maxwell (Figure 176). The Jicarilla were allowed to remain but increasing Anglo intrusion led to strife.

American period, 1846–present

From the outset of the American occupation the native people had to deal with a new mind-set quite foreign and very dangerous to them. First, the American government subscribed to the concept of Manifest Destiny, a tenet that supported domination of the continent to the Pacific Coast. Second, the philosophy of Social Darwinism gave justification to the vanquishing of the weak by the strong. The Indians never really had a chance.

In 1850 the American military built Ft. Union on Llanero land (Figure 176), and by 1855 incursions by whites increased. The limited resources of the land made the tribe dependent on food rations. In 1853 a group of 250 Jicarilla were settled on the Rio Puerco del Norte (on the north side of the Jémez Mountains), but a treaty with the governor of New Mexico was not honored by him and war again broke out. In 1854 the Jicarillas were defeated by the United States Army. During the Civil War the American military was preoccupied elsewhere and raids by Navajo and Mescalero Apaches intensified, but the Jicarillas were mainly peaceful during this time. In 1864 the Navajo and Mescalero Apache were both interned at Bosque Redondo at Ft. Sumner in east-central New Mexico. The government had originally wanted to include the Jicarilla with them.

In 1869 the British bought the Maxwell Land Grant (Figure 176) and the Llaneros were kicked out. The Jicarilla thus became a people without a home. Between 1872 and 1873 the U.S. tried to move the Jicarilla to Ft. Stanton, New Mexico, but the Jicarilla stiffly resisted. The year 1873 began a decade of uncertainty during which the tribe was forced to wander about over rugged country of uncertain ownership where they attempted to settle down and grow crops. The two separate bands centered around the Indian Agencies: the Ollero at Abiquiu, New Mexico, and the Llanero at Cimarron, New Mexico. In December of 1873 the two bands met together with American officials to consider a new homeland. An executive order was issued the next year to establish a reservation on the north side of the San Juan River and east of the Navajo Reservation, and the Jicarilla found the idea attractive (Figure 177A).

However, during the next few years the government made no effort to move the Jicarilla to the new reservation. It was no coincidence that at this same time the San Juan Mountains north of the proposed reservation were proving to be rich in minerals. In 1876 President Ulysses S. Grant voided the executive order of 1874 and returned the lands north of the San Juan River to the public domain, thus opening up the San Juan River Valley to Anglo settlement.

In 1878 an Act of Congress was passed to force the Jicarilla south to the Mescalero Apache reservation, but this time the Jicarilla dragged their feet. Arable land on the Mescalero reservation was scarce and the remaining land was more suitable for grazing. The Jicarilla were spread out, and rounding them up for removal to Mescalero land was like herding cats. However, later that year a caravan of Jicarilla under military escort began the journey from Abiquiu to the Mescalero reservation, but many people began slipping away and returning to their homes. Only 33 arrived at their destination. The effort was a dismal failure.

In early 1880 a delegation of Jicarilla traveled to Washington 1880 to petition for a reservation in northern New Mexico, and found sympathetic ears. Later that year the 1878 act mentioned above was repealed and a new reservation proposed astride the Continental Divide (Figure 177A). Before they could move to their new home the bands began to quarrel among themselves. The Llaneros wished to return to the Cimarron area and hunt buffalo, even though the herd no longer existed, but they were convinced that they could live better off the new reservation than on it. In 1882 the Llaneros slipped away and returned to their former home, but they were soon forced to return.

At about this time politics began to raise its head in earnest. A group of "interested parties," including the New Mexico governor, complained to the Indian Office and urged that the proposed reservation be open to settlement and that the Jicarilla be moved to Indian Territory (Oklahoma). At the same time the Indian Office was being petitioned by people in southern New Mexico to move the Mescalero Apaches out of their area. The Indian Office was not in favor of abolishing the new Jicarilla reservation, and so they ignored the northern complainers, but they did listen to

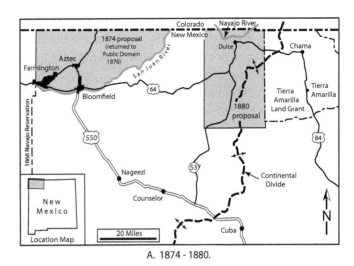

A. 1874 - 1880.

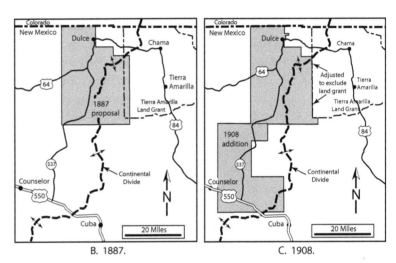

B. 1887.   C. 1908.

Figure 177. Jicarilla Apache Reservation, 1874–1908.
(Modified from Tiller 1983.)

239

the southerners and entertained the notion of moving the Mescalero north to join the Jicarilla. However the Jicarilla and Mescalero were not on good terms and the proposal was dropped.

Most of the new reservation occupied the high country astride the Continental Divide, and with the exception of a strip along the Navajo River there was little arable land. Accordingly the Indian Office worried that the Jicarilla would not be able to support themselves. It was therefore decided that the Jicarilla would after all be moved to the Mescalero Reservation, contingent on them living on land apart from the Mescalero. In August 1883 the Jicarilla bitterly began their southward 350-mile "Trail of Tears."

At Mescalero the Jicarilla gave farming life a try, but the land, as mentioned, was just more suited for grazing than agriculture, and the best arable land had already be assigned to the Mescalero Apaches. The rift between the Olleros and the Llaneros deepened as competition for the remaining arable land intensified. Things got so bad that in early 1886 the Olleros argued that they be allowed to return to the Rio Arriba area and homestead on the Public Domain like any other citizens. While the bureaucracy dabbled, several groups of Jicarilla escaped and made their way back north.

In 1884, the 1880 reservation had been returned to the Public Domain but closed to homesteading. During the three years that the Jicarilla were on the Mescalero Reservation many squatters had claimed land on the new reservation, but did not file with the Land Office because the reservation had not yet been surveyed. In 1886, when it was announced that the Jicarilla would return, the squatters arrived in a flood to preempt the best arable lands. It was politically impossible for the government to forcibly evict them. Meanwhile about 25 settlers had in good faith claimed about 4,000 acres of prime agricultural land. If at all possible their claims had to be respected. To make matters worse, a group of wealthy New Mexican ranchers controlled much of the grazing land and resisted the return of the Jicarilla. The ranchers ran some 10,000 head of cattle on the reservation and stripped the grass cover. The situation was truly a mess. The government realized that something had to be done about the settlers and ranchers if the Jicarilla were to make a living on the reservation.

But then the tide turned in favor of the Jicarilla. On February 11, 1887, an Executive Order signed by President Grover Cleveland formally created the slightly modified, 416,000-acre Jicarilla reservation with boundaries modified only slightly from those of 1880 (Figure 177B), and the next month the Jicarilla left the Mescalero lands for good. They finally had a permanent home, but they were yet not free from problems.

It quickly became even clearer that the reservation was definitely not suited for agriculture, but it did have two other resources: timber and grass. A plan was advanced for the Jicarilla to waive the rights to the timber, which would be cut and sold, the proceeds going into a fund to provide the funds to purchase livestock. Due to the high elevation and harsh climate of the reservation, as early as 1896 it was suggested that a southern range would be required for year-round grazing of the livestock. This came to pass in 1908 when a southern area of lower elevation and less rainfall, covering 15 townships (about 350,000 acres), was added to the reservation by Executive Order. The reservation henceforth consisted of a northern range for summer grazing and a southern range for use in winter (Figure 177C).

Logging began in 1908–1909 and a sawmill began operation in 1910 to produce railroad ties for the Rio Grande and Southwestern Railroad, which was building a spur to the reservation to move the timber out. However, the proceeds were not turned over to the tribe but rather impounded by the government, ostensibly for the benefit of the Indians. Some of the funds were used to purchase a herd of sheep and cattle that remained under government control. The southern addition was leased for a pittance to non-Indian ranchers, with the proviso that the lessees develop the water resources. Instead the ranchers severely overgrazed the range and degraded its value. Funds accumulated in the federal treasury while the Jicarilla wallowed in poverty. Not until 1920 was the livestock distributed to the Indians. The Jicarilla were in business at last, although the federal agencies continued to siphon off funds from timber sales to finance government overhead. By the mid-1930s the timber had all been harvested.

In 1934 Congress passed the Indian Reorganization Act (IRA). It was enacted in part to improve the economic condition of American Indians by protecting their natural resources and increasing their land base, and to give the Indians a heightened level of political autonomy over their own affairs. The Indians would hold their reservations by charter. Ironically the Jicarilla vehemently opposed the act. Many of them held allotments, to which they were to receive patented titles as per the Dawes Act of 1887 (more about this later). The Jicarilla feared that they could easily be deprived of ownership of the community lands on the reservation given to them via the Executive Orders of 1887 and 1908, whereas patented lands were guaranteed. In short they justifiably feared that "what one lawyer givith, another would taketh away."

During the next several years the government bought out most of the privately-held arable land on the reservation. Gradually the Jicarilla came around and realized that acceptance of the IRA was in their best interest, and in 1937 they overwhelmingly approved the Jicarilla Apache Constitution and Bylaws. The reservation was divided into six representative districts, all of which were in the northern part of the reservation because that was where the Indians had their permanent homes.

One of the first orders of business was to purchase the trading post and buildings in Dulce that had been run and owned by Emmet Wirt, an extremely influential figure in the area since 1889. The store was renamed the Jicarilla Apache Cooperative Store. A single branch store was set up, the Otero Store in the southern addition, about four miles northeast of the ranch of Mariano S. Otero, Jr. The main store at Dulce became the center of Jicarilla economic life. The southern addition continued to be communally owned by the tribe. Livestock numbers increased as did the Jicarilla population. The tribal constitution recognized the existence of a common, inalienable land base, the communal ownership of all natural resources, a sound federal trustee relationship, and, at least until the 1940s, a stable agrarian way of life.

During the post-war recession the demand for agricultural products declined and individual income began to decrease. At the same time mineral exploration on the southern addition began, slowly at first, and became outright exuberant in the 1950s. Income from oil and gas increased about tenfold before the end of the decade. A severe drought in 1950 crippled the Jicarilla livestock industry and many stockmen gave up and moved to Dulce where the Indian Agency was located. The population of the tribal capital swelled.

With increasing tribal income from oil and natural gas production and declining personal income from agriculture, the tribe in 1952 began per-capita cash payments as a social stabilizer, and this allotment became the principal source of income for most Jicarillas. However, the cash payments had the opposite effect, and proved to be socially destructive because it undermined the willingness of some Jicarilla to work.

At the same time, mismanagement at the cooperative store led to a hemorrhage of funds due to the overextension of credit to a small number of scalawags. Recognizing the problem, in the mid-1950s the Tribal Council agreed to have the store run by private management on a cash-only basis. As people moved to Dulce they found inadequate housing, little employment, and few community services. The per-capita payments often were the people's only source of sustenance. By 1955 the population of Dulce reached about 1,000 and included the vast majority of the tribe's members.

In 1957 the Council approved a contract with the Stanford Research Institute (SRI) of Menlo Park, California, to make recommendations for long range economic and social development on the reservation. The objective was to increase tribal and individual income via gainful employment, and to replace reliance on depletable oil and natural-gas resources and per-capita cash payments. In 1958 the tribe began a five-year program to phase out the allotments while at the same time embarking on an economic development effort. In 1960 the tribe adopted a new constitution which gave more autonomy to the Tribal Council and reduced that of the BIA. The constitution was amended several times until by 1968 it took on a form modeled after the U.S. Constitution.

In the early 1960s the Master Plan prepared by SRI to develop the Dulce community was implemented. New housing was constructed, modern water and sewage systems were installed, and Southern Union Gas Co. constructed a natural-gas line to the town. At this same time the New Mexico Highway Department constructed 32 miles of fence along NM-44. Throughout the 1960s the Master Plan focused on employment, outdoor recreation, and tourism to replace revenue from diminishing oil and gas production. By the late 1960s annual tribal income was over $1 million.

The tribe made an unusual investment in 1970 when it produced a movie, shot in part on the reservation, called *Gunfight*. The film told the tale of two over-the-hill gunfighters (staring Kirk Douglass and Johnny Cash in his screen debut) who sell tickets for one more gunfight to take place in a bull ring. The tribe shared in the net profits and the music and television rights. Also in 1970 the Jicarilla leased about 400,000 acres of land to oil and gas companies, mainly in the "southern addition," and shortly afterwards El Paso Natural Gas Co. and Northwest Pipeline Co. constructed a 500-mile gathering pipeline system.

In 1972 a long-standing dispute involving the Jicarilla Apache land claim came to a conclusion. This monumental event had its roots in the Indian Claims Commission Act of 1946. The act's motive was to settle Indian claims against the U.S. so that the government could with a clear conscience terminate the guardianship relationship with the tribes. The Jicarilla made their initial claim in 1948 and an amended claim in 1950 for 14 million acres in northeastern New Mexico and southeastern Colorado. The first hearings began in 1958. By 1963 it was determined that the Jicarilla Apache had title by virtue of long use to certain lands, that the U.S. had taken these lands without consent of the tribe, that the U.S. had not compensated the tribe for the taking of these lands, and that the area encompassed by this claim (less Spanish and Mexican land grants made before 1848) amounted to about nine million acres (Figure 176). It was determined that the value of this land in 1883 (the year the bison herd disappeared) was almost $10 million. Less costs and other deductions, the tribe finally in 1972 received a cash award of $9,150,000 after a quarter of a century of litigation (Nordhaus 1995). About 50% of the settlement was used for investment and capital improvement and the other 50% was designated for per-capita payments (Tiller 1983).

In 1976, for the first time the tribe entered into a a joint oil and gas-drilling venture with Palmer Oil Company of Billings, MT, resulting in seven success-ful wells. The tribe then instituted a tribal severance tax (a tax to remove or "sever" a natural resource from the ground) in addition to the negotiated royalties (a percentage share of the production). The next year the Jicarilla bought out Palmer's interest and thus became the first tribe in the U.S. to own and operate oil and gas wells on its reservation (Nordhaus 1995).

In the mid-1980s the Jicarilla began a serious effort to enlarge their land base by acquiring new land with undisputed title. By the mid-1990s the tribe had ac-quired about 137,000 acres to the north-east and east of the northern sector of the reservation (Figure 178). Several of these acquisitions were ranches that had fallen into bankruptcy and had ceased to pay property taxes.

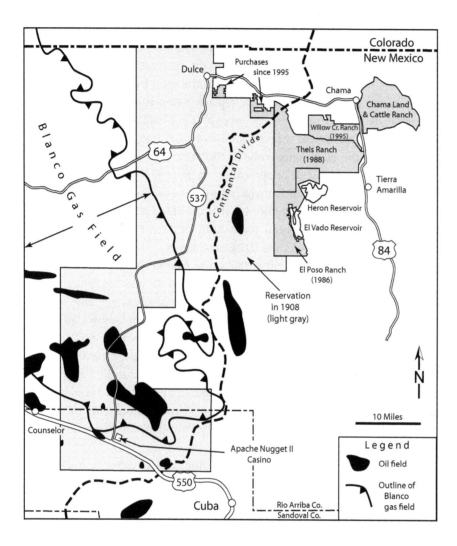

Figure 178. Jicarilla Apache Nation (shades of gray), 1980–present.
(From Jicarilla Apache Nation website.)

24

Indian Country II—the Navajo Nation (*Diné Bikeyah*)

The Navajo word *diné* (pronounced DIN-eh) denotes "the people." The Navajo Nation, or *Diné Bikeyah* ("Land of the People") is the largest Indian reservation in the United States. (The term *Dinétah*, on the other hand, used in Chapter 21 and below, refers to the ancestral Navajo homeland in the Gobernador/Canyon Largo country). It encompasses about 27,400 square miles (about 17.5 million acres) in New Mexico, Arizona, and Utah, and is larger than ten of the 50 states in the U.S. (Figure 179A). Tribal members number almost 300,000, about 107,000 of whom live in northwestern New Mexico (2000 U.S. Census).

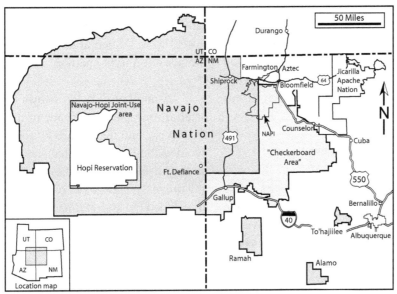

A. Present Navajo Nation and Checkerboard Area (light gray).

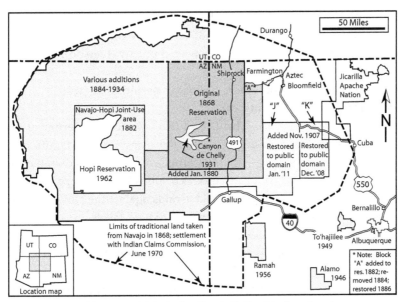

B. Additions and deletions to Navajo Nation (shades of gray).

Figure 179. Navajo Nation. (Modified from Goodman 1982.)

The word "Navajo" has an interesting genesis. It was first used by the Spanish in the early 17th century to refer to the homeland of a group of Apaches living in northwestern New Mexico. A Spanish friar, Fray Alonso Benevides, referring to these people in 1630 wrote [in Spanish]: "these of Navajò are very great farmers, for that is what 'Navajò' signifies—'great planted fields'" The Tewa-speaking pueblo Indians of northern New Mexico claim that their word *navahú* means "the large area of cultivated lands." The Tewa noun *náva* means cultivated field, and *húu* denotes canyon, so the combined word designates a canyon with cultivated fields (Reeve 1946). The English spelling of the word is "Navaho," but the Spanish, "Navajo," is generally used today. A region characterized by "canyons with cultivated fields" is clearly the ancestral Navajo homeland in the area of Gobernador Canyon and Canyon Largo and the present-day Navajo Reservoir. The Navajo people call this place *Dinétah*, which means "homeland."

Pre-Spanish and Spanish periods to 1821

The Dinétah was occupied by the ancestors of the Navajo at least as early as about 1500, although some sources claim an earlier date. When the Spanish arrived to stay in 1598 they founded a colony, *San Gabriel del Yunque*, at the site of a former pueblo, on the west bank of the Rio Grande opposite Ohkay Owingeh (formerly San Juan) Pueblo. The Navajo considered this the eastern edge of their turf. They accordingly raided and generally harassed the Spanish until by 1610 the Spanish were forced to relocate the colony to present-day Santa Fe.

By the early 17th century the Spanish recognized the Navajo as a group distinct from the Apaches. The Spanish at this time referred to two Navajo bands. One occupied land west and northwest of Jémez Pueblo and south of the San Juan River. This was the Province of Navajo. The second band occupied land in western New Mexico in the vicinity of Mt. Taylor (Kelly 1968).

The next 70 years were marked by periods of peace interrupted by periods of incursions and insidious slave raids by both the Navajo and the Spanish. In 1680 the pueblos rose up and drove the colonists out of New Mexico. Fearing the return of the Spanish, many of the pueblo people fled westward from their homes along the Rio Grande and its tributaries and mingled as refugees with the Navajo in the Dinétah. The two cultures fused and the Navajo adopted the puebloan practices of weaving, pottery, masonry architecture, and farming (Linford 2000).

During the chaos following the uprising of 1680, the Navajo (and pueblos) acquired a large number of stray sheep and horses abandoned by the Spanish and began their gradual shift to a pastoral economy. The Spanish returned in 1692 and quickly overwhelmed the pueblos. A new wave of refugees fled west to join the Navajo. The Spanish resumed their slave raids on the Navajo and the Navajo reacted in kind until about 1710. During this time the Navajo built fortified pueblitos up in the Largo and Gobernador canyons to protect themselves from the Spaniards as well as from marauding bands of their traditional enemies the Utes (Chapter 21). These *casas fuertes*, or pueblitos, were positioned on mesa tops overlooking the canyon floor where their crops were grown. Observation towers were located at strategic line-of-sight positions to provide a connected network of observation points (Reeve 1956).

A number of bold Spanish raids deep into the Dinétah during the first decade of the 17th century stunned the Navajo, and for the next 50 years the Navajo generally avoided the Spanish settlements and focused their raids on the pueblos. Increased pressure from the Utes eventually drove the Navajo out of the Dinétah. By 1760 they had abandoned the area and had moved west (Roessel 1988).

Almost continual contact with pueblo people up to this time gave the Navajo a distinct culture. The Navajo became less nomadic, less dependent on hunting, and adept at weaving. Contact with the Spanish since the mid-1600s gave them sheep and horses. Thus a new Navajo world emerged centered on domestic animals. To augment their herds the Navajo became raiders vs. warriors. By the late 18th century livestock had become an essential block of the Navajo economy (Iverson 1981).

Mexican period, 1821–1846

The Mexican government encouraged commerce with the Americans. Manufactured goods, including firearms, were brought down the Santa Fe Trail into New Mexico. Guns increased the ghastly efficiency of Mexican slave raids against the Diné and tensions gradually ratcheted up to peak levels.

American period, 1846–1868

By the beginning of the American period in 1846, the Navajo frontier was in a general state of war. For the next 14 years the frontier was marked by frequent and nasty skirmishes, by the Navajo for livestock, and by New Mexican "volunteers" for slaves. The year 1860 began an eight-year period the Navajo ruefully call "being chased time." It was marked by fear, despair, and exile. In early 1861 the outbreak of the Civil War resulted in the removal of regular American troops from New Mexico Territory, and their replacement by the same New Mexican volunteers who had plagued the Navajo with their slave raids. In the latter part of 1861 all-out war between the Diné and the whites broke out (Linford 2000).

In 1863 Colonel Christopher "Kit" Carson was ordered to totally subjugate or exterminate the Navajo. Carson, somewhat reluctantly, eliminated the Navajo's ability to survive by killing livestock and destroying homes and crops. In early 1864 the Navajo, starving and freezing, surrendered. For a year, beginning February 1864, groups of Navajo were marched from western New Mexico to Bosque Redondo near Ft. Sumner on the Pecos River in eastern New Mexico. Some 8,500 finally arrived and perhaps about 1,000 died on the march. Several thousand Navajo refused to join the march and hid out in the wilds wherever they could, out of sight of white intruders (Kelly 1968, 1970).

The Navajo at Bosque Redondo, along with about 400 Apaches, were supposed to be taught how to settle down and become productive farmers. The combination of poor soil, small plots, shabby treatment by their military guards, and harassment by the ever-present Comanche caused the experiment to fail miserably. By 1866 even the U.S. military realized that the experiment had been a huge mistake. In early 1867, control of the Navajo was taken from the army and placed under the Indian Service. In mid-1868, at long last, a treaty was negotiated with the Navajo for a reservation in the Four Corners area. A few weeks later a long column of Indians left the hated Bosque Redondo and headed northwest to their new home.

Navajo Reservation of 1868 and later additions in New Mexico

The Navajo Indian Reservation was created in 1868. The rectangular block, defined by lines of latitude and longitude, astride the north-south line separating the Arizona and New Mexico territories, encompassed about 3.4 million acres—less than 10% of their former range (Figure 179B). This area soon proved to be woefully inadequate in size because it was generally barren and useless. The government purposely had made the reservation small and compact to encourage the Navajo to give up the nomadic

life and to adopt an agricultural and pastoral economy. The government had especially encouraged the raising of sheep, but such an enterprise required that the flocks be moved to usable range and water. The best land was the well-watered area along the San Juan River to the north and along its tributaries to the east, outside of the reservation, and also coveted by white settlers. Therefore the Navajo tended to drive their flocks to these areas.

Before the late 1870s the Navajo had grazed their livestock in the lush San Juan River valley bottomland despite being harassed by their traditional enemy, the Utes. In 1876 an Executive Order restored a huge swath of land north of the San Juan River, which had been proposed for a Jicarilla Indian reservation, back to the public domain (Figure 177A). This opened the area up to Anglo settlement. Beginning in 1877, settlers began to trickle into the Animas and San Juan River valleys. By the end of the decade the irrigatable bottomland between the Navajo reservation and the mouth of Canyon Largo was occupied by non-Indians. Conflict between the Navajo and the settlers soon flared up over grazing rights in the valley east of the reservation near Farmington and Bloomfield. The Navajo considered the region not only their traditional right but a downright economic necessity. In essence, the bone of contention between the Navajo and non-Indians was not land, but water.

In January 1880 the reservation south of the San Juan River was extended 15 miles to the east (Figure 179B). This included a strip of land, the portion of Township T29N R14W lying on the south side of the river ("A" in Figure 179B). Intense lobbying by settlers along the river resulted in this strip being returned to the public domain in May 1884. However, without access to the waters of the San Juan River, the land in the southern part of T29N R14W was worthless. The government belatedly realized this and in April 1886 restored the strip to the reservation (Reeve 1946). All this controversy indelibly established the ironclad bond existing between land and its water.

Checkerboard Area

The so-called "Checkerboard Area" is a huge swath of territory flanking the Navajo Nation on its east and southeast sides (Figures 179A and 180). Today about two-thirds of this area is Navajo land. Before their exile to Bosque Redondo many Navajo had lived in this area. When they left the Bosque Redondo they had no idea where the new reservation was so many returned to their old familiar haunts (Kelly 1968).

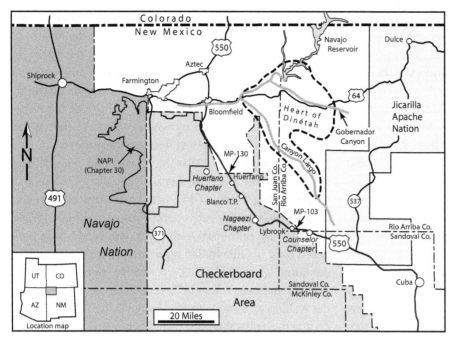

Figure 180. Checkerboard Area and vicinity.
(Modified from BLM 1994.)

Many of the Navajo holdings are only a quarter mile square (160 acres) in size and are non-contiguous, hence the name. The pattern is in part the legacy from the deal struck between the federal government and the railroads. The infamous Dawes Severalty Act of 1887, sometimes known as the Indian Allotment Act, was not generally applied to the Navajo area due to the poor quality of the land and the existence of only one living stream—the San Juan River. The act provided for reservation lands to be allotted to individual Indians in 160-acre parcels (generally clustered by family) under the tenets of the Homestead Act of 1862. It was an effort to assimilate them into the U.S. population as "responsible farmers" and to protect the traditional use areas from non-Indian stockmen (Linford 2000). However, the reality on the ground is that a minimum herd of 250 to 400 head of livestock are required to supply a family of five and the arid land allocated to the tribe could not support anything near that number.

By about 1885 the carrying capacity of the land had been reached. To make matters much worse, after about 1890 the railroad and non-Indian settlers had laid claim to all the arable land surrounding the reservation, and the Navajo felt compelled to sell wool to augment their income. The expanding market for wool led to the establishment of trading posts by the 1890s to handle the new commodity. The growth of the Navajo economy was limited, however, especially in the off-reservation Checkerboard Area to the east. Large-scale livestock operators there gained control of huge regions of range by buying or leasing sections with water and refusing the Navajo access (Kelley 1977). The Navajo were pinned in. This situation persisted fully until after the Great Depression when stock-raising became less profitable.

After 1900 artesian water was discovered in the Checkerboard Area. The main and shallowest aquifer was the Ojo Alamo Sandstone. With the drilling of wells and installation of windmills and water-collection tanks hauled in by railroad, the once worthless terrain became very attractive as grazing land. The railroad land companies began leasing the land to Anglo cattlemen, and with these leases in hand the cattlemen had a vested interest in these lands. The non-reservation Navajo living in the area became alarmed and requested that this region be added to the reservation by presidential executive order.

In November 1907 President Theodore Roosevelt, a friend of the Navajo, issued two Executive Orders to restore much of the tribe's traditional habitat east of the reservation into the reservation proper as the two executive-order reservations, "J" and "K" (Figure 179B). This region was collectively called the "Eastern Extension of the Navajo Reservation." The intent was to prevent Anglo entry into those areas inhabited by these non-reservation Navajos for a period of time, during which the Indians would be given allotments from the public domain under a section of the Dawes Act. After the allotment process the remainder of the land would be restored to the public domain for Anglo entry (Kelly 1968).

A gigantic political outcry from the powers-that-be pressured the president to rethink his orders, but he refused to back down. He recognized that the Eastern Extension is a desert country and cannot be used effectively in small tracts. To return it to the public domain and make it free open range for Anglo ranchers' livestock would be a cruel injustice to the Navajo, who needed it all (Counselor and Counselor 1954). Anglo opposition to expansion of the reservation was relentless, and pressure was then brought to bear on the succeeding Taft administration. Block "K" was returned to the public domain in late 1908.

That same year the northeastern part of "K" became the southwestern corner of the Jicarilla Apache Reservation. In 1911, with New Mexico on the verge of statehood, the territory's delegate to Congress had acquired considerable influence and thus block "J" too was restored to the public domain (Iverson 2002). By 1918 Congress had had enough of executive-order reservations and mandated that all additions to reservations be authorized only by Congress. The era of expansion of the Navajo reservation had come to an end (Kelly 1968).

When New Mexico became a state in January 1912, school lands had been carved out of the public domain via two acts of Congress. This further aggravated the checkerboard pattern of land jurisdiction in the region. The Ferguson Act of 1898 had stipulated that sections 16 and 36 of every township be granted for the benefit of the "common schools" (now called the "public schools"). If these sections were mineral lands or lands already granted for mining or homesteads, the territory would make alternative choices. As the political gears turned slowly in the direction of statehood, President Taft signed the Enabling Act of 1910, which provided for a constitutional convention and granted the additional sections 2 and 32 in each township to supplement support for the schools. Thus, four sections in each township (commonly colored blue on land-ownership maps) are today school lands administered by the State Land Office.

In 1946 Congress passed the Indian Claims Commission Act. This legislation provided for monetary compensation of legitimate Indian claims against the U.S. for lands taken from the tribes in order to create the reservations. (In the previous chapter we saw how this issue was adjudicated for the Jicarilla Apache Nation.) The Navajo tribe filed a claim for 23 million acres of traditional homeland taken from them in 1868 when the tribe was relocated to the reservation. In June 1970 the U.S. compensated the tribe for the taking of about 12.3 million acres of ancestral lands, all of which lies in the Checkerboard Area (Figures 179B and 180).

US-550 traverses the Checkerboard Area from the tiny community of Lybrook at MP-103 to MP-130 (Figure 180). Within this domain there are no towns, and the only visible human settlement consists of scattered Navajo dwellings. This pattern of habitation—so different from that of the pueblos—are the products of the checkerboard ownership, the ubiquitous pick-up truck, the scattered distribution of trading posts, and the low carrying capacity of the range.

Navajo Chapters

The Navajo Nation is divided into political entities called agencies, and the agencies are in turn subdivided into chapters. The chapter system was begun in 1927. The Navajo word for chapter means "Group of Three," referring to the chapter's three officers. The chapters are roughly equivalent to counties, and today the chapters number 110. Each chapter has a meeting place—the Chapter House—where local issues and problems are discussed. The members elect delegates to represent them in the Tribal Council at the Navajo capital at Window Rock, Arizona. The format is similar to that of a town-hall meeting, and discussions are conducted in Navajo. New chapters can be formed when a single chapter's membership exceeds 1,000.

The Checkerboard Area along US-550, part of the Eastern Navajo Agency, contains three Navajo chapters (Figure 180). The Counselor Chapter House is located across from the Counselor trading post on the south side of the highway, and the Nageezi Chapter House is behind the trading post on the southwest side of the highway. The Huerfano Chapter House, the site of the former Carson Trading Post founded about 1917, is located 7.5 miles west of US-550 on route N-44.

25

Side-Journey down to Chaco Canyon

In this book I refer to the gap through Crow Mesa at US-550's crest near MP-104 as the "Lybrook Gate." It is a gate in several senses. First, geologically, from this point westward we exit the "inner" San Juan basin behind us and penetrate the "outer" basin (Figures 162 and 181). Second, hydrologically, this is the divide between the Canyon Largo and Blanco Wash drainage basins (Figure 170). The road bed through the gate is only about 64 feet lower than that over the Continental Divide (Figure 13B). From here on it's generally downhill to Bloomfield and the valley of the San Juan River. Third, visually, the gate is the divide between two major viewsheds. The broad back of the Sierra Nacimiento suddenly disappears behind us as we enter the viewshed of Huerfano Mountain. We are just entered a new visual world (Figure 182).

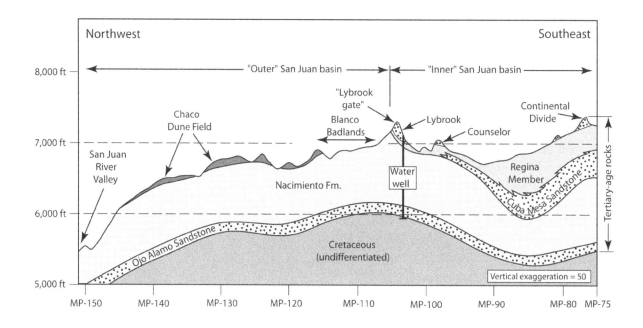

Figure 181. Schematic geologic profile along the western segment of US-550.

Along this reach the highway very closely follows the divides between a number of individual drainage sub-basins of the San Juan River basin (of which the Blanco Wash subbasin is one). For ease of movement the early travelers kept to the high ground along the divides. Later the highway engineers did the same thing to avoid road-washout problems and excessive construction costs. The course of the early NM-55, and its successor NM-44, was preordained by the landscape (Figure 183).

West of the Lybrook Gate, the most important place easily accessible from US-550 is clearly Chaco Canyon and its fabulous ruins. Therefore I'll dedicate the rest of this chapter to this exquisite place, and flesh out some of its geologic story in the next (Chapter 26). But first, let's deal with an annoying little detail—the mileage bust, and tell the short story of another trading post.

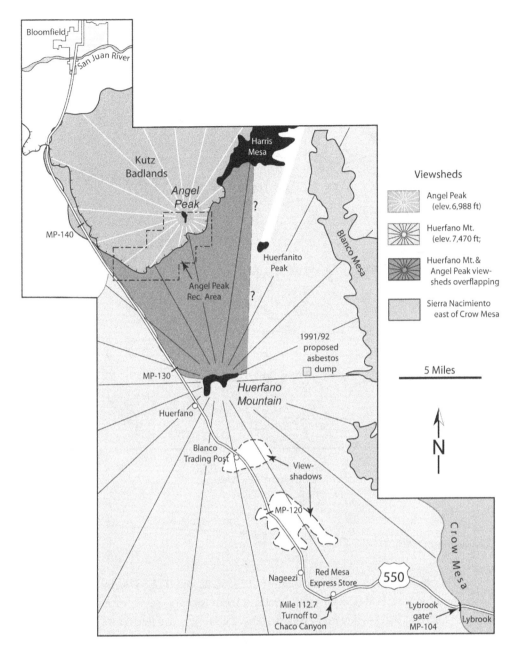

Figure 182. Major viewsheds.

Mileage bust

One of the best ways to study a highway and to organize its scenic features is to construct a detailed road log. Such a log should be geared to your vehicle's odometer, and the points of interest on the ground can then be accurately positioned on the log. I made one in the course of researching this book. I carefully plotted the location of the official highway mileage signs (mile posts or MPs) on the USGS topographic maps, and spot-checked only a few select ones. The distance between MPs of course should be exactly one mile, right? Most of them are, but one is way off, and it took me seemingly forever to locate it. It seems that the distance between MP-104 (at the crest of the hill in the Lybrook Gate) and MP-105 is 1.3 miles. However, MPs are not about to be moved about or corrected any time soon because they're at points fixed by the NMSHTD. So there they remain. This entire narrative therefore keys off the official MP signs as they exist on the ground, despite this annoying glitch.

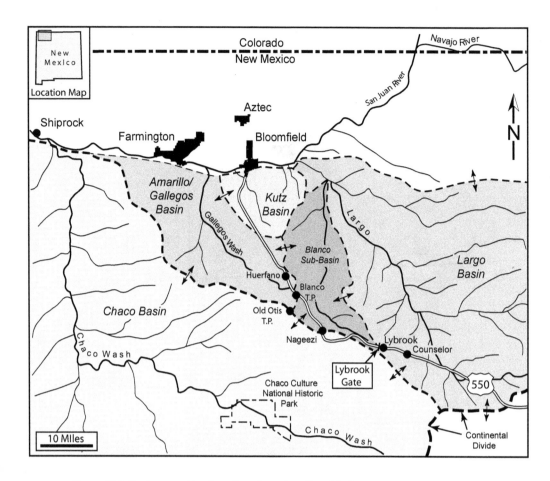

Figure 183. Drainage subbasins of northwestern New Mexico.

Red Mesa Express store

This store located on the north side of US-550 at mile 112.5 was formerly called the Forty Four Store. It was established by Woodrow Stiles in the early 1960s. It went through several hands until the Thriftway convenience store chain bought it about 1980 (Kelley and Francis 2006). Today it is the Red Mesa Express store and it provides gas and supplies for visitors to Chaco Canyon, turnoff located 0.2 mile farther down the highway on the south.

Chaco Canyon

One of the very most important places in New Mexico is the Chaco Canyon Culture National Historical Park. Access to this special place is from mile 112.7 a short distance beyond the Lybrook Gate at the intersection of US-550 and CR-7900 (Figure 184). The park is the crown jewel of the San Juan basin (if we discount the mineral wealth). The literature about this special place is enormous and well beyond the scope of this book. I'll instead focus on a few select geologic subjects.

The word "chaco" is probably derived from the Navajo word, *Tségai*, meaning "White Rock," or "Home" to many Navajos who consider the canyon their ancestral home. The word was later Hispanicized to both "Chaca" and "Chacra" (Linford 2000). Miera y Pacheco's 1776 map refers to the Chaco Canyon area as "Chaca," but by the middle 19th century the name had become "Chaco." Today the sandstone ridge bordering the north flank of Chaco Canyon is variously shown on maps as Chaco Mesa, Chaca Mesa, or Chacra Mesa. We'll use the former.

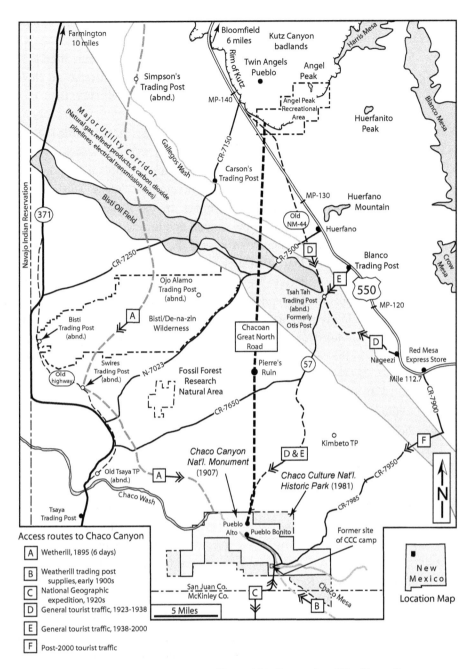

Figure 184. Geography of the Chaco Canyon area.
(Modern geographic features shown for reference.)

Early accounts of Chaco Canyon

The south-facing Chaco Mesa formed the northern rim of a natural military corridor to the San Juan country. In 1823 José Antonio Viscarra led a military expedition to Chaco Canyon. His official records include numerous Spanish place-names that evidently were not being used for the first time (Brugge 2004). After the Mexican War, the exploration and development of the West became items of high importance to the U.S. government. In 1849 Colonel John M. Washington led his expedition from Santa Fe to the San Juan country. He followed a course essentially northwest from the Rio Salado Pass directly to the canyon via the south face of Chaco Mesa. Lt. James H. Simpson, a member of the party (Chapter 7), documented his trek in an obscure 1852 report. He was enthralled by the ruins at Chaco Canyon. He assigned names to the sites suggested by his able Mexican and Indian guides, who apparently were quite familiar with the place. His report, however, attracted little attention.

In 1877 a government exploratory party, the U.S. Geological and Geographical Survey of the Territories, a.k.a. the Hayden Survey (Chapter 14), paid a visit to Chaco Canyon. The group's photographer was William H. Jackson, who fully expected to be the first to photograph the site. He hauled his heavy 8 x 10-inch camera up and down into every nook and cranny of the canyon and took almost 400 exposures. Later, when he processed the glass plates, he was horrified to learn that they were all blank due to his use of faulty materials. He went on to become famous as the "Pioneer Photographer of the West," but his reputation was certainly not based on his work at Chaco! In 1888 Victor Mindeleff of the Smithsonian Institution spent six weeks in the canyon and—unlike Jackson—successfully produced the first photographic record of the ruins (Lister 1984). Slowly the public took note and interest grew, but the canyon remained virtually inaccessible to all except the very intrepid.

Richard Wetherill—cowboy archeaologist

The name Richard Wetherill (1858–1910) is indelibly linked with the subject of Southwestern archeology and especially with Chaco Canyon. He lived an interesting life. In 1881 Richard's parents had moved the family of six kids from eastern Kansas to their new homestead in southwestern Colorado. There, along the Mancos River near the base of Mesa Verde, they established their Alamo Ranch. The family eked out a living by ranching and farming (Lister and Lister 1985). Probably due to their Quaker background they valued education and soon took an interest in the area's Indian ruins, especially the wonderful nearby cliff dwellings at Mesa Verde. They soon realized that the quest for antiquities and the guiding of the occasional tourist to the ruins could be turned into a profitable business. Richard, the eldest son, took the lead. Via these tourists, Richard had by the late 1880s developed a network of "clients" more than willing to purchase his finds of artifacts (Snead 2001).

In 1892, the Hyde brothers of New York, Talbot and Frederick, Jr., wealthy heirs to the Babbitt Best Soap Company fortune, stopped by the Alamo Ranch on an early stage of a world tour. After Richard took them up to see the cliff dwellings the dazzled brothers continued on their way. The following year, having returned home, the brothers attended the Columbian Exposition in Chicago, and renewed their acquaintance with Richard Wetherill, who was there on the prowl for clients (Snead 2001).

For a very long time collection of relics played a central role in the study of history. The simple reason—as any teacher would tell us—was that most people are best able to respond to concrete objects rather than to abstract ideas. During the 1880s and 1890s the American upper classes were enthusiastic collectors of art. This promoted the establishment of museums, whose mission was to provide venues for exhibiting such objects, but not to study or interpret them. The 1893 Columbian Exposition in Chicago focused this growing interest as never before. The exposition was to celebrate the 400th anniversary of Christopher Columbus' arrival in the New World, but it also coincided with the closing of the American frontier. This capsule-depiction of the 1893-America featured displays to champion advances in commerce, industry, and science (Snead 2001).

The Anthropological Building provided displays of ethnology and archeology. Many entities and organizations participated in the exposition, including individual southwestern states and territories. The state of Colorado hired Richard Wetherill to collect and display cliff-dweller relics for its exhibit. The displays of southwestern artifacts generated enormous interest in the attending public. The Southwest suddenly was elevated from an unknown, vague abstraction to a real, exotic destination (Snead 2001).

The Anthropological Building was run by Dr. Frederic Ward Putnam, of the American Museum of Natural History in New York. Putnam was bound and determined to establish archeology as a respected profession. The expanding field of archeology required a constant supply of artifacts, and that required the services of professional collectors. The problem was that the individual collector's objective was only to acquire artifacts for sale, and the historical context of the items was far down on his list of priorities. He needed only a shovel, and a wagon and team to haul off the finds. Soon, relic-hunting became an important economic activity in the southwestern states and territories, especially after the arrival of the railroad in 1879–1880. Meanwhile the archeologists bitterly lamented their lack of control over the resource (Snead 2001).

After their meeting at the Columbian Exposition, Richard Wetherill and the Hydes began to talk about the Hydes possibly funding a collecting expedition. The result was the First Hyde Exploring Expedition (HEE) in 1894 to the Grand Gulch in southeastern Utah. The Hydes, apparently not at all financially challenged, were impressed with the expedition's results (Snead 2001). The Hyde boys were hooked.

In 1895, Richard Wetherill caught wind of the ruins at Chaco Canyon and visited it for the first time with the Palmer family (an intrepid family of traveling musicians who had visited the Wetherill's ranch), which included his wife-to-be, Marietta Palmer. From the river crossing at Bloomfield they traveled south cross-country by wagon for six difficult days. At this time there were no roads, and they used Huerfano Mountain on their left as a guidepost (Figures 182 and 184). They stayed in the canyon about a month. During this time Richard became convinced that the ruins would provide "relicts in quantity" (Lister 1984). Chaco Canyon was located on public land and there were no restrictions against collecting. Next they traveled to Albuquerque from where Richard wrote a letter to the Hyde brothers.

By this time Talbot Hyde had become acquainted with Frederic Putnam of the American Museum. Putnam's need for collections of artifacts coincided with Talbot's interest and ability to finance another expedition, to be led by Richard Wetherill. The Hydes would provide the funds and Putnam, through his representative on the site, a young archeologist named George Pepper, would provide the intellectual control. In 1896 the Second Hyde Exploring Expedition (HEE) made its way into Chaco Canyon to commence excavations that would last through 1899 (Frazer 2005).

The Weatherills settle in

In 1897 Richard and his now-wife Marietta set out from Durango, Colorado with their household items on the way to their new home in Chaco Canyon. Once on the scene a degree of animosity rapidly developed between Wetherill and Pepper. Wetherill's ambition to be accepted as a full-fledged professional archeologist was continually and effectively thwarted by Putnam. The two sides represented two opposing philosophies—the collector vs. the archeologist.

Meanwhile the needs of a large number of Navajo and Hispanic laborers and for the members of the expedition required a steady stream of supplies. Teams of mules hauled things such as flour, coffee, sugar, and hardware from the rail centers of Albuquerque, Thoreau, and Farm-

ington. A typical wagon haul from Albuquerque through Cabezón normally took 12 days via the old trail along the south face of Chaco Mesa (southern part of Figure 184). Large volumes of feed, lumber, fruits, and vegetables came in from the Mormon communities along the San Juan River to the north.

During the first year, 1896, the precious and fragile artifacts were carefully hauled out by wagon over primitive roads to the narrow-gauge railhead in Durango, Colorado. An entire freight car filled with artifacts was taken to Alamosa, Colorado, where the material had to be transshipped onto standard-gauge stock for the long trip back east. In subsequent years the Hydes hauled the artifacts south over the dirt road to the standard-gauge Santa Fe railhead at Thoreau for more efficient transport back east.

The natural offshoot of supplying the expedition was trade with the Navajos, including the buying and raising of livestock (Schmedding 1951). Expedition members began purchasing woven items from the Navajo women and encouraged them to produce more. The Navajo men were given wage jobs and at first were paid via checks. When they encountered problems cashing them at nearby trading posts it became necessary to pay the men in groceries (Brugge 2004).

By 1897 Richard and Marietta Wetherill were mired in debt and sought to establish a stable source of income. Richard believed that trading could become a profitable venture. He approached Fred Hyde and requested financial backing (Snead 2001). In 1898 the Hydes and Richard Wetherill opened a small, one-room trading post in Chaco Canyon. The post was built against the north side of Pueblo Bonito and shared the wall of the ruin. Eventually they enlarged it to three rooms—one room as a residence for the Wetherills, one for Pepper, and one for the trading post. Recognizing the far-reaching opportunities, the Hydes and Richard expanded the merchandising business and soon had 17 trading posts in the area dealing in livestock, wool, hides, and woven rugs.

By 1901 the Wetherills had built a hotel next to their residence located immediately west of Pueblo Bonito. They applied for a post office under the name "Pueblo Bonito." The U.S. Post Office complained that the name was too long, so the name "Putnam" was chosen to honor Wetherill's benefactor, Dr. F.W. Putnam of the American Museum, who had planned the first excavations (Julyan 1996).

From 1898 the name Hyde Exploring Expedition (HEE) became associated with the trading enterprise headquartered at the ranch in Chaco Canyon. The business soon became the mainstay of the entire Wetherill family. By the time the family's Alamo Ranch failed in 1902, all the Wetherill brothers and their wives worked for the HEE (Snead 2001).

In addition to the growing friction between the archeologists and Wetherill the pot-hunter, local resentment toward the expedition was festering. In 1900 the Santa Fe Archaeological Society was formed, and an educator by the name of Edgar Lee Hewitt was recruited to head it. Hewitt argued that the ruins and artifacts of the Southwest were cultural resources and should be protected from "despoilers" (translation=HEE) looting the area and shipping the resources "back east" (Snead 2001). Hewitt's constant agitation let to the passage by Congress of the Antiquities Act in 1906, which prohibited unauthorized excavation and collection of artifacts on public land. In 1907 the canyon was established as a National Monument, and Richard Wetherill was forced to concentrate his efforts on his trading and livestock businesses.

In 1910 Richard was gunned down and killed by a disgruntled Navajo, and Marietta was forced to flee the canyon with the remainder of her family. The Wetherill era was over. The Cuba cattle baron Epimemio Miera (mentioned in Chapter 15) purchased the falling-down Wetherill buildings left behind at Pueblo Bonito. Richard was buried a short distance east of Pueblo Bonito. His lonely grave marker occupies a tiny cemetery plot at the foot of the sandstone cliff, and provides the only commemoration of this sturdy pioneer and early archeologist. How that latter appellation would have pleased him! He would not have been pleased however to learn that the National Park Service (NPS) would later obliterate all traces of his ranch in the canyon.

Marietta went on to live a long, adventurous life. As mentioned earlier (Chapter 15), she and her children settled in and around the town of Cuba. Later she moved to eastern Arizona to operate a small ranch and store at Chambers, near Petrified Forest National Park, and then she worked in stores in Utah. In old age she moved to a little house on Peach Street in Albuquerque. While there, in 1953, she dictated the recollections of her full life to a reporter from Silver City named Louis Blachly. Her story filled up 75 reels of tape, and the transcriptions occupied more than 2,000 single-spaced typed pages! Blachly had hoped to make a living by recording the stories of historical figures in New Mexico and publishing them under the auspices of the Pioneers Foundation, but the notion never

caught on with his investors. The tapes and transcripts later became the core material for a biography, entitled *Marietta Wetherill: Reflections of Life with the Navajos in Chaco Canyon* (Gabriel 1992). Nine months later, in 1954 she died peaceably in her sleep at the age of 77, her memories safely preserved. Her ashes were buried in Richard's grave in Chaco Canyon.

Years later the efforts of the Hyde Exploring Expedition would be belatedly appreciated by the archeological community. The name Talbot Hyde, the elder benefactor of the first major expedition at Chaco Canyon, lives on at Hyde Memorial State Park, formed in 1938 in the mountains northeast of Santa Fe on land donated to the state by his widow (Julyan 1996).

The tourists arrive

Slowly but surely members of the public were drawn to the canyon. As one visitor remembered a trip taken in 1921 via a wagon and team: "In those days there were no roads to Bonito . . . for uncounted miles around us was a maze of tracks made by Navajo wagons. We could follow any one, go any direction or add our track to the network." And after a storm, "The rain had obliterated every animal and wagon track; the earth was as clean as a washed slate" (Wattles and Wetherill 1977). Even though there were no real roads, the network of wagon tracks connecting the far-flung trading posts did provide a crude way to get around in the early part of the 20th century (Figure 9).

During the years 1921–1927 the National Geographic Society carried out serious work at Chaco Canyon. This large, well-funded, and prestigious expedition had to be supplied from the railhead to the south at Gallup, and from there via truck over a capricious 106-mile dirt road. If the weather was dry the trip would take maybe seven hours (Judd 1925). By 1923 one of the first state highways, NM-35, was built south from Bloomfield, through Chaco Canyon, and on to Gallup (Figure 10). The southern part of this route was likely the same route used to supply the National Geographic Society. By 1929 NM-35 was restricted to that stretch south from NM-55 (see below), through Chaco Canyon and on to Thoreau to the south (Figure 11). The designation NM-35 existed at least to as late as 1934.

As we've seen, the first "highway" connecting the ruins of Chaco Canyon with the Rio Grande Valley was the dirt track constructed in 1929 between Cuba and Bloomfield, designated NM-55. The turnoff from NM-55 onto NM-35 south to Chaco Canyon was at the so-called

"Otis turnoff" at the Otis Trading Post, established in 1929 or 1930. The Navajo name for this place is *Ts'ahtah* (Sagebrushland?).

In 1929, Emer E. Otis, with his wife and two kids, established residence here and applied for a homestead (he received a patent in 1936). In 1934 the homestead consisted of a one-room stockade house (probably the trading post), a three-room lumber house, two cellars, a machine shop, a cow shed, chicken house, corral, and a well. The store building contained a small lunch counter. Interestingly, in 1930, prohibition officers raided the Otis store and found cases of beer under cases of soda pop. A post office was operated out of the store from 1930 to 1932. Eliza W. Otis was its postmistress (Kelley and Francis 2006).

Otis had been a blacksmith on the Denver and Rio Grande Railroad, and surprisingly was a little guy of only about 125 lbs. He drove an old 1933 Chevy truck, and when he would drive to Cuba and load the truck up with supplies he would amuse the locals by climbing through the windshield to get inside. After Otis died the place was sold (Hernández 2000b). The building has been demolished.

In 1938, NM-55 was moved north to the present alignment of US-550 (New Mexico State Highway Department 1939). Otis was, therefore, no longer located on the main highway and it gradually withered away. The 1966 USGS 7.5-minute topographic map (Blanco Trading Post quad) shows the site of Otis as occupied by the "Tsah Tah Trading Post," but this may be confusing the Navajo name for the site for the defunct trading post. Today the site is occupied by a small assortment of ranch structures (Figure 184).

In the late 1940s the northern approach to Chaco, the northern part of NM-35, was redesignated NM-56. Also in the 1940s, NM-55 disappeared from the maps and NM-44, which until then had run north from Counselor down the Canyon Largo to the San Juan River Valley, replaced it. Around 1970, NM-56 was given its present number NM-57 (Riner 2004). Blanco Trading Post had been established at the junction of NM-56 and old NM-55, most likely in the late 1930s after the realignment of the main highway.

The National Monument was enlarged in 1980, and in 1981 was redesignated the Chaco Culture National Historical Park (Figure 184). Sometime shortly in the late 1990s the critical access turnoff to NM-57 at the Blanco post at mile 123.3 was discontinued. A sign at the trading post now clearly says that this is not the way to Chaco, but a few poor souls still try it anyway. Today the real turnoff from US-550 is at mile 112.7 onto County Road 7900

from the Red Mesa Express store. Five miles of paved road are followed by 16 miles of a bone-breaking "washboard" gravel road. Starting in 2005, after many years of heated discussion, there began to be serious talk about finally paving this stretch. The purists are opposed because, they say, the gravel road is a desirable deterrent to the dilettantes and, besides, the monument's capacity to host additional visitors is now at its maximum. Others maintain that an improved road would allow more people to visit and enjoy this special place. Of course both are correct. The debate continues. Whatever the condition of the access road the trip is well worth the effort. So buckle down your teeth and secure your eyeballs and let's make a side journey down to Chaco.

What causes those awful washboard roads?

The 16 miles of unpaved road mentioned above provides quite an experience and takes the motorist back to yesteryears. It is a classic washboard road (Figure 185). For those too young to remember, before roads in the U.S. were routinely paved, the dirt or sandy roads typically developed an awful corrugation that caused the vehicle to bounce about almost uncontrollably. Frustrated motorists likened the bouncy surface to a washboard — a primitive device (my mother used one when I was a boy) — that was placed one end into a wash basin filled with soapy water and used to knead dirty clothing across its horizontal ridges to work the grime out. Everyone then knew what a washboard was and the term became fixed in our vocabulary, even though its source is now obscure.

The washboard road leading to Chaco Canyon can be traversed in one of three ways. First, full tilt so that the tires fly across the bumps and only intermittently make contact with the road. Heaven help you if there is a sudden need to change direction or to stop. Tires have no traction on air! Or one could slow down to about four mph to ride up and down the bumps. That would make a four-hour drive to the canyon! A third way is to drive a "normal" speed, say 30 - 40 mph, at which rate the bouncing is at a maximum and one has to hold on tight to his/her teeth and sanity. Indeed you have to really want to visit the ruins at Chaco, and that fuels the paving debate.

What causes washboarding? One of the first scientific studies to determine the cause of this bane of travel in dry regions was done by an Australian named Keith Mather (Mather 1963). He was keenly aware that hundreds of miles of the Australian arid outback were "beautifully

Figure 185. Washboard road on the way to Chaco Canyon
(bumps are 22 to 24 inches apart). (Photo by author 2006.)

corrugated." He soon learned that this weird type of "beauty" was prevalent in many other arid regions of the world, especially in Third World nations, but even in more advanced nations where low traffic density often combines with prohibitively high paving costs to keep the roads from being surfaced.

Mather learned that washboarding was in fact the natural result of "suspended" rolling wheels interacting with a dry, loose surface. When a sandy or gravelly surface is dry it lacks cohesion and the particles are free to move about. As a wheel rolling on such a surface surpasses a critical speed, usually about four mph, it eventually will hit some irregularity such as a pebble. As the wheel bounces upwards the suspension then pushes it back down. The wheel impacts the loose surface and splashes material forward and outward, creating a little valley, and begins to move for a distance in short hops. So far no problem. But then the wheel on the next vehicle comes along, hits the same valley, becomes briefly airborne, comes down in the first valley, becomes airborne again, and then comes down and digs a second valley up ahead. Each new vehicle contributes to the process. The washboarded stretch becomes longer and more accentuated. The spacing, or "pitch" of the valleys is mainly dependent on the velocity: the faster the speed the greater the pitch. Continued passage over the surface slowly extends the entire pattern of bumps and valleys forward.

Today's harder tires and tendency for greater speeds create higher stresses and aggravate the situation. Interestingly, steel railroad tracks can also become slightly washboarded. The pattern develops slightly ahead of the connections between rail segments. The pitch is only a few inches but it leads to a phenomenon known as "singing rails" (Mather 1963).

What to do? One could reduce the stress by deflating the tires. I can just cannot see tourists taking the time to do this, and then somehow figure out how to re-inflate. Last time I checked, few vehicles today carry air pumps around. Or one could slow down to about four mph or less, as mentioned above. Or one could avoid dry unpaved roads and skip Chaco Canyon altogether.

In this chapter I will introduce some geologic subjects that normally don't accompany discourses about Chaco Culture National Historical Park. Hopefully these notes will provide an interesting compliment to the enormous literature that presently exists about this renowned place.

The development of a sophisticated stone-age culture that flourished for about three centuries (ca. 850–1130 CE) in Chaco Canyon was no accident. This special setting was provided by the area's unusual geology, especially the split-up of the so-called Chaco Mesa. Let me explain my use of the phrase "so-called." A mesa is defined as an isolated, nearly level landmass looming above the surrounding landscape. Chaco Mesa is not really a mesa at all, but rather a ridge of sandstone. The sandstone body dips gently down to the north, thus offering a sharp, south-facing escarpment (Figure 186). Geologists call such a landform a "cuesta," but we'll continue to call it Chaco Mesa.

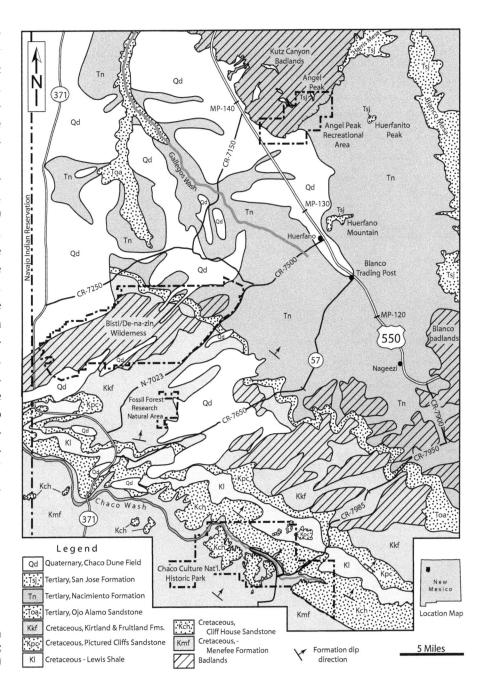

Figure 186. Geology of the Chaco Canyon area. (Modified from McFadden et al. 1983; Wells 1983; NMBG&MR 2003.)

Legend

Qd	Quaternary, Chaco Dune Field
Tsj	Tertiary, San Jose Formation
Tn	Tertiary, Nacimiento Formation
Toa	Tertiary, Ojo Alamo Sandstone
Kkf	Cretaceous, Kirtland & Fruitland Fms.
Kpc	Cretaceous, Pictured Cliffs Sandstone
Kl	Cretaceous - Lewis Shale
Kch	Cretaceous, Cliff House Sandstone
Kmf	Cretaceous, - Menefee Formation
⫽	Badlands

Formation dip direction

5 Miles

Location Map

New Mexico

Chaco Mesa is formed by the stacking of three distinct segments of the Cretaceous-age, Cliff House Sandstone. For most of its length the "mesa" forms a single ridge, but in the Chaco Canyon area it splits up due to its composite nature (Figures 187A and 187B). This fact has created an inhabitable valley within the ramparts of the composite Cliff House Sandstone.

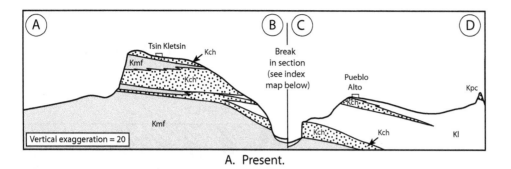

A. Present.

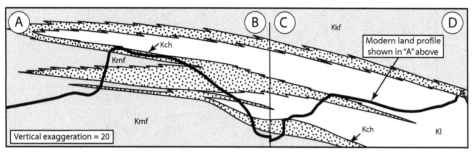

B. "Restored" to before development of modern landscape.

Legend

Kkf Kirtland/Fruitland Formations (non-marine, locally coal-bearing delta-plain mudstones & sandstones)

Kpc Pictured Cliffs Sandstone (regressive beach sandstone)

Kl Lewis Shale (marine shale & mudstone)

Kch Cliff House Sandstone (transgressive beach sandstone)

Kmf Menefee Formation (non-marine, locally coal-bearing delta-plain mudstones & sandstones)

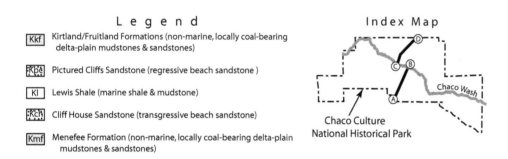

Index Map

Chaco Wash

Chaco Culture
National Historical Park

Figure 187. Generalized south-to-north geologic cross section across Chaco Canyon. (Modified from Mytton and Schneider 1987.)

To understand the geology — and the human habitation — of Chaco Canyon it helps to understand the origin of the Cliff House Sandstone. The unit is the product of deposition along the edge of a north-facing coastline bordering a shallow sea during the Cretaceous Period, some 75 to 80 million years ago (Figures 4, 188A, and 188B).

This region was then of course at sea level. During this time the land was slowly subsiding in little pulses and the sea was advancing over the newly-submerged terrain from north to south in a series of fits and starts. The shoreline itself was characterized by a zone of sandy beaches located just above the water line. The top surface of the sand — the "shoreface" — extended for some distance north and downward beneath the sea (Figure 188C).The shoreface is a zone of wave surge, and oceanic, "marine" critters burrowed down into it. The effect of this burrowing was a churning of the sand, destruction of any bedding structure, and therefore creation of a massive body of unbedded sand. Storm surges periodically sloshed sand far down the shoreface into deeper water where the sand interfingered with oceanic, marine muds.

Streams flowing across the low-lying coastal plain from south to north delivered their sand to the sea through river-mouth gaps in the beach zone (Figures 188A and 188B). Longshore currents then transported the sand parallel to the shoreline as the waves and surf pounded the shoreface and accreted the new sand onto it. If subsidence of the land had stabilized and the relative elevation of sea level remained constant, the only place that this new sand could accumulate was where there was space to accommodate it, and that was out in the sea, north of the shoreface (Figure 188C).

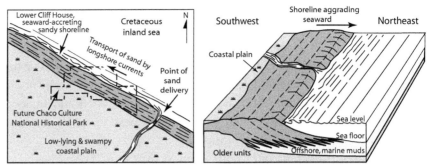

A. Geography during lower Cliff House time., ca. 78 Ma.

B. Block diagram of lower Cliff House shoreline.

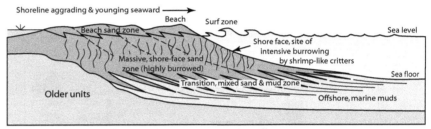

C. Closer look of some essentials of shoreline-sand deposition (expanded front panel of "B" above).

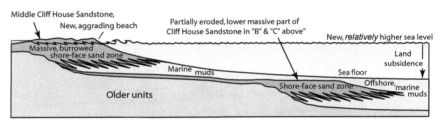

D. Subsidence of land, "transgression" of sea, and establishment of new, relatively higher shoreline in middle Cliff House time.

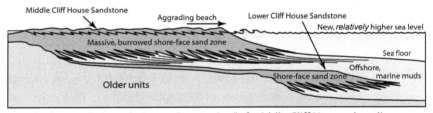

E. Seaward aggradation, or "regression" of middle Cliff House shoreline.

Figure 188. Schematic development of multiple "stacked" Cliff House Sandstone bodies (dark gray) at Chaco Canyon. (Not to scale, vertical dimension exaggerated.)

The resumption of land-subsidence upset the delicate balance. As the land subsided beneath the level of the sea, the old shoreline was "drowned" and a new one established landward and relatively higher than the old one (Figure 188D). During the process of temporary sea advancement to the south, the upper parts of the earlier, drowned shoreline sands, the beach and upper part of the shoreface, were typically eroded away on the sea floor before they could be buried and preserved. The new shoreline then aggraded seaward just like before and the cycle repeated itself (Figure 188E).

With this understanding of ancient depositional environments we can now begin to speculate how and why the Chacoans exploited the natural environment of rock in the canyon as they did. Two of the wonders of the ruins at Chaco are the huge size of its major buildings—the multi-storied "great houses"—and the exquisite masonry of their walls. These two characteristics are of course what attracted the attention of travelers and archeologists alike. Prior to the middle 9th century CE, home construction in the canyon was typically of *adobe*, or *jacal*—vertical poles tied together and slathered over with adobe. This material was structurally weak and tended to dissolve when wet. The Chacoans desired large buildings—for reasons that are beyond the scope of this book—and had to invent new masonry systems to support the enormous weight of multiple stories.

Several distinct masonry styles evolved over time. The earliest was rather crude, but in time it consisted of a careful stacking of rather thin, slabby sandstone blocks facing outwards from a core filled with rubble (Figure 189). Sometimes the sandstone slabs exhibit internal laminations, indicative of slow deposition of sediment in protected waters. Where did the builders find these slabs? Looking around from the canyon floor one sees little but massive Cliff House Sandstone cliffs lining the sides of the valley. The slabs certainly didn't come from these things. The inquisitive mind naturally wonders, "Where were the quarries?" And how much rock was removed from them?

Figure 189. Slabby, laminated sandstone blocks in masonry wall of Wajiji Ruin, Chaco Canyon. (Photo by author 2006.)

Let's do some back-of-the-envelope calculations. It has been estimated that one of the great houses, Chetro Ketl with about 500 rooms, contained about 50,000,000 chunks of rock in its walls (Frazier 2005), or about 100,000 pieces per room. The largest great house, Pueblo Bonito, contains about 800 rooms. The other seven great houses in the canyon are smaller, so to be conservative we'll assume that the average number of rooms per remaining great house is a mere 100 (=700). That gives a total of 2,000 rooms (500+800+700), or perhaps 200,000,000 (2,000 rooms x 100,000 pieces/room) pieces of rock. Of course as we continue along this line of speculation the margin of error compounds, but let's be fearless and forge ahead. Assume that the average size of a single piece of rock is something like 2 x 2 x 3 inches, or 12 cubic inches (12 in³). If this is correct, a single cubic foot (ft³), with its 1,728 in³, would contain 144 pieces (1 piece/12 in³ x 1,728 in³/ft³). Let's be even a little more conservative and assume 150 pieces per cubic foot, so our 200 million pieces of sandstone slabs would occupy a volume of 1.3 million ft³ (200,000,000 pieces/150 pieces/ft³). Now, hold this pregnant thought while we look for the quarries.

We should be able to find the quarry areas by doing a little geologic detective work. As we've seen, the Cliff House Sandstone was formed by deposition of sand along an ancient shoreline (Figure 188). As we've also seen, the three environments occupying the transition zone between the shore and the open sea were 1) the beach-sand zone, 2) the submarine shoreface zone seaward of the beach, and 3) the offshore marine environment beyond the influence of the shoreline. In the transition zone between #2 and #3, slabby beds of sand are intimately interbedded with marine muds. Here we should look for our discrete sandstone slabs. However, if this mixed type of sedimentary rock consists of perhaps 50% sandstone and 50% mudstone, the volume of rock quarried to obtain the needed sandstone slabs would about double from 1.3 million ft³ to approximately an astounding 2.5 million ft³ (1,300,000 ft³/0.50).

Can we express this volume in terms we can comprehend? Threatening Rock, the huge piece that collapsed on the back part of Pueblo Bonito in 1941 (more about this below) had a volume of roughly 450,000 ft³ (97 feet high x 140 feet long x 34 feet thick; NPS 1993). So, the total volume of rock quarried to build the great houses of Chaco Canyon was equivalent to about 5.5 Threatening Rocks (2,500,000 ft³/450,000 ft³). Quite a feat performed by stone-age people wearing sandals without machinery or beasts of burden!

Earlier I asked where were the quarries located. I have walked tops of the cliffs above the ruins on the north side of the canyon in search of the quarry sites, to no avail. I conclude that most if not all of the favorable rock material has been completely quarried. Could this gradual dearth of optimal building material have been a factor in the progressive, terminal decline in the sophistication of masonry construction at Chaco Canyon?

Collapse of Threatening Rock

In 1901, S.J. Holsinger, a special agent of the U.S. General Land Office, was sent to Chaco Canyon to investigate rumors of irregularities involving Richard Wetherill and the Hyde Exploring Expedition. His report cited a huge slab of rock threatening to fall onto the back (north) side of Pueblo Bonito (Figure 190). He called it the "Elephant." Later it was dubbed "Braced-Up-Cliff" from the translation of its Navajo name. The NPS called it "Threatening Rock." It was a 30,000-ton slab of lower Cliff House Sandstone that had separated from the main sandstone mass along a nearly vertical fracture. The slab sat on about 15 feet of Cretaceous-age, Menefee Shale that is partly exposed above the level of Quaternary-age alluvium on which Pueblo Bonito is built (Schumm and Chorley 1964). It was to become the most famous rock in the state of New Mexico.

Figure 190. West view of Threatening Rock behind Pueblo Bonito, 1901. (Photo courtesy of Palace of the Governors Photo Archives (NMHM/DCA), negative #6153.)

The NPS people weren't the first to be concerned. When the Chacoans were building Pueblo Bonito they belatedly realized that they had a problem on their hands. In the 11th century they had jammed a number of timbers into the rock's Menefee Shale footing and built up some platforms and buttresses in an attempt to stabilize it (Lister and Lister 1981). Clearly there must have been signs of movement to spur this work.

In 1929 the aviator Charles Lindbergh and his wife Anne, only two years after his solo flight across the Atlantic in his iconic plane, the Spirit of St. Louis, photographed Chaco Canyon from the air. One of his photos is a north view of Pueblo Bonito, showing Threatening Rock in the process of detaching itself from the 100-foot cliff of lower Cliff House Sandstone behind it (Figure 191A). This fascinating image shows Threatening Rock before its 1941 collapse onto the back of the ruin (Figure 192B). The NPS clearly understood that Pueblo Bonito was in serious jeopardy. Something had to be done—or at least tried. And fast.

A. Charles Lindburgh's 1929 northeast, high-oblique aerial photo of Threatening Rock behind north wall of Pueblo Bonito. (Photo courtesy Palace of the Governors Photo Archives (NMHM/DCA), negative #130232.)

B. *Google Earth* image of Pueblo Bonito after January 1941 fall of Threatening Rock.

Photo 191. Pueblo Bonito and its Threatening Rock.

In 1933 the NPS conducted a careful survey of Threatening Rock. At that time the slab was completely detached from the cliff behind it, but it had also settled into the soft underlying shale about eight inches. In 1935 the NPS installed steel bars set in concrete to measure the movement. Measurements were taken from 1936 until the rock's fall in January 1941. The result was a fascinating record of the movement and final demise of Threatening Rock, and documented one of the major geologic processes that form canyons. It was learned that the movement wasn't steady, but rather was concentrated in the winter months. Due to less evaporation, winter precipitation had more opportunity to soak into the supporting Menefee Shale. It was also found that the movement was proportional to the cumulative amount of precipitation during all seasons, and that the movement was accelerating (Schumm and Chorley 1964).

The winter of 1940–1941 was one of the wettest on record. It rained and it rained. By the end of 1940 measurements indicated that the rock had moved outwards 12 inches and had settled downward a few inches. In January 1941, before the fall, it moved outward another ten inches to the southwest and settled four inches. Things were definitely out of control. On the evening of January 21, 1941, a large chunk of rock fell from near the top of Threatening Rock. The NPS estimated that the rock had moved nine inches that night, probably at the time the chunk broke loose. The next morning NPS officials could hear the rock "popping and cracking"(Schumm and Chorley 1964).

All that morning Monument Custodian, Lewis T. McKinney, had been taking photographs and measurements of the shifting rock to record the big event for his report to the regional office. By midafternoon, he ran out of film and rushed to the old Hyde Exploring Expedition boardinghouse at the nearby Pueblo del Arroyo for a fresh supply. While inside he felt the ground shake. It was 3:24 PM.

Prior to the fall, the Civilian Conservation Corps (CCC) had played an important role in the saga of Threatening Rock. In 1937 a small "CCC-ID" (Indian Division) camp was set up at Pueblo Bonito for Navajos workers involved in restoration activities. Later, in the summer of 1939 the full-blown CCC camp NP-2-N (the name comes from National Park Service, camp No. 2, New Mexico) had been established in the national monument. It was manned by a company of about 200 "boys." The camp operated for 2.5 years. The boys must have been deeply impressed by what they saw on their first trip away from home, but what about the impression they left on the landscape itself? Such camps were beehives of activity with many trucks and people coming and going. The camp's "footprint" should remain visible today, but one needs to know where to look. It is odd that the location is not indicated on the current maps or in the Visitors Center literature. So just where was the camp? Two sources place it "at the [east] foot of South Mesa" (Brugge 1980; McNitt 1966). However, a newly available Internet research tool, *Google Earth*, allows one to search the area via high-resolution satellite images. The images reveal nothing up against the east side of South Mesa, but they do show rectangular grid of disturbances west of the intersection of the historical park's one-way loop road and NM-57 where the latter exits the park to the south (Figure 184). I am convinced that this is the site of the old CCC camp.

The existence of this camp is an unheralded piece of Chaco Canyon history, and this is unfortunate considering the amount of conservation work the CCC accomplished here. This work included the planting of 100,000 cottonwood, plum, willow, and (unfortunately) tamarisk trees throughout the canyon, and the construction of many earthen berms for erosion control. The CCC also initiated a project to construct a road for vehicles to the top of the cliff directly behind Pueblo Bonito. Perhaps fortunately, this project was scrapped upon the onset of World War II.

The senior foreman for the CCC camp was Claire J. Mueller. Many years later, in 1990, looking back fondly to his CCC days, he wrote down his account of the futile attempt to control Threatening Rock. The text below is taken from his story (Mueller 1990):

"Through the years I have lost track of other members of the Technical Personnel staff. As far as I know I may be the last living member. As I recall the camp was established as a 200 man camp. The initial group came from the East and was from Pennsylvania and New Jersey. After about a year the men came from the Southwest and were mostly Spanish-Americans.

"When the camp first opened the emphasis was on soil conservation. There was a crew which had drawn the assignment to stabilize Threatening Rock. This was the group I supervised. For many years Threatening Rock had attracted the attention of National Park Service personnel. Many ideas were received and considered in an effort to find the most practical way to prevent its toppling onto Pueblo Bonito.

"Long before a CCC camp was established at Chaco Canyon, Monument Custodian Lewis T. McKinney had installed two simple gauges on the top of Threatening Rock to facilitate determining that the rock did 'breathe' and to what extent. Each gauge consisted of a length of halfinch steel pipe and a steel rod. The pipe had been set in the wall of the mesa and the rod had been set in Threatening Rock. Before setting the rod in place a notch was cut into it. The rod was placed into the pipe with the notch up and then affixed to the rock. By measuring the distance from the end of the pipe to the notch the amount of movement of Threatening Rock could be determined. Two CCCers had been assigned to work with Custodian McKinney and it was their responsibility to climb the crude prehistoric stairs behind Pueblo Bonito at least twice a day to read and record the measurements of the gauges.

"In the 1938–39 timeframe McKinney's records showed the rock had been moving in and out at almost the same measurements. Suddenly a change was noticed—the rock was moving out but not returning the same distance.

Something had to be done in a hurry to prevent the rock from falling. It was concluded the debris and especially the geodes [sic] that had been eroded from the mesa wall were preventing the rock from returning to its morning position. Several ideas for preserving Threatening Rock were considered:

1. "Drill several vertical holes deep into the rock and into the mesa wall into which heavy steel rods would be inserted leaving 34 feet of rod protruding. These rods would then be tied together with used cable from the elevators at Carlsbad Caverns.

2. "Use the rock as a quarry. Take off the first 30 or 80 feet to lower the center of gravity. This rock would be used for the projects to be constructed by the men of the CCC.

3. "Regardless of what was to be done later it was decided the debris and geodes [sic] behind the rock had to be removed. If that could be accomplished it might be possible to use Plan 1 and pull the top of the rock towards the mesa wall and leave it resting in that position. Even if the bottom did move out the damage to Pueblo Bonito would be slight.

"It was decided to initiate Plan 3. The project was assigned to me. It was a dangerous job but the men liked the challenge and worked hard. At first we used wheelbarrows and buckets, but we soon found they were unsatisfactorily slow. The camp blacksmith, Eli J. Anderson and mechanic, Clyde L. Fletcher, came up with an idea that really got the work going. The engine from a disabled Ford truck was converted into a donkey engine which pulled an especially designed sled in and out from behind the rock. We found very few artifacts in the debris . . . We were unable to remove several fairly large geodes [sic] from the mesa wall and several which were lodged between the wall and the rock because they were out of reach. Work on the project began to lag because of the weather. We had rain almost daily for the month of December 1940.

"I should tell you we were now having some very anxious moments. To appreciate the problem you have to know that Threatening Rock was resting on a six to eight-inch layer of lowgrade coal or perhaps oil shale. It had been that way for centuries and nothing had happened. But, we had been having too much rain and water had found a way to penetrate the mesa wall and reach the layer of lowgrade coal or shale. A black substance began to ooze from the layer of coal or shale.

"We watched the situation grow steadily worse.

The gauges were indicating the worse possible condition: the rock was moving out more than it was receding. And the rain continued! The large geodes [sic] which we had not been able to remove were now steadily moving downward. We had hoped the pressure would break them but that was not to be. Work was stopped on the removal of debris and all equipment removed. The fall of Threatening Rock was inevitable.

"On 20 January 1941 the gauges showed the rock was moving out and not receding at all. On the 22nd McKinney knew the time had come. It had rained all night but was clearing a bit by midmorning. He set his tripods and cameras at the most advantageous sites and he and his CCC assistants set by awaiting the big fall. It was almost noon and nothing had happened so the vigil crew went to lunch. Just before they returned to their positions they heard a load noise and looking towards the rock saw a cloud of dust. When it cleared it was found that Pueblo Bonito had survived better than had been anticipated."

At the time of the collapse three Navajos woodcutters had witnessed the big event. Afterwards many of them feared greatly that the world was about to come to an end because Braced-Up-Cliff had come down. The Park Service pieced together their oral accounts as follows:

"The slab leaned out about 30 of 40 feet from plumb, settled sharply, and when it hit solid bottom, rocks from the top of it were broken loose and propelled into the ruin. The lower two-thirds pivoted on its outer edge and fell down the slope toward the ruin. The whole mass broke into many fragments and an avalanche of rock catapulted down the slope and into the walls of the back portion of Pueblo Bonito" (Schumm and Chorley 1964).

Let's now return to Mueller's (1990) account:

"The Navajos got the news of the fall in a hurry . . . in the morning it was found that every new exposed break of the rock had been sprinkled with corn pollen and turquoise dust. For several days I was the bad guy at Chaco Canyon. Every time I got near a Navajo he would change his course and utter 'chindi!' " [Note: this term is from the Navajo word meaning "ghost" or "spirit." The Navajo have a deep respect for the spirits of the dead. In this case they evidently felt that some kind of death was associated with the collapse of Threatening Rock and that Mueller somehow housed such a "spirit."] "A couple of days after the fall I went to Arthur Tanner's Trading Post to get my mail. The bench in the building was filled with

Navajos. When I opened the door I heard "chindi". As I walked past them they looked the other way and slid down the bench and out the door. When I turned around the store was empty and none were to be seen outside. In a few weeks things got back to normal and I didn't hear 'chindi' when we met."

Post mortem analysis of the measurements revealed some interesting conclusions. Taking these data and projecting backwards in time suggested that Threatening Rock had begun its southwestern journey at about 550 BCE. It had taken about 1,000 years to move the first foot, 650 years to move the second, 200 years the third, 60 years the fourth, eight years the fifth, and two and one half months to move the sixth (Schumm and Chorley 1964).

Valley-parallel fracturing

The peeling off of huge slabs of sandstone, like Threatening Rock, from the canyon walls along fracture surfaces is a primary mechanism of canyon widening called "mass wasting." The Lindbergh photo (Figure 191A) also shows the debris from an earlier rock-fall just to the west (left) of Threatening Rock These rockfalls have been going on for a long time. But where are all the others? They should be everywhere. The answer is hidden below the alluvium of the Chaco valley floor. The alluvium is up to 125 feet thick (Patton et al. 1991) and was fully in place by no later than 900 CE (Force et al. 2002). Many earlier rock-falls are probably hidden below its surface. In fact, excavators have found evidence of one earlier fall, in the form of large blocks of sandstone, lying beneath the floor of the earliest rooms of Pueblo Bonito built in the 9th century. A later fall buried another part of the structure built in the 11th century (Schumm and Chorley 1964). Someday the remainder of the larger slab, from which Threatening Rock broke loose will come tumbling down and likely wipe out at least the entire rear of the ruin.

The near-vertical fractures that make these falls possible are a recurrent theme throughout Chaco Canyon. These are sometimes called "valley-parallel fractures" because they follow the topographic contours of the valley (Figure 192). They develop in response to the imbalance of forces created by the rapid incision of a valley into a thick body of massive, somewhat brittle sandstone such as the lower Cliff House Sandstone. Gravity is always pulling the sandstone downwards, building up internal pressure. Because the rock is very slightly elastic it would squash downward and expand slightly sideways if it could. At the valley wall, however, the sandstone is unconfined. A steep gradient of diminishing pressure is established from the interior of the sandstone body outward toward the vertical valley walls. The imbalanced stress is relieved via fracturing (Whitehead 1997). The fractures gradually widen as water enters, and plant roots penetrate and make themselves at home. The cracks expand and form detached slabs, *a la* Threatening Rock, which eventually peel away and thus widen the canyon (Figures 193A and 193B). The process was much more important during the ice ages of the Pleistocene Epoch (ca. 800 ka to about 12 ka). Drying out of the climate during the past 10,000 years or so has greatly decelerated the process. Today the process of slab development and failure is—except for Threatening Rock—in a state of suspended animation and is "fossilized" in place.

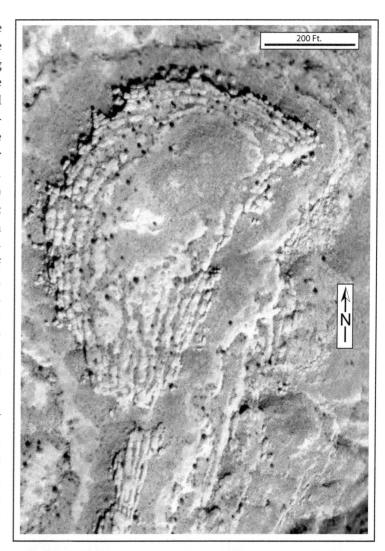

Figure 192. *Google Earth* image of small mesa at west end of Chaco Canyon showing valley–parallel fractures.

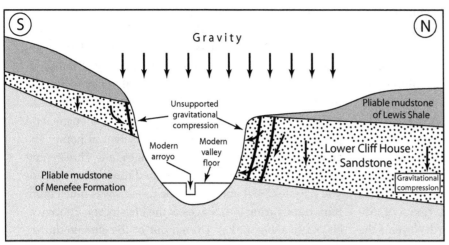

A. Unbalanced forces (curved bold arrows) at canyon walls in Cliff House Sandstone causing development of valley-parallel fractures (heavy black lines)

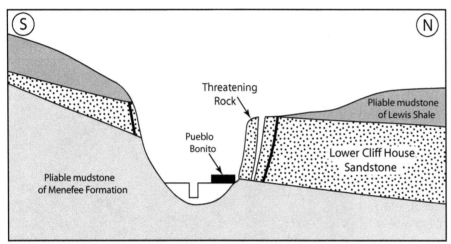

Figure 193. Development of valley-parallel fractures.

B. Separation of rock slabs along widening valley-parallel fractures

The dam

A small number, perhaps no more than about 2,000, Ancestral-Puebloans, or in this case "Chacoans," lived and farmed on the valley floor of Chaco Canyon during its heyday (Frazier 2005). The valley floor had changed its character over the centuries, however, and the people had to cope the best they could with the changing conditions. Four constructional periods of Chacoan "great houses" between about 900 and 1130 CE danced an exquisite tango with the evolving geologic situation (Figure 194A).

Sometime before about 10,000 years ago the ancestral Chaco River cut a valley and channel into the Cretaceous-age bedrock (Wells et al. 1983). Mass wasting widened the canyon and steepened the walls, especially on the north side where the lower Cliff House Sandstone is thickest (Figure 187). At about 200 BCE a large sand dune became established across the mouth of Chaco Canyon and cut it off from Escavada Wash, the canyon's natural outlet to the west (Figure 194B). Studies have determined that such a sand-dune dam could have been formed extremely rapidly, perhaps in as little as a single windy year. It is difficult to realize that a pile of wind-blown sand could form an effective dam. Such a dam, however, would be able to impede water behind it while at the same time leaking enough to prevent dam erosion (Force et al. 2008).

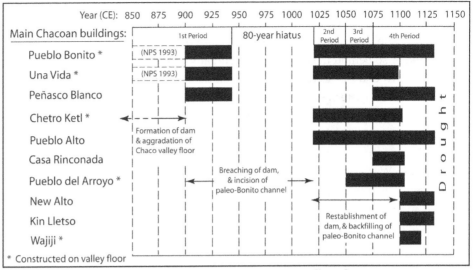

A. Four periods of great-house construction in Chaco Canyon
Modified from National Park Service 1993; Force et al. 2002; Frazier 2005.

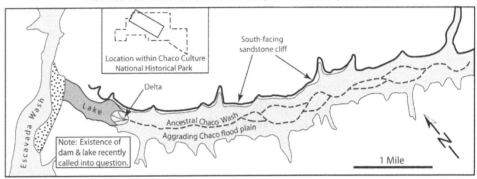

B. Sand-dune dam (dotted), Chaco playa lake, and aggadation of Chaco flood plain, ca. 850 CE.
Modified from Force et al. 2002.

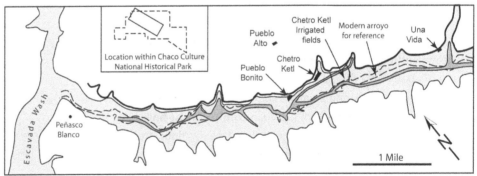

C. Incision and early back-fill of paleo-Bonito channel (dark gray) vs. construction of Chacoan "great houses" (black), ca. 1050 CE. Modified from Force et al. 2002.

Figure 194. Relationship between big-house construction in Chaco Canyon and development of dam at western end of canyon.

For the next 1,100 years, until about 900 CE, sediment accumulation along the drainage upstream from the sand-dune dam built up the level of the valley floor higher than that of Escavada Wash downstream. The water table rose with it, creating an unusually wet bowl that was attractive for human habitation. An interesting yardstick for the rate of valley filling is provided by an old ruin exposed near the bottom of the 15-foot-deep modern arroyo—a Basketmaker structure dated at 585 CE. A thickness of 13 feet of sediment had been deposited in the 500 years (585–1100s CE) after the site's abandonment (Frazier 2005). For perhaps 0.5 mile or so behind the dam a lake eventually formed in the western end of the valley and a small delta built out into the lake's eastern end (Force et al. 2002, Figure 194B). Chaco Canyon had lake-front property!

About 900 CE the dam broke, the lake drained, and the drainage once again flowed directly down into the Escavada Wash, some 15 feet lower. (Note: recent studies have called into question the existence of a sand-dune dam and its lake; the issue remains controversial.) At first the well-watered, still-intact flood plain was now ripe for increased human habitation. For the next 40 years the Chacoans conducted their first period of construction of great masonry houses (Figure 194A). However, because of the 15-foot drop from the Chaco flood plain to Escavada Wash, the water table dropped, and a new channel, the so-called "Bonito channel," progressively cut downward and headward (eastward) into the old flood-plain valley fill (Force et al. 2002).

By 940 CE arroyo downcutting and headward advancement reached an advanced state, and for the next 80 years there was a hiatus in great-house construction as the Chacoans struggled to cope with the challenging situation (Figure 194A). The segmentation of the valley floor by the arroyo compelled the Chacoans to limit their farming to those areas conducive to flood irrigation, especially on the north side of the valley (Figure 194C). At about 1020 or 1025 CE a new sand-dune dam was established and downcutting of the Bonito channel ceased. From then until about 1090 CE the Bonito channel filled back in, thus slowly re-establishing a smooth flood plain. Thus was introduced the golden age of Chacoan culture. Great-house construction extended onto the restored valley floor and over the filled-in old Bonito channel (Figure 194A).

Finally, beginning about 1130 CE a prolonged drought set in and by about 1300 CE the Chacoans left the canyon for good. The abandoned great houses suffered the elements in silence until their "discovery" by José Antonio Viscarra in 1823. The modern arroyo that we see today only partly follows the course of the old filled-in Bonito channel, and for the most part developed between the 1880s and about 1920 (Figure 194C).

27

Badlands to the Chaco Dune Field

As the title indicates, this chapter deals with two sharply-contrasting landscapes, each with its own unique history. The following three chapters (Chapters 28, 29, and 30) will deal with subjects that emerge spacially as we traverse US-550 to the north and northwest. In the penultimate chapter (Chapter 31) I'll put these two landscapes into their regional context. But first, badlands and a big dune field.

Badland country

For the six miles beginning at mile 108.6, the point at which the highway department has excavated a deep roadcut through a bluff of Nacimiento Formation (Figure 195A), we traverse the badlands carved into the soft mudstones of the Nacimiento Formation (Figures 181 and 195B). Here the "dueling" and opposing drainage systems of the Chaco basin to the southwest and the Blanco basin to the northwest have clashed head-on in their competing aims of eating up the mudstones. The badlands display a distinctive profusion of red, green, and black beds. The hues are mainly due to the variable content of the mineral bentonite, which is an iron-rich dust that erupted from volcanos far to the west and settled onto, and mixed with, the muddy sediments of the Nacimiento Formation between 64 to 61 Ma (Cather 2004). This chaotic zone presents one of the most interesting landscapes along the entire US-550.

A. Southwest view of deep roadcut in Nacimiento Formation at east entrance to badland country. Location in "B" below. (Photo by author 2005.)

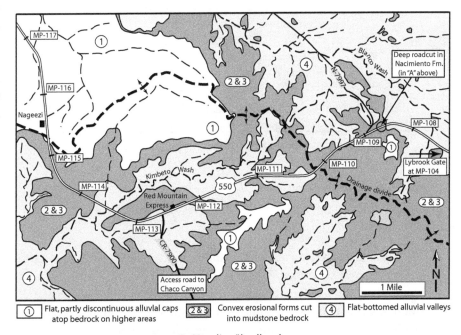

Figure 195. Badland country west of the "Lybrook Gate."

B. "Dueling" badlands

The term "badland" was first applied to an area in South Dakota, which the early French traders called *mauvaises terres* (Love 2002). The French and English terms refer to an intricately eroded, hilly landscape developed on soft, clay-rich rocks, with little or no vegetative cover. This landscape is typically difficult to cross on foot or horseback and almost impossible by wheeled vehicle. It is useless for agriculture. Therefore they're "bad." The Spanish word *malpaís*, in contrast, refers to broken-up volcanic terrain.

The Nacimiento Formation is a specific sequence of sedimentary rocks that is prone to badland-development. The unit is dominated by a rock type called "mudstone," a consolidated mixture of silt and clay, i.e., mud. The silt fraction consists of extremely fine-grained particles of quartz, i.e., finer than sand. The clay fraction, usually about a third of the total volume, gives the badlands its odd character due to its ability to absorb water and swell. With alternating wet and dry spells (winter vs. summer) the mudstones that are exposed to water alternatively swell and contract.

Badlands have a distinctive geometry, consisting of four geomorphic zones (Figure 196). Above the badlands is zone 1, a surface of undissected bedrock often with a discontinuous alluvial cover. Zone 2 of the badlands is a steep, concave outer rim, sharply incised into the surrounding undissected terrain. Downslope from the outer rim of Zone 2 is Zone 3, a broader rim of convex, rounded forms developed in the mudstone (Figure 197A). The clay-rich rocks absorb water and hold onto it so tightly that there is none left over for the roots of plants. With no plant cover, erosion is rapid. As the convex surfaces weather, an intricate series of closely-spaced shallow cracks develop. Wetting and drying of the thin upper layer via the entryways provided by these cracks cause expansion and contraction. The weathered and broken-up mantle, perhaps less than an inch thick, detaches itself from the underlying unweathered material. The mantle soon becomes a mosaic of loose chunks, sometimes called "popcorn terrain" (Figure 197B). Eventually the popcorn chunks become detached and are able to tumble down the slope (Figure 197C). Torrential summer downpours next wash the chunks into an accumulating deposit of alluvium in the central Zone 4 (Guttierez 1983).

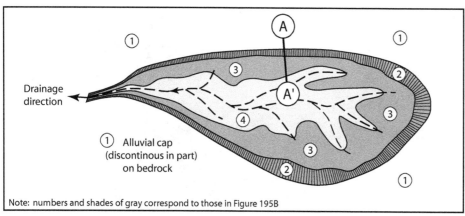

A. Schematic map view of typical badland area. (Modified from Gutierrez 1983.)

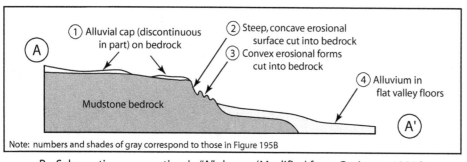

B. Schematic cross-section in "A" above. (Modified from Gutierrez 1980.)

Figure 196. Geometry of badlands.

The key characteristic of badland terrains is the rapid rate of erosion. By way of example, in the badlands of eastern Arizona's Petrified Forest National Park about three inches of weathered mudstone is eroded away in only ten years—that means a 25-foot hill is removed in 1,000 years (Wiewand and Wilks 2001). A beneficial side effect of this easy erodibility is that for over a century the badlands of the San Juan basin have been an extremely rich source of vertebrate fossils because such fossils are continually being exhumed. Geologically, badland terrains are extremely young, generally having been developed only during the past 5,000 years (Wells 1983).

The geographic distribution of badlands is controlled by the aerial distribution of the requisite mudstone. Badlands in the San Juan basin were preferentially formed not only on the Tertiary-age Nacimiento Formation (Tn), but also on the Cretaceous-age Kirtland and Fruitland Formations (Kkf) such as those exposed in the spectacular Bisti/De-Na-Zin Wilderness (Figure 186).

Nageezi

Lying between the Blanco Badlands and the Chaco Dune Field (see below) at mile 115.7 is the tiny, non-incorporated Navajo community of Nageezi, which does not even quite live up to the status of a wide-spot-in-the-road. The name comes from the Navajo word *Naayízí*, meaning "Squash," possibly in reference to a nearby Navajo pumpkin field. An earlier store was located several miles to the northeast on the old down the Blanco Wash. It was built in about 1915 and operated by a certain Barela (Figure 9).

In 1938 Jim Brimhall built the present store. In 1950 he sold it to Jim McEwan, and in 1970 McEwan sold it to Harry Batchelor. In 1974 the post was an L-shaped one-story building with board-and-batten facade and portico. The original structure, built of left-over highway construction materials, has been extensively modified (Kelley and Francis). In 1990 Batchelor, his son and daughter-in-law opened an adjacent bed and breakfast (Eddington and Makov 1995), but it has since been closed. The little place, anchored by a post office established in 1941, is really a hub for some 800 or 900 scattered families. The people are isolated from the main highway, US-550, by at best washboarded gravel and at worst by non-graveled dirt roads. And for police or medical emergencies, Bloomfield is an hour away (Nobis 2001). The post office has special significance because it serves as the official post for Chaco Culture National Historic Park.

A. Badlands developed on Nacimiento Formation mudstones near MP-111.

B. "Popcorn terrain" developed on Nacimiento Formation mudstones (wallet for scale.)

C. Chunks of "popcorn" tumbling downslope.

Figure 197. Development of badlands. (Photo by author 2005.)

Blanco Trading Post

The name of this trading post at mile 123.3 comes from Blanco Wash, located about three miles to the east. The original post was located two miles east. It was built in 1916 by W.G McClure and was operated by him until at least 1931 and possibly until 1941 (Figure 9). Also in 1916 the site was used by Wilfred "Tabby" Brimhall as headquarters for his sheep ranch located to the south by Chaco Canyon. In 1941 or 1942, Tabby and his brother built the present Blanco Trading Post along the Cuba-to-Bloomfield highway. In 1952 a Myrl Harper bought the store and in 1957 sold it to Leonard Taft. After 1960, Taft's partner Harold McDonald, ran the store and after 1974 bought out Taft's interest. In 1988, Harold McDonald was robbed and killed at the store (Chapter 20). Then Bruce and Julie Burch bought the store. In 1995 Bruce was killed in a highway accident. In 1997 the owner was Grant Savage (Kelley and Francis 2006). From sometime after 1970 until sometime after 1995 this was the principal turnoff to Chaco Canyon via highways NM-56/57.

Chaco Dune Field

At about MP-124, just beyond the Blanco Trading Post, we climb up onto what is called the "Chaco Dune Field" (Figure 198). The terrain on this now-inactive sand-dune field is gently rolling to nearly flat and reminiscent of western Kansas. It persists to MP-145 — nearly to Bloomfield. What's this almost-flat expanse doing here? To take a look we have to go back in time to the end of the ice ages of the Pleistocene Epoch.

The Pleistocene Epoch came to an effective end about 13,000 to 12,000 years ago (Figure 4; more about the Pleistocene ice ages in Chapter 31). From that time onward the climate became warmer and drier. The terrain then consisted of a broad surface, sloping gently downward to the west and southwest from the foot of the original outcrop of the Tertiary-age San José Formation (remnants of which form the tops of Huerfano Mountain and Angel Peak). Rivers flowed southwest across this surface of beveled bedrock down to the level of the ancestral Chaco River, the bed of which was about 160 feet higher than at present, and covered the bedrock surface with a veneer of river-borne sand.

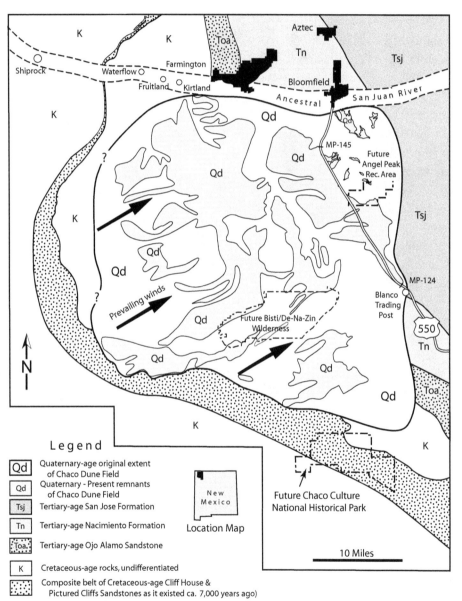

Figure 198. Original extent of Chaco Dune Field. (Modern geographic features shown for reference; modified from McFadden et al. 1983.)

After a long period of stability, two stages of regional lowering of base level (defined in Chapter 1 and discussed more fully in Chapter 31) rejuvenated the ancestral Chaco River and its tributaries to produce a composite, slightly carved-up land surface (Smith 1983). On this surface were located extensive outcrops of weathered, south-facing *cuestas* (asymmetrical ridges with a gentle slope on one side and a steep slope on the other) of Cretaceous sandstones, especially the Cliff House and Pictured Cliffs Sandstones. And then there were the prevailing southwesterly winds. These three factors combined to produce the perfect storm: the necessary conditions were in place for the development of a large sand-dune field downwind (northeast) from the sandstone outcrops. What was once a composite bedrock surface became the floor of the dune field.

Three periods of dune development ensued. The first operated from about 13,000 to about 7,000 years ago. A volume of about 75 million cubic yards of sand was blown off the friable outcrops and spread out by the prevailing southwest winds across the landscape to the northeast. The result was a dune field with a pronounced northeast-southwest "grain." A long period of stability followed. Dune development resumed about 5,000 years ago and continued for the next 1,500 years. During this time some 14 million cubic yards were stripped from the sandstones and heaped onto and across the older dune field. Finally, mostly during the past 1,000 years, an additional 37 million cubic yards were deposited. This time the main source of the sand was from the bed of the Chaco Wash itself and secondarily from the Cretaceous sandstones (Schultz 1980 1983). In all, a volume of some 126 million cubic yards of sand was spread across the landscape (Figure 198). For perspective, this equates to almost eight million 16-cubic-yard dump-truck loads. Subsequent erosion has sliced the dune field itself up into discrete segments. The surviving remnants are inactive because the dunes have been stabilized by vegetation, principally big sagebrush.

Main viewsheds

The semi-circular Huerfano mesa (elevation 7,474 feet, unfortunately called "Huerfano Mountain" on the USGS topographic map, Huerfano Trading Post 7.5-minute quadrangle), is a major landmark in this part of the San Juan basin and casts an expansive viewshed (Figure 182). The mesa is a geologic anomaly. Its height is maintained by an isolated hard cap of Tertiary-age San José Formation crowning and protecting the softer Nacimiento Formation below. Clearly, the San José Formation once covered a more extensive area, but its limit has been since eroded back to the east, except here and at a few other scattered places.

Huerfano Mountain, *Dzil Ná' oodilii* ("People Encircling the Mountain") is one of the sacred mountains of the Navajo, and the Indians frown on efforts to climb it. Because of its prominence it figured large in the line-of-sight network that linked places of importance to the Ancestral Puebloans of Chaco Canyon. A few years ago a high school student named Katy Freeman discovered that Huerfano was in a direct line of sight between Pueblo Alto (a great house above and on the north side of Chaco Canyon), 25 miles to the south, and Chimney Rock Pueblo in southwestern Colorado near Pagosa Springs, 60 miles to the northeast. The Chimney Rock site (ca. 900–1125 CE) was a religious center dedicated to the moon, and the calculations and observations made there were important to the Chacoan leaders to the south. Freeman also found the ruins of many fireboxes on the top of Huerfano. Messages conveyed by flame at night, or by mirrors of polished obsidian or pyrite during the day, could have been communicated in real time from Chimney Rock to downtown Chaco in minutes by using the Huerfano "repeater station" (Leckson 2004).

In 1991–1992 a company proposed an asbestos disposal site on a quarter-section of private land four miles east and in plain view of Huerfano Mountain, probably in NE/4 Sec. 1-T25N R9W (Figure 182). The Navajo of the Nageezi Chapter went ballistic, and their efforts to block the project were successful at public hearings the following year. Not only is the site considered sacred ground, called *T'iistah Diiteeli* ("Spread-Out Cottonwoods"), but it is the site of an old trading post. Sometime in the 1910s, a José Romolo Martínez had settled with his family along the wagon road running down the Blanco Wash to the San Juan River, and soon applied for a homestead. After Romolo (known to the Navajo as *Atsoo'i* or, for some inscrutable reason, "Tongue") died, his widow proved up the tract. The store complex consisted of a four-room rock house, a two-room picket house, a one-room frame house, outbuildings, stable, picket corrals, and two wells. A one-room building served for a while as a school. The 1992 archaeological record identifies only one large building, a masonry rubble mound (the store) and a foundation about the same size as the one-room frame house. The family closed the store in the middle or late 1920s and sold the homestead in 1939 (Kelley and Francis 1994, 2006).

The second important viewshed in this area is that of Angel Peak (Figure 182). The peak is quite a prominent feature from the north, but its viewshed is surprisingly limited to the south. The mystery disappears when it is realized that the top of Angel Peak is about 500 feet lower than Huerfano Mountain (6,988 vs. 7,474 feet). Angel Peak sticks up as a "peak" only because hundreds of feet of rock have been removed from its base in Kutz Canyon (more in Chapter 31).

28

Along the Chacoan Great North Road

One of the exciting archeological discoveries made during the late 1970s and early 1980s was that of the so-called Chacoan roads. A remarkable characteristic of these features is their linearity. They radiate out from Chaco Canyon and maintain straight segments regardless of topographic obstacles. If they must turn, they do so in dogleg fashion. A great deal has been written about the Chacoan roads, and to avoid getting swallowed up in the subject I will focus on only that segment relevant to US-550. This is the so-called "Great North Road." It connects Chaco Canyon on the south with Kutz Canyon on the north across about 31 miles of rather bleak terrain. From Pueblo Alto, the great house on the north rim of Chaco Canyon, the road bears north 13 degrees east for about 1.5 miles, then within 0.5 degree of true north for ten miles to a collection of structures called "Pierre's Ruin," and finally north two degrees east for 19 miles to the south rim of Kutz Canyon (Figure 184; Gabriel 1991).

The Chacoan roads were constructed during the late fluorescence of Chacoan culture from 1050 to 1115 CE, and the Great North Road was one of the last of them. Most illustrations tend to depict the road as a continuous trace (e.g., Figure 184), but in reality, today at least, it consists of discontinuous segments. Construction of the road mainly involved removal of vegetation and soil to produce a smooth, level surface. It exists as two to four closely-spaced parallel traces, and its linearity suggests that it was laid out as a single project (Sofaer et al. 1986). After nine centuries the road has been nearly obliterated by blowing sand and historic human activities. It is difficult and often impossible to recognize on the ground, but in places its trace can appear quite striking on aerial photographs or high-resolution satellite imagery (more below).

What was the purpose of the Great North Road? To answer this we must remember the significance of "north" to the ancient people. The North Star is the sole fixed point in the night sky, and the Chacoan sky-watchers must have considered the north as an enormously important and fundamental position. Although some of the Chacoan roads may have had a utilitarian purpose, such as for transportation of food and timber, others had purely ritualistic purposes. The Great North Road seems to be one of the latter for there is no evidence that it had any utilitarian purpose at all.

The early part of the 12th century marked the decline of the culture at Chaco Canyon and the rise of the new Ancestral-Puebloan community near present-day Aztec (today's Aztec National Monument), about 50 miles due north of Chaco. The Great North Road may have been constructed to commemorate the shift of power to this upstart community on the Animas River. The "Aztec" community was perhaps legitimized by its position astride the symbolically important north axis emanating from Chaco Canyon (Frazier 2005).

More insight comes from modern Puebloan beliefs. Pueblo cosmology incorporates the four cardinal directions. To them north—the upward direction—represents the upper world, and south—the downward direction—the inner world. The passageway between the upper and inner worlds is the *sipapu*, or *shipapu*. Often the sipapu is viewed as a hole or canyon. Perhaps the Chacoans—the Ancestral Pueblo people—considered Kutz Canyon as the sipapu. This may explain why they purposely deviated the road two degrees east of true north from Pierre's Ruin (see below) so as to encounter the sipapu provided by Kutz Canyon. In short, the Great North Road is an over-engineered feature and lacks a practical destination. It clearly seems to have been constructed by the Chacoans to establish harmony between their world and the spiritual landscape (Sofaer et al. 1986).

The places associated with the Great North Road are Pierre's Ruin and two sites in Kutz Canyon. None is on a par with the magnificent structures at Chaco Canyon, nor even with those to the north such as Salmon Ruins west of Bloomfield and Aztec National Monument near Aztec. However they are far from the thundering crowds and provide a wonderful sense of solitude in the open spaces of the beautiful San Juan basin. They are therefore fun places to search out and visit.

Pierre's Ruin

The trajectory of the Great North Road passes through a group of archeological structures collectively called Pierre's Ruin (Figures 184 and 199A). The site was discovered in 1974 by E. Pierre Morenon, a young archeologist who had been working along the San Juan River and who had set out to find roads running southward. In 1976 the site was inventoried by the Bureau of Land Management (it is located on BLM land). It consists of two massive structures facing one another atop a mesa of Ojo

Alamo Sandstone. The two are together referred to as the Acropolis. A third structure is located at the base of an adjacent conical hill, on top of which there are traces of a hearth. The hill is located on the centerline of the Great North Road. Evidently the hill served as a beacon of sorts, perhaps to guide travelers, and it is appropriately named *El Faro* (Spanish for "the Lighthouse"). No traces of the road can be found on the ground at Pierre's Ruin, but segments show up on aerial photographs both to the north and south (Gabriel 1991). The site has not been excavated, so it remains essentially in an "as-found" condition.

Today the site is protected by the BLM. It is rather remote and quite unheralded, but it is easy to reach, open to the public, and very much worth the effort to get there. To reach Pierre's Ruin (Figure 199A), exit west from US-550 onto dirt highway NM-57 at the Blanco Trading Post (mile 123.3). Drive southwest for ten miles until a short spur takes off to the right (west). Take the spur across the cattle guard and immediately take a left (west) on CR-7650. Proceed four miles. Take the primitive dirt track (preferably with a high-center vehicle) north about 0.25 mile to an oil-field pumpjack. Park a short distance beyond the well amidst the sagebrush. A hike of a little more than a mile due north across several drainages and a fence takes you to the base of the mesa on the north side of the main dry wash. The mesa can be recognized by its trademark "smiley face" (Figure 199B), and by a BLM perimeter fence.

Kutz Canyon Chacoan sites

At mile 136.7 on the east side of US-550 is county road (CR) 7175. A BLM sign indicates that this is the main access road to the BLM's Angel Peak Recreation Area, featuring Kutz Canyon. Located in the recreation area are two fascinating Chacoan sites: the terminus of the Great North Road known as the Upper Twin Angels Mound, and the ruins of Twin Angels Pueblo.

Upper Twin Angels Mound

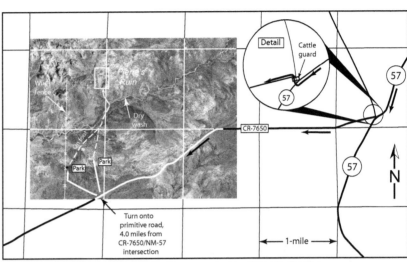

A. Access to Pierre's Ruin augmented by *Google Earth* image.

B. North view of "smiley face" mesa on which Pierre's Ruin is located. (Photo by author 2005.)

Figure 199. Pierre's Ruin.

County Road 7175 leads northeast for 0.85 miles through the mess of the Envirotech land farm, and then makes a sharp bend to the southeast to hug the canyon rim (Figure 200A). About 1,000 feet past the bend on the canyon (north) side is a picnic area with a restroom. Paradoxically the john is the terminus of the Great North Road. Until recently the road was detectable only on certain aerial photographs, but the amazing satellite imagery recently introduced by *Google Earth* has provided another source of information (Figure 200B). The remarkable images clearly show the old road bee-lining north toward the restroom. Despite the road's clarity on the satellite images, the trace of the road is not at all recognizable on the ground.

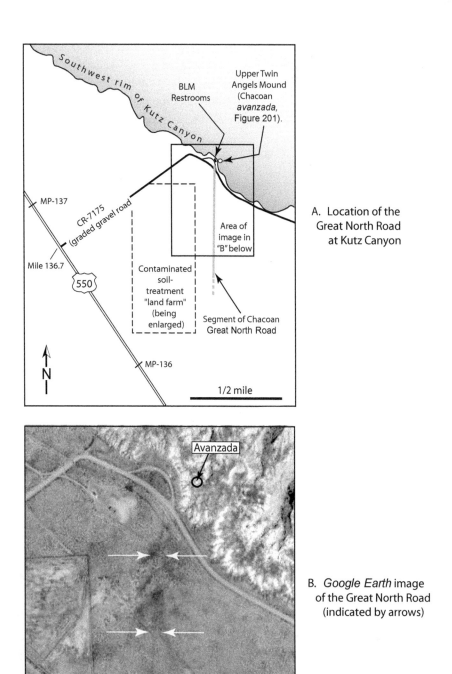

A. Location of the Great North Road at Kutz Canyon

B. *Google Earth* image of the Great North Road (indicated by arrows)

Figure 200. Chacoan Great North Road.

About 500 feet due east of the restroom, north and slightly detached from the canyon rim, is a dome-shaped mound of mudstone of the Nacimiento Formation. This is the Upper Twin Angels Mound. On top of the mound are the remains of an outpost, an *avanzada*, which may have been a small shrine or perhaps a signal post (Figure 201). A wooden stairway was discovered on the north side of the mound descending to the canyon floor. The stairway, since collapsed, consisted of a number of platforms supported by posts and crossbeams packed with soil (Gabriel 1991). Whether the road was ever extended beyond this point is questionable. This is likely the terminus of the Great North Road. Why is the mound slightly offset from the road's trajectory. Might the use of the mound have been an afterthought?

Chacoan *avanzada*

Figure 201. Northeast view of Upper Twin Angels Mound (location in Figure 201). (Photo by author 2005.)

Twin Angels Pueblo

The ruin of Twin Angels Pueblo is another one of those off-the-beaten-track places that are fun to visit. For good reasons, the BLM doesn't go out of its way to advertise such sites, and this little-visited site normally requires some research to locate (Figure 202). The ruin occupies the top of a narrow mesa that juts out on the west side of the inner Kutz Canyon rim and perches about 100 feet above the flat canyon floor. A trip to the site is not difficult, and is highly rewarding, not only for the incredible solitude but for the impressive views of the Kutz Canyon below.

Twin Angels Pueblo was first excavated in 1915 by the young Earl Morris, one of the first archeologists to work in the area. In his notes he referred to the place as both the Twin Angels Ruin and the Kutz Canyon Ruin (Carlson 1966). Earl Halstead Morris (1889–1956) was a native of the Farmington area. In 1916, under the auspices of the American Museum of Natural History, he conducted excavation and restoration work at the Aztec Ruins until 1923 when the site was declared a National Monument. He became the monument's first custodian. Later that year he moved on to work at Canyon de Chelly in eastern Arizona and remained there until 1929. He went on to a establish a respected reputation for himself in Mexico and Guatemala.

To reach Twin Angels Pueblo I advise use of a four-wheel-drive, high-clearance vehicle. Access from US-550 is east from mile 139.2, just south of the Giant Service Station on the opposite (west) side of the highway (Figure 202). The following unmarked mileages are cumulative values measured from this turnoff from US-550. Take the turn off to the east of the highway, cross the cattle guard and go a few feet to the unpaved pipeline road that parallels US-550. Take the

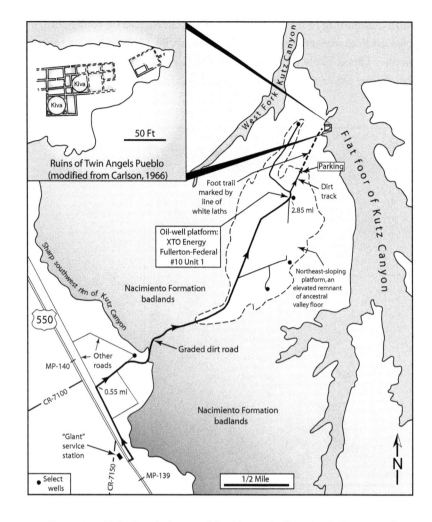

Figure 202. Twin Angels Ancestral Puebloan ruin (location in Figure 208).

pipeline road north to mile 0.55 and then take a right. The road curves to the right around where a gas well used to be and then begins a descent across the badlands northeastwards into the canyon. The badlands are carved into the soft mudstones of the Nacimiento Formation. At mile 1.0 the road crosses a cattle guard, and at mile 1.45 levels off considerably and traverses a broad surface that slopes gently down to the northeast. This old rather smooth surface is a remnant of the ancestral valley floor that existed before streams incised it and developed the present rugged topography around it. Resist the temptation to take the occasional side road and instead continue straight ahead downhill to the northeast.

At mile 2.85 the road seems to end at a well site consisting of a oil-well pump jack and an enclosed compressor. The sign identifies the well as the XTO Energy Fullerton-Federal No. 10, Unit 1. From the well site take a sharp left onto the dirt road that passes the small metal shed. Just a few tens of feet past the shed is a rutted, very subtle track striking off northeast. Follow the track on the slight downhill grade for 0.2 miles, for a cumulative of 3.05 miles from US-550. The track ends at a little sign that says "Twin Angels Pueblo." Park there. A gentle foot trail follows a prominent line of white plastic laths (pipeline trace?) for about 0.5 mile to the enclosed site.

The Twin Angels Pueblo was built by the Ancestral Puebloan about 1100 CE and abandoned sometime between 1150 and 1300 CE. The site consists of two separate structures (inset in Figure 202). The main one is a block of 17 rectangular rooms and two kivas. There is no evidence of a second story. The pueblo was clearly a defensive structure. To the east out on the end of the promontory is a smaller block of six rooms. Much of both blocks has come loose and tumbled down the slope.(Carlson 1966).

The site must have been quite important to justify the effort to build and maintain it. Some questions naturally arise such as where did these people get their food? Where did they find the flaggy, slabby sandstone beds to build the pueblo because there is no obvious source of this material anywhere to be seen. What about water? The people may have trapped rain water in a *tinaja* (depression or tank) in the massive sandstone lip at the top of the big drop-off next to the pueblo on the north side, or perhaps they found a nearby spring. One might conclude that the site wasn't continually inhabited but rather was manned on special occasions or on an as-needed basis. Wonder about these things as you soak up the ambience of this little jewel.

29

Mormon Settlers, Cattle Companies, Trading Posts, and Kutz Canyon

We are now at the point where it is appropriate to stir some human history into our stew of geology and geography. That history is rich, rather complex, and utterly fascinating.

First settlers

The San Juan River is joined by the La Plata River and the Animas River, all of which have their headwaters to the north in southwestern Colorado. About 70% of the total surface water supply of the state of New Mexico is here. It was therefore very well-watered, lush, and potentially extremely rich agricultural land — if it could be successfully irrigated. However it was also Indian country and it remained off limits to settlers.

After the Civil War, prospectors and miners began to filter into southwestern Colorado. When the Utes ceded their claims to that area in 1873, miners surged into the valley of the upper Animas River, north of present-day Durango. Durango itself was founded in 1880 by the Denver and Rio Grande narrow gauge railroad, and settlement accelerated afterwards. The valley of the San Juan River, located well to the south of all this activity, was the land of the Navajo, the Utes, and the Jicarilla Apache. In 1874 a huge swath of land north of the San Juan River had been set aside as a reservation for the Jicarilla (Chapter 23), but the tribe preferred not to get that close to the other two and avoided the area. Congress withdrew the reservation in 1876 and opened the area for white settlement.

Colorado cattlemen were the first to be attracted to the valley as winter range for their herds. In the late 1870s Anglo farmers from southwestern Colorado and Hispanic families from the Tierra Amarilla area of New Mexico began to filter into the valley (Duke 1999). The Anglos tended to concentrate near the junction of the Animas and San Juan Rivers. This place was naturally first referred to as "Junction." The Navajos called it *Totah* — "land where the waters meet." This place would eventually become Farmington. The Hispanic farmers clustered upstream to the east near the mouth of the Largo Wash.

Many of the settlers were not stockmen, but rather farmers, and irrigation of the San Juan River's flood plain was the key to growing crops in this arid land. This was a time before powered earth-moving equipment, and the digging of irrigation ditches was backbreaking and often futile work. Periodic floods would wipe out the ditches and force the settlers to begin all over again. Until the 1950s the history of this area revolved around the struggle to control the river. I'll talk about the history of the Bloomfield area in the final chapter of this book (Chapter 32), but right now let's take a look at one group of newcomers to the San Juan country: the Mormons.

Enter the Mormons

The Mormons began to trickle into New Mexico's lower San Juan River Valley in 1878 (Figure 203). In fact, the end-point of our voyage, Bloomfield, is named after an early Mormon settler. The Mormon communities were centered on the towns of Olio and Burnham (later renamed Kirtland and Fruitland, respectively) (Julyan 1996). The Mormons went on to become very influential in the economic development of the lower San Juan River Valley. However, non-Mormons (like me) tend to know little about this extraordinarily industrious group. What is Mormonism? Their story is best told in context, so let's take a brief look back at their tumultuous history, their core beliefs, and their arrival in the western U.S. Most of the history below comes from Arrington and Bitton (1992).

Beginnings of Mormonism

The movement that was to become Mormonism has its roots in western New York state during the early 19th century. This frontier region was at the time wracked by the religious turmoil of what was to be later called the "Sec-

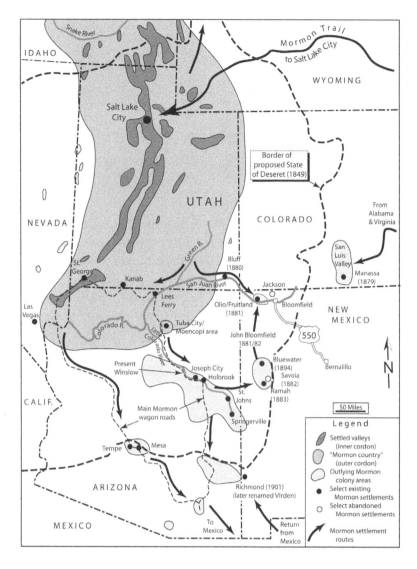

Figure 203. Mormon pioneering in the Southwest. (Modified from Walker and Bufkin 1979; Tietjen 1980; McTighe 1984; Arrington and Bitton 1992.)

ond Great Awakening" (1790–1840). This was a period of great religious upheaval, with widespread Christian evangelism and conversion. It was named for the "Great Awakening," a similar period that occurred almost a century before in the 1740s. Many people sought a return to what they believed were the original fundamental concepts of New Testament Christianity, and they rejected the new doctrines and practices that had developed over the centuries from European religious traditions

Into this cauldron arrived the Smith family. The Smiths had earlier attempted to put down roots in New England, but were driven from the rocky fields to western New York by the crop failures of 1816, the "year without a summer." The Smiths settled in Palmyra, near Rochester. One of their many sons was Joseph Jr., born in 1805.

In about 1820, when Joseph Jr. was about 14, he reported his "first vision," in which Jesus Christ appeared and spoke to him. In 1823 he had another vision. This time an angel called "Moroni" announced himself as last prophet of a vanished race that had lived in the ancient New World. Moroni told Joseph that he, Moroni, and his father "Mormon," had engraved a history of their people many centuries ago on a set of gold leaves. He instructed Joseph to retrieve them and told him where to look. Joseph obeyed and went to a hill near his home. He reported that he found a stone box with the plates and a strange instrument to be used for translation. The instrument consisted of two transparent stones, which Moroni had identified as the biblical *Urim* and *Thummim* used by ancient seers. But Smith was unable to remove the plates. Moroni told him to return in a year for another try and repeat the exercise each year until he was successful. On his fifth try, in 1827, he retrieved the plates. He was now 21 years old.

He told only his family about it, but the story of the plates leaked out. The text on the plates was in "Reformed Egyptian" and needed translation. Smith needed solitude and peace to begin the large task ahead so he and his new wife, Emma, moved to her parents' home in Pennsylvania. Using the transparent stones, Smith, hidden behind a screen, translated the plates and dictated the text to his scribe seated outside. In 1829 he finished the 275,000-word manuscript, the "Book of Mormon."

Book of Mormon

The Book of Mormon is the Mormons' holy book, and they consider it a complement to the bible. The crux of what the Mormons believe is summarized below, and is taken from an amalgam of sources that include Arrington and Bitton (1992), Tietjen (1980), and Remini (2002).

The book is purported to be a chronicle of several peoples named after Moroni's father, Mormon, an ancient American soldier-historian. The story begins ca. 600 BCE in the Holy Land. The Hebrew prophet, *Lehi* (a descendent of *Manasseh*, one of the twelve sons of the prophet Joseph, son of Jacob and founder of one of the 12 tribes of Israel), was instructed by God to flee with his family just prior to the Babylonian invasions. Lehi's sons, under the leadership of the youngest, *Nephi*, gathered up their people's history recorded on brass plates, and led the family across the Arabian peninsula to the Indian Ocean. There they constructed a boat, and sailed supposedly east across the Pacific Ocean some 15,000 miles to a promised land on the west coast of Central or South America.

Once there they founded a great civilization. However they soon began to quarrel and split into two factions: the rebellious *Lamanites* (after *Laman*, Nephi's older brother), and the more obedient *Nephites* (after Nephi). In time the two brothers became the patriarchs of two diametrically opposed civilizations.

Mormons believe that Christ, after his crucifixion, made an appearance in the Americas and gave the people the authority to carry on his work. Some 200 years later, discord set in once again among the two factions, and in about 421 CE the Lamanites defeated the Nephites in battle. Mormon, the last prophet of the Nephites, took all the engraved plates and abridged them into a form that would become the Book of Mormon. Moroni, Mormon's son, buried the records of his people in a hill, where they remained until found by Joseph Smith Jr. some 14 centuries later.

The victorious Lamanites mixed with other groups from Europe and Asia to become the Indians of South, Central, and North America. God eventually "cursed" the Lamanites by giving them dark skin. Thus was laid the foundation for the Mormons' affinity toward the dark-skinned native people of the Americas.

After showing the golden plates to several of his friends, Smith, as instructed, returned them to Moroni. Throughout 1829 Smith had other revelations regarding the proposed organization of his new church. He issued "A Revelation on Church Government," which would become the church constitution. The Book of Mormon was first printed in 1829–1830. In 1830 in Fayette, New York, Smith and others officially organized the "Church of Christ," later referred to as the "Church of the Latter-Day Saints" (emphasizing that this was Christ's Church in the last days), and in 1838, as the"Church of Jesus Christ of Latter-day Saints."

Initial resistance

After 1830 word circulated of a new gospel that aroused ridicule and opposition from many quarters, but belief from a few. Word spread abroad as well. Mormonism had become a movement. The Smith family, as well as a few friends and neighbors, rallied to the cause. In 1831 Joseph Smith and some New York Mormons moved to Kirtland, Ohio, which for nearly seven years Kirtland was the center of Mormon activity. The locals, however, felt threatened because the new religion in their midst challenged accepted Christian values. And then there were the Indians. The Mormons believed that the Indians—the Lamanites—were destined to rise and assume their rightful heritage in the Americas, and the Mormons' suspected agitation of the Indians was deeply suspect. The Mormons were viewed also as economic competitors. To boot they were also seen as political competitors because they typically voted as a bloc. Finally, the Mormon belief that all the other religions were false and offensive to God, and that their leaders were corrupt was hardly a way to "make friends and influence people," especially during this emotionally-charged period of American history.

The resulting persecution convinced the Mormons that they were the heirs of the early Christian saints. They moved to Missouri and then, in 1839, to western Illinois. Within a year the town of Commerce, Illinois (renamed "Nauvoo") became the Mormon hub. From Nauvoo, Smith sent several trusted advisors, including Brigham Young, to England to preach, and by 1841 there were several thousand Mormons in England, 400 of whom emigrated to the U.S. the first year. Over the next five years Nauvoo grew to a town of 15,000 and became the second largest city in the state after Chicago. Incredibly, as many as 35,000 converts joined the movement in the first 14 years. Mormon preaching in the U.S., as well as in Europe, especially England and Scandinavia, brought a steady stream of converts and the migration to Nauvoo continued unabated. In 1844, Joseph Smith was killed by an angry mob, and Brigham Young stepped into the breach.

Appeal of the new faith

How to explain the enthusiasm for this new movement? The Mormons presented a "primitive gospel" theme, which claimed that religion should be made more personal and acceptable to the common man. They felt that the established churches had abandoned their primitive and pure Christian faith. The Mormons believed that a mass conversion to Christianity would prepare the way for the Second Coming of Christ and the imminent end of the world. They felt themselves charged to establish the moral, social, and political framework necessary before the Second Coming. They felt that the here and now was of vital importance and were not preoccupied quite as much with the hereafter. To them, the acquisition of material wealth, via farming or commercial activities, was a desirable objective to advance the work of the Church.

Most of the early converts were people of the lower economic classes, with no specter of an inheritance and little hope for better times. The new faith appealed to their deep-seated emotional needs. Also Mormonism encouraged escape from present oppressive and discouraging conditions to a place of safety and security. Such "gathering" together was a natural screening device because many had to give up property, money, family, and friends. After making such emotional decisions, reversal became very difficult. They were committed. They quickly became involved with the movement and their fellow converts, and the collective activity created psychological bonds that became reinforced by financial bonds. Mormonism brought the downtrodden and lonely into a society where they were called "brother" and "sister" and made to feel part of an extended family.

The great migration

By 1845 it became clear to Brigham Young that the Mormons had to leave Nauvoo. But to where? Fortunately the American frontier provided an "escape hatch." Young acquired a book written by explorer John Frémont about his 1843–1844 expedition to the Great Basin of present Utah, and Young liked what he read. He decided to move the saints west. In 1846 the Mormons abandoned Nauvoo—the "City of Joseph." The mass exodus involved about 15,000 people, 3,000 wagons, 30,000 head of cattle, and many head of sheep, horses, and mules, all strung-out into a moving chain of loosely-linked groups.

Late in 1846 and in 1847 the Mormons arrived in the Great Salt Lake Valley. The valley was attractive to them simply because it was unattractive to everyone else. They didn't wish to compete for land again. Their Salt Lake organization was considered a "stake in the tent of Zion." In 1848 Young organized the community into 19 territorial wards, or congregations, of about 100 families each. Each ward was led by a bishop, who saw to it that the ward built a school, fences, roads, and canals.

Through 1855 the approximately 22,000 emigrants up to that time were about 90% English, 9% Scandinavians, and a smattering of French, Italians, and Germans. During the second wave of Mormon settlement in the last quarter of the century the number of Scandinavians eventually totaled more than 30,000. Few converts came from Spain, Portugal, Italy, Greece, the Balkans and eastern Europe due to the fact that some of these countries prohibited Mormon missionary work. Between 1846 and 1887 European emigration to Utah totaled more than 85,000. By 1870 British-born immigrants made up nearly 25% of Utah's total population. Most had urban and industrial backgrounds and tended to locate in the main cities. They were easily absorbed and soon moved into leadership positions. Scandinavians, however, encountered unforeseen problems. Arriving behind the English, most were forced to go farther afield to find arable land and water, and of course there was the language problem.

Colonial outreach and proposed State of Deseret

Soon after arriving in the Great Salt Lake Valley, Brigham Young instituted a vigorous colonizing effort. During the period 1849–1854 colonies were established in the arable valleys within Utah, the so-called inner cordon (Figure 203). An outer cordon reached well beyond Utah and encompassed an enormous area. Mormon colonization was based on the unique relationship they understood to exist between themselves and the Indians. Conversion of the Indians and teaching them effective methods of agriculture were therefore primary obligations. Furthermore, Brigham Young dreamed of a self-sufficient community that could supply its own staples. The Mormons believed in isolation and separation, but often out of necessity they had to deal with the gentiles (non-Mormons), mainly through the accepted Mormon occupations of freighting and selling of produce. Colonization was necessary both to relieve population pressures on the irrigated valleys of the inner cordon and to extend the productive area. Settlement toward the north was somewhat inhibited by a short growing season and low agricultural production. Settlement toward the warmer south offered more promise.

Brigham Young intended the Mormon lands to be accepted by the U.S. as a territory, carved out of acreage ceded by Mexico in 1848. Before colonization was complete, and realizing that California and Colorado were also questing statehood, in 1849 he proposed that this vast area be organized as the State of Deseret (Figure 204). The term "deseret" comes from the word "honeybee" in the

Book of Mormon, and the bee and beehive—symbols of industriousness—are today Mormon icons. Congress was aghast at Young's territorial overreach and instead, via the Compromise of 1850, created a much-reduced Utah Territory. The territory extended from the Continental Divide in the Rocky Mountains to the Sierra Nevada in California, and included the southern California coast, which at that time was barely inhabited.

Figure 204. Brigham Young's 1849 proposed Mormon state of Deseret (gray) vs. scaled-back 1850 Territory of Utah (heavy solid line).

In 1854, missionaries were sent to southern Utah to organize this outreach program. Between 1858 and 1862 they crossed the Colorado River into the Hopi lands of Arizona and eventually arrived at Moencopi, where the Hopi encouraged the Mormons to settle (Figure 203). From their base in Moencopi, in the early 1870s the Mormons explored the valley of the Little Colorado River in northern Arizona to investigate the possibility of settlement there. The following year they established four colonies near present-day Winslow, only one of which

(Joseph City) survives today. Internal bickering in that area resulted in a group of Mormons going on a mission to place called "Savoia," near present-day Ramah (originally called Navajo) in western New Mexico. One of these settlers was John Bloomfield, who arrived there in 1882 (Tietjen 1980). He would later appear in the San Juan River Valley (Chapter 32).

Mormons in the San Juan River Valley

In 1880 a group of 250 Mormons, with 80 wagons and a large contingent of livestock, headed toward the valley of the San Juan River in southeastern Utah and settled in the area of present day Bluff (Figure 203). Soon conditions became overly crowded and a small group continued on upstream and settled in the area of present-day Fruitland, New Mexico, on the San Juan River. Pleased with what they saw there, the pioneers in 1881 convinced others to join them (Tietjen 1980).

The first order of business was to install a small community ditch to irrigate their crops. In 1882–1883 the Mormons attempted to settle on the lower La Plata River a few miles upstream from Fruitland, called Jackson. Conflict with non-Mormon cattlemen and water shortages doomed the effort. A second colonizing effort was launched in 1898, upstream on the San Juan River above present-day Bloomfield near Blanco. This colony was named Hammond after Francis A. Hammond (d. 1911), whom the Mormon Church had authorized to preside over the San Juan Stake. By 1911 this colony had failed as well due to the inability to construct a durable irrigation system, and the colonists returned to their base in the Fruitland area (Duke 1999). Today the Fruitland/Kirtland area remains the center of Mormon influence in the San Juan River Valley. Only an "echo" of the Mormon movement—the name adopted for the City of Bloomfield—remains at the end of our US-550 (more in Chapter 32).

Three diverse Mormon beliefs

One Mormon belief has been a source of grief to the Mormons, and two have been a boon to Mormons and non-Mormons alike. Polygamy, the doctrine of plural marriage, was introduced by Joseph Smith as early as 1841. The Mormons placed great emphasis on marriage and having children. Plural marriage (at least for those few

283

who could afford this expensive hobby) was part of that endeavor and, indeed, part of their faith. But they also considered themselves part of the United States, and the practice alienated them from the political mainstream. Anti-Mormon "crusades" raged during the 1870s and 1880s and many polygamists wound up in prison. Only in 1890, bowing to the overwhelming pressure, did the leader of the Mormon church issue a decree, called the Woodruff Manifesto, banning the practice. With this hurtle behind them Utah acquired statehood in 1896.

However, not all Mormons were happy with the Manifesto and with outsiders "meddling" in their religion. Many voted with their feet and, beginning in 1885, fled to the Casas Grande Valley in the state of Chihuahua, northern Mexico for refuge. During the next 20 years there was a constant flow in that direction and the Mormons established a number of successful colonies. While there they were also free to continue the custom of plural marriage. When the Mexican revolution flared up in 1912 the Mormons found themselves harassed alternatively by the two opposing sides. Many Mormons returned to the U.S. via El Paso, but many others opted to remain in Mexico, where they prosper today.

The second Mormon belief is that of postmortem baptism. This holds that baptism is a necessary ritual to erase sins and to find salvation in the hereafter. Because the faith had been non-existent for so many centuries, untold past generations lacked the benefit of baptism. This explains the Mormon focus on genealogy—to save as many deceased family members as possible via postmortem baptism. The Mormon Church maintains a huge database of genealogical data available to Mormons and non-Mormons alike on the Internet, called [familysearch.org], and no genealogist can afford not to access it.

Finally, an overarching Mormon characteristic is the belief in hard work, education, and cooperation. This explains the extraordinary success of this extraordinary group.

Carlisle Cattle Company and Gallegos Wash

In 1883 the Carlisle brothers, Edmond and Harold, and their Kansas and New Mexico Cattle and Land Company purchased 7,000 head of cattle from ranchers on the east side of the Blue Mountains near Monticello in southeastern Utah. By 1885 the herd had increased to 10,000 head. The Carlisles established an enormous operation in the Gallegos Wash area, south of the San Juan River in New Mexico (Figure 184). The wash was astride the main wagon route and later the main "highway" between the San Juan River and Gallup.

The Gallegos Wash, a tributary of the San Juan River, is often misleadingly referred to as Gallegos Canyon (e.g., on the revised 1979 USGS Gallegos Trading Post 7.5-minute quadrangle topographic map). It's a stretch to call this a canyon. Anywhere else it would be called simply a valley, but in this arid country the term "wash" will do. The Gallegos Wash is unable to incise a rugged topography, unlike areas to the east such as at Kutz Canyon. A schematic cross section reveals the reason: the floor of the wash is underlain by the relatively hard Ojo Alamo Sandstone, whereas the Kutz Canyon area is underlain mainly by the soft mudstones of the Nacimiento Formation (Figure 205). This fortunate event guaranteed that downcutting of the Gallegos drainage would be impeded and that a gentle valley would initially develop. This provided a natural passageway that attracted travelers, and the travelers established wagon roads. Last but not least, the extensive grass cover on smooth terrain made it prime ranching country.

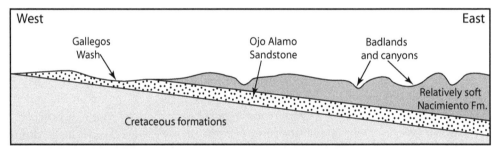

Figure 205. Schematic cross section showing geologic contrast between broad Gallegos Wash and rugged badlands to east.

The Carlisles had chosen their ranch site well. They invested a considerable sum for stock tanks and windmills to water the several thousand head of cattle they ran there. Twice a year the herd was moved up north through Mancos, Colorado, in the fall it was driven down to Pueblo Bonito in Chaco Canyon, and in the spring it was moved back up to the high country to graze (Kutac 2001). The company established its headquarters 14 miles up the canyon to the south (McNitt 1962).

Most of the cowboys who rode for the Carlisle outfit were from Texas. They were tough, wild, and reckless. Butch Cassidy and others of the "Robbers Roost" band worked at the Carlisle Ranch at one time or another. But others were good and decent. Harry H. Green later went on to become mayor of Moab, Utah, and Emmet Wirt established a trading post on the Jicarilla Reservation and was highly respected (Kutac 2001).

During the time the company operated (1883–1900) its cowboys managed to become equal-opportunity offenders. They had serious run-ins with Ute Indians in Colorado and Utah, and Mormon settlers and peaceful Navajos in New Mexico. The Gallegos ranch was located only 15 miles to the east of the Navajo Reservation, and when 50 Navajo families moved off the reservation and settled along the Gallegos, the Carlisles were furious. They griped so loudly that the Department of the Interior ordered the cavalry to move in and remove the Indians. Even though the Navajo were then no longer living in the Gallegos they still continued to visit a trading post run by a certain W. B. Haines. The Carlisle foreman accused Haines of encouraging the Indians to come to the Gallegos. The Haines' store soon went out of business (Kutac 2001).

The Carlisles had less success with intimidating Hispanic sheepherders. A gunfight on the Gallegos broke out between one sheepherder and one cowboy, but before long others joined. In the aftermath several sheepherders lay dead. For several days the cowboys were under attack, hiding in a pump house. Finally the sheepherders withdrew with their dead and wounded. A court trial followed. While the sheepherders who had survived the shootout gave their testimony in the courtroom, the cowboys pointed loaded rifles and pistols at them. The cowboys were acquitted. The New Mexico territorial governor caught wind of the dubious proceedings and ruled that the testimony was obviously given under duress and should be discounted. He went on to offer a reward for the capture of the men who had killed the sheepherders. He also accused Edmond Carlisle of seizing control of portions of the public domain that had been used by the sheepherders for a generation or more (Kutac 2001).

The Carlisle Cattle Company's days were numbered. More and more sheep moved into the Four Corners area. In the 1890s there was a drought and the price of beef fell. Beset on all sides the Company began to sell off assets and ceased operations in 1900. They were not missed (Kutac 2001).

Area trading posts

In 1868, when the Navajo were allowed to return from their internment in Bosque Redondo, many returned to their former homes in the Checkerboard Area east of the reservation that had been set up for them (Figures 179A and 180). They were destitute. There were no towns or roads. The army was charged with supplying the Navajo with food, clothing, and tools from Ft. Defiance at the southern end of the reservation—a formidable distance for many to travel. The next year the army began giving out sheep, and eventually sheep became the basis for the Navajo economy, augmented by goats, some cattle, and horses. With the wool the Navajo weavers produced rugs, clothing, and saddle blankets.

It didn't take long for peddlers to realize that the Navajo had something of value to sell and during the 1870s and 1880s they slowly moved into the area. Prerequisites for siting a post were that it be a natural gathering place for the Navajo, near water if possible, and accessible by horse-drawn wagon. In the beginning they peddled their wares out of tents or the back of wagons, but by the 1880s and 1890s their places of business became, fixed, well-known points. They provided groceries such flour, sugar, and coffee, and took in trade wool, rugs, and sheep. The number of posts multiplied in the early 1900s (Roberts 1987). Three are relevant to this segment of our story.

Simpson's

As mentioned above, Gallegos Wash was a natural corridor south from the San Juan River to the Checkerboard Area and points south and southeast. A dirt road—later to become State Road 35—led up the canyon from the San Juan River Valley and until the 1920s was the main road south to Gallup and to Albuquerque via Chaco Canyon (Figures 9 and 10). It was also a natural location for a trading enterprise and Dick Simpson heeded the call. Richard "Dick" Simpson was born in England in 1863. At the age of 16 he left school to become a bank clerk. He resigned in 1890 and moved to London, where he worked as a stockbroker. He didn't do well, so in 1892 he moved to the U.S. and the following year made his way to the San Juan country. Like Will Lybrook, he evidently was a "remittance man"—one whose family paid to go off somewhere out of sight and harm's way—because occasionally over the years he received large sums of money from his family back in England (McNitt 1962).

He farmed and raised sheep near Kirtland west of Farmington for a few years and then sold out. In 1896 he moved south up to the junction of the west and east forks of the Gallegos Wash (Sec. 20-T27N R12W) next to the dirt road highway, where he acquired the land and abandoned former headquarters of the Carlisle Cattle Company (Figure 184). He built an L-shaped adobe trading post across the road from the Carlisle buildings, a well house, a three-room home for himself and his Navajo wife, a granary, blacksmith shop, and guest house (McNitt 1962). His wife *Yanabaa'* was a skilled weaver from a local Navajo family, and Simpson's Navajo name was *AyEhE*, or son-in-law (Kelley and Francis 2006).

By about 1910, the store was the largest one south of the San Juan River. It served as a supplier of wholesale goods to some of the smaller stores farther south (including Carson's, see below) and was the major supplier of wagons and buggies to the region's Navajo residents (Kelley and Francis). In addition to running the store, Simpson raised large numbers of sheep. The money he received from his family in England allowed him to dabble in other ventures as well. In about 1921 he organized the Simpson Mercantile Company, a wholesale trading firm in Farmington. He became a naturalized American citizen and only once, in 1919, returned to England.

After Yanabaa' died, Simpson sold the post in either 1927 (McNitt 1962) or 1929 (Roberts 1987) to the Progressive Mercantile Company and moved to Farmington. He married a white woman and—true to form—operated a small grocery store there until his death in 1945 (McNitt 1962). In about the late 1950s the new owners of the Gallegos store abandoned Simpson's buildings and located a new post a little more than a mile northeast (Sec. 16-T27N R12W) near a more reliable source of water. They named the new enterprise the Gallegos Sheep and Mercantile Company (Roberts 1987). The post was operated until about 1972 when it was closed and sold to the Navajo Agricultural Products Industries (Chapter 30; Kelley and Francis 2006). Today the latter site has been scraped almost clean, leaving only the vestige of perimeter markers made up of cobbles of quartzite hauled in from the San Juan River Valley to the north. Simpson's is gone.

Carson's

The Navajo name for this place is *Hanáád*, for "It Flows Back Out." The name "Carson" is synonymous with trading in northwestern New Mexico. John Christopher, or "Kit" Carson (no relation to the famous scout and In-

dian fighter) came to the U.S. in the late 1860s or 1870s. For a while he worked for the Union Pacific Railroad as a stoker. When he reached Colorado he gave up his wonderful life shoveling coal to work in the mining towns near Lake City. Kit staked a claim, and his brother Will joined him from Canada. They did quite well and sold their claim for a nice profit. A mining district, Carson, was named for them. The brothers moved to the Farmington area and homesteaded on adjacent plots just east of the present town. Kit and his brother married sisters, the daughters of a fellow homesteader who lived across the river—the proverbial girls next door.

Will and his new wife moved on, but Kit stayed and became a fixture in San Juan River life. He and his wife had seven children. One of those was Orange J., nicknamed "Stokes" from a cartoon character, born in 1886 (MacDonald and Arrington 1970; Roberts 1987). In 1916 he took a job at a store at Star Lake, about 30 miles southeast of Chaco Canyon. He wanted his own store and so kept his eyes and ears open for an opportunity, and in 1917 found it on the Gallegos Wash.

By 1907 a Shultz Trading Post, a wood-frame affair, had been located near the present site of the post. In 1914 Joe Hatch and George Kentner took over that store, but soon sold it to the Reidner Brothers. By about 1917 the store burned down and was never rebuilt. Later that year Nick and George Maher, who were then operating out of a tent, started building a store at the current site, but ceased after a disagreement. In 1918 they sold the partly-finished store to Stokes and his wife. The Carsons operated out of tents until they finished building the new store. The Carsons filed a homestead claim on the tract, and proved it up in 1923 (Kelley and Harris 2006).

Some of Carson's first wares were supplied to him on credit by Dick Simpson, his nearest neighbor down Gallegos Wash. At this time the store's Navajo customers negotiated a web of unmarked roads leading from home to home and from water well to canyon (Figure 9). The best, most traveled roads were dirt or "improved," meaning graveled (Roberts 1987).

The Navajo chapter system was introduced in 1927 to give the spread-out Indians a voice in their local affairs. In 1931 Stokes donated a few acres to the tribe behind the store for a chapter house, the Huerfano Chapter. After Dick Simpson sold out his enterprise on the Gallegos in the late 1920s, Carson acquired some of his customers. In 1936 he bought a tiny post called *Huerfano*, near the foot of Huerfano Mountain. In 1938 state route NM-44 was paved and the Huerfano store found itself on the main drag leading to the San Juan River. Over the years he acquired additional stores, mostly on the Navajo Reservation, some of which were operated by his kids. Farmington was the wholesale supply center for the stores. In 1952 Stokes sold the Carson's store to his daughter and son-in-law, put the Huerfano store up for sale, and with his wife moved to Farmington. He was no longer a young man.

The late 1960s and early 1970s were times of great political and social unrest in the U.S. In 1972 the federal government clamped down on the trading establishment for a long list of real and perceived abuses toward its Navajo customers. The long-held practice of taking in pawn was terminated, and new regulations were installed. The endless and often acrimonious proceedings were a special agony for Carson. He died in 1974 at age 88. Various members of the Carson family ran the store until 1986, when they closed it. In 1990, the new owners were Robert and Lorraine Garlinghouse. The present facility consists of a multiroom trading post with living quarters, lumber barn; lumber utility shed, and corrals.

Today few real "trading posts" remain today. Most have been replaced with modern, far-less interesting, chain convenience stores (Roberts 1987). But Carson's Trading Post is still there, off the main highway, US-550 (Figure 206A). For an interesting side trip to Carson's turn west from US-550 at mile 130.25 onto County Road 7300, and continue 7.5 miles (Figure 206B). The sprawling post is on the north side of the Gallegos Wash at the intersection with CR-7150. The latter road is paved and leads north to the Giant service station and store back on US-550 at mile 139.3.

A. North view of Carson's Trading Post and old bridge over Gallegos Wash. (Photo by author 2005.)

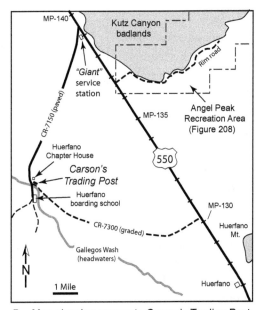

Figure 206. Carson's Trading Post.

B. Map showing access to Carson's Trading Post.

287

Huerfano

Two defunct Huerfano posts were located near the intersection of US-550 with CR-7500 (at mile 127.3), that leads to the Bisti/De-Na-Zin Wilderness. Around 1918–1919, an A.M. Smith had a store in the badlands just south of Huerfano Mountain, and there was perhaps another, later store. In 1936 Stokes Carson bought one of these stores from Glen Swires. In the late 1930s one account refers to an "El Huerfano Trading Post" located "near the foot of the peak" (Workers of the Writers' Program of the Work Projects Assocation 1989), and this was probably Carson's. Later (ca 1940?) Carson moved to a new building two miles west of Huerfano Mountain. In the 1940s, when the highway was paved, he moved to a new building on the west side of the highway. Carson sold the store in 1952. In the 1960s the new owner built a store across to the east side of the highway. It was a multi-room cinderblock building (a ruin by 1997). In 1974 partners Jerry Apodaca (soon to be New Mexico governor), William Marchiondo, and Jack Richards bought the store. By the late 1970s they sold the property for the Navajo Indian Irrigation Project (Chapter 30) and the store was abandoned. The Carson store on the west side of the highway by 1997 had been reduced to a rubble mound (Kelley and Francis 2006).

Kutz Canyon Badlands

The colorful badlands of Kutz Canyon lie off to the east side of highway US-550 via County Road 7175. The official name for a good part of the canyon is the Angel Peak Recreation Area. The recreation area was established by the Bureau of Land Management (BLM) in 1964. The event was the outcome of ten years of prodding by the Bloomfield Lions Club (Reager 1965). The recreation area features an unpaved (often washboarded) road along the south rim, a wonderful view point for a cliffy feature called Castle Rock, two picnic areas, and a small, dry, but very nice campground with a very impressive view of Angel Peak.

The origin of the name "Kutz" is unknown. Early on the area had

been called "Coots" Canyon by some (Taylor 1981). It may be more than coincidence that a prominent merchant from Aztec was named John Koontz, who with his wife in 1892 donated the land on which the San Juan County courthouse would be built (Duke 1999). Or, more likely, the name "kutz" might be an unsuccessful attempt to pronounce, and spell, the Navajo word, *Tsé Gizhi*, meaning "Rock on Top of Two Prongs" (Reager 1965), for the area's signature landmark Angel Peak? This area has gone by several nicknames, including the "Enchanted Desert" (Woods 1962) and the "Nacimiento Badlands" (Reager 1965).

Kutz "canyon" is not a discrete canyon at all. Rather, it is a large swath of broken, badland terrain carved into the Nacimiento Formation of Paleocene age. The area is bounded by a sharp rim on the south and southwest and opens out to the San Juan River Valley to the north (Figure 207). The cliffy rim consists of the upper part of the Nacimiento Formation, which is rich in channel-sandstone bodies. The typical badlands in the interior lower parts are developed on the soft mudstones of the lower part of the Nacimiento Formation, which is the product of deposition of muds in river flood plains and lakes.

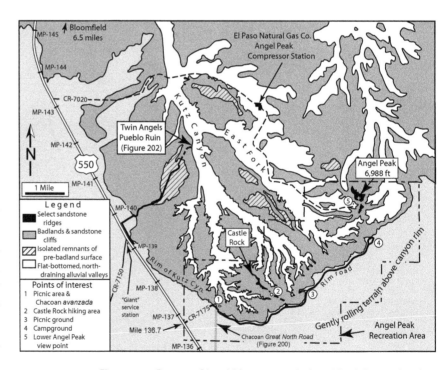

Figure 207. Bureau of Land Management's Angel Peak Recreation Area and Kutz Canyon.

The canyon's principal landmark of course is Angel Peak, so named because of its vague resemblance to an angel with outstretched wings (Figure 208). The "angel" is propped up by two little resistive, remnant caps of the San Jose Formation that protect the relatively softer Nacimiento Formation below (*a la* Huerfano Mountain). However, the angel's days are numbered. The little caps will eventually weather down to insignificant nubs and the final chunks of sandstone will come tumbling down.

Figure 208. North view of Angel Peak (No. 5 in Figure 207) in Kutz Canyon. (Photo by author 2004.)

The canyon's second principal landmark is the more obscure Castle Rock, a prominent prow of cliffs that juts north from the southern canyon rim. It is easily accessible from the rim. At a point 2.3 miles east along the rim road from the first picnic ground (No. 1 in Figure 207) is a small gas-field installation on north side of the road on the canyon rim. Behind it is a steep, narrow dirt road that plunges 260 feet north down into the canyon. At the end of the road is a graded platform containing a gas well (No. 2 in Figure 207). Gas-field trucks make it down to this point so you can too, albeit with a four-wheel-drive vehicle and steely composure. From the well, with a little care one can hike to heart's content amidst some very imposing cliffs and overlooks (Figure 209).

Figure 209. North view of Castle Rock (No. 2 in Figure 207) in Kutz Canyon. (Photo by author 2004.)

30

Indian Water Rights—the Essentials

As US-550 approaches Bloomfield it passes by the edge of a huge, irrigated farm. At a lofty elevation of 6,200 feet this appears to be a highly unlikely place for an irrigation project. Where's the water coming from? Well, the story behind this project has a significance that reaches far beyond the borders of US-550, and even beyond New Mexico. This farm is an exemplar of the festering issue of water rights and the senior claimants of those rights—the Indians. This issue is enormously important for New Mexico and indeed much of the Southwest and it needs to be understood.

The American Nile—doling out the Colorado River Basin's water

In 1912, Joseph B. Lippincott, a tireless advocate of the City of Los Angeles' water interests, famously said, "We have in the Colorado [River] an American Nile awaiting regulation" (Carrier 1991). His comparison of the Colorado with the Nile is apt. Both rivers are "exotic" streams in that they acquire most of their water from a relatively small upstream watershed and then channel their flows great distances through arid lands. In the case of the Colorado River Basin these arid lands are potentially arable, if only they could avail themselves of the water necessary for irrigation.

A decade after Lippincott's pronouncement the process of regulation and control formally began with the Colorado River Compact of 1922. Delegates from seven western states met in negotiations led by Commerce Secretary Herbert Hoover and divided the Colorado River drainage basin up into an "upper basin" and a "lower basin," with the dividing point arbitrarily set at Lee's Ferry located just below today's Glen Canyon Dam (dedicated in 1966) in northern Arizona (Figure 210).

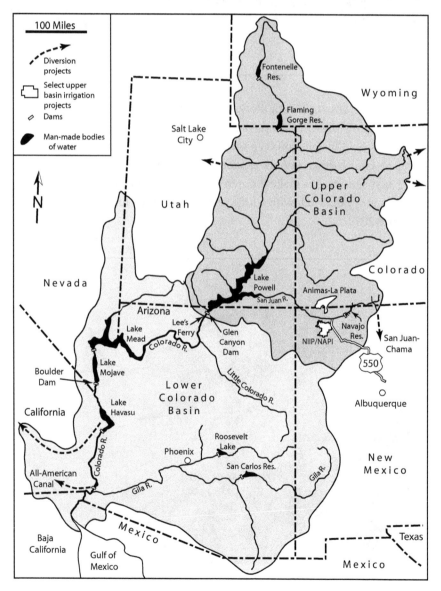

Figure 210. Colorado River Basin and main water projects.

As a benchmark for planning, the compact accepted the Bureau of Reclamation's estimate of 17.5 million (M) acre feet for the river's total annual flow. The acre-foot (ac-ft) is the most useful unit for quantifying the large water volumes of stream flows and irrigation projects, vs. the standard "gallon" used to measure the liquid volume of such things as gasoline tanks, water cisterns, etc. An acre-foot is the volume of water required to cover an acre of land (43,560 square feet) to a depth of one foot, and is equivalent to about 326,000 gallons. The infamous river-flow figure of 17.5 M ac-ft turned out to be an error of epic proportions. That number had been based on about 18 years of steam flow measured by instruments primitive by today's standards. Also, during these years the river's flow was prodigious and not at all a representative average rate that would be useful for long-term planning.

Only about 85 years later was it learned that a see-saw type of relationship exists between precipitation in the Southwest (including the Colorado River Basin) and the temperature of the Atlantic Ocean. A warmer Atlantic brings on increased hurricane activity in the western Atlantic and drought in the Southwest; a cooler Atlantic diminishes the hurricane activity and gives the Southwest its wet years. The see-saw has been tipping for centuries, and although the exact mechanism is still unknown, global warming does not appear to be a suspect. The tips of the see-saw occur rather rapidly, but then stay fixed that way for decades (McCabe et al. 2007). Although it could not have been known at the time, the period from about 1900 to 1920—the period used to generate that 17.5 M ac-ft number—was very wet in the Southwest. Other tips of the see-saw affecting the Southwest were the drought in the 1950s (warm Atlantic), the wet years of the 1960s (cool Atlantic), and mid-1990s to the present (warm Atlantic, major hurricanes).

The Colorado River Compact apportioned 7.5 M ac-ft to both the upper and lower basins (2 x 7.5M=15M) and left it up to each basin to work out between themselves the distribution of its proper share. A flow of 1.5 M ac-ft was reserved for Mexico, as per treaty (15 M+1.5 M = 16.5 M), and—amid great controversy—the final 1.0 M ac-ft were apportioned as a bonus to the lower basin (16.5 M+1 M = 17.5 M). The legislatures of the seven states upset the plan and flatly refused to ratify the compact (Reisner 1987). In 1928 Congress intervened and authorized two big projects for the lower basin: 1) Boulder Dam between Nevada and Arizona, and 2) the All-American Canal from the Colorado River to southern California (Figure 210). The lower basin's constituents, led forcefully by Los Angeles, was

therefore first out of the gate in putting the water of the Colorado River to beneficial use.

Slowly, especially after the Great Drought of the 1930s, suspicions grew that perhaps the venerable 17.5 M ac-ft figure was not what it was cracked up to be. The upper basin, which lagged behind the lower basin in water development, gradually became alarmed. It had its own river-dam projects on the books that would require the basin's full 7.5 M ac-ft apportionment to be viable for electrical generation. If the lower basin were to be fully put to use, and thereby establish a claim to, its 7.5 M ac-ft, and if the total flow were to be something less than 17.5 M, the upper basin might have to do with less than its supposed 7.5 M ac-ft. Would the shortage be shared equally? Would the earlier projects claim seniority on the water?

Then the bombshell: the Upper Colorado River Basin Compact of 1948 determined that the volume of water available for non-Indian use was contingent on the volume actually put to Indian use (Clark 1987). Meanwhile the aim of the Bureau of Reclamation was to develop water that was "surplus" to senior claims. But who could say what volume of water was surplus?

In 1953 a respected hydrologic engineer announced at a conference in Washington, D.C. that the river's flow at Lee's Ferry had averaged a little less than 12 M ac-ft since 1930. Despite this, at the very same time the Colorado Basin states were furiously planning and building as if the total were actually 17.5 M. By 1965 even the Bureau of Reclamation estimated that the total flow was perhaps only as high as 15 M ac-ft. After delivering the 1.5 M ac-ft to Mexico, and losing about 1.5 M to evaporation, that left only about six M ac-ft for each basin. By the late 1960s the total figure was revised downward to 13 M ac-ft (Reisner 1987). After satisfying Mexico and allowing for evaporation, each basin might be faced with a nominal share of only five M ac-ft. A shortfall of 2.5 M ac-ft in an arid region represents a loss of colossal proportions. The time to panic had arrived.

In 1968 Congress enacted the Colorado River Basin Storage Act. This was the largest authorization of public funds in history. As the political price for support and passage of the act, the California delegation held out for, and got, a guarantee of 4.4 M ac-ft, regardless of shortages. The shortages would therefore come out of someone else's hide.

Finally, on December 13, 2007, after a monumental effort, the seven states in the Colorado River drainage basin signed a sweeping 20-year agreement to work out their future controversies in the Colorado River Basin via consultation vs. litigation. The pact eliminates the furious competition between the upper basin served by Lake Powell and the lower basin served by Lake Mead. This is the most significant agreement reached among the seven states since the original 1922 compact (Ritter 2007).

Navajo water rights—the Winters Doctrine

During the 1850s and 1860s the U.S. based its Indian policy on treaties. These pacts were intended to allow for safe passage of people through lands controlled by the nomadic tribes. By the late 1860s this policy was replaced by the reservation system. The intent of the system was to pacify the frontier once and for all. With the adoption of permanent reservations, the U.S. accepted the responsibility to protect the land and the waters supplying those lands.

Despite being confined to the lands of their reservation, the Indians clearly had a claim on the waters flowing into the reservation because that water was essential for them to adopt a settled existence—the very purpose of the reservation system. This overriding principle was to be enshrined in the statutes by an infamous U.S. Supreme Court case dealing with a contentious issue that had flared up far to the north. The Ft. Belknap Indian Reservation had been established in 1888 in north-central Montana along the Milk River (a tributary of the Missouri). Settlers soon flocked to the land upstream from the reservation and by the late 1890s began to appropriate the river water for irrigation. The U.S. brought suit against one of the settlers, Henry Winters, and his neighbors. At the nub of the issue was the implied reservation of a volume of water sufficient to satisfy the irrigation needs of the Ft. Belknap Indians, and by extension, other Indian tribes in similar situations. The Supreme Court ruled in the so-called Winters decision of 1908 that the Indians indeed held a senior claim to waters adequate to irrigate their land. Unfortunately the court did not rule on what an "adequate" volume of water was (Clark 1987). This ruling was quietly forgotten but it would reside in the corpus of court precedent like a piece of buried unexploded ordinance ready to go off when probed hard enough.

The Winters decision was later upheld and reaffirmed in 1963 by another landmark Supreme Court Case, Arizona vs. California. This interminable case had been plodding along since 1953. At issue were the respective water rights of the states in the Lower Colorado River Basin, but the court added a zinger heard loud and clear upstream as well. The judgement bolstered Winters by

ruling that the states had rights only to that water free from claims predating the formation of the Colorado River Compact of 1922. Furthermore, Arizona vs. California linked the total amount of water reserved for Indians to the total irrigatable acreage on their reservations (Clark 1987). The land used to construct the Navajo reservation had been land that for the most part no one else wanted. Its agricultural potential was limited by poor, sandy soils and a short growing season, but there was a lot of it and it therefore potentially at least required a great deal of irrigation water. The Navajo had implied rights to about 600,000 ac-ft of water—about 20% of the total runoff of the state of New Mexico. Arizona vs. California affirmed that the Navajo tribe could use every drop of that water even in the event of a crushing drought (Reisner 1987). These vague, implied rights to this amount of water were fully legalized on March 25, 2009, when the U.S. Congress formally ratified the tribes rights to the 600,000 ac-ft. Under terms of the act the Navajo Nation would give up its rights to sue for an even larger share of the water—which they possibly could win. The years of legal wrangling and uncertainty have thus come to an end. (Fleck 2009).

Upper Basin irrigation projects

With this legal background in mind, let's take a look back to the origins of irrigation projects on the Navajo Reservation. With the creation of the reservation in 1868 the government had encouraged the raising of sheep. An expanding population required an ever-growing amount of sheep. Overgrazing on the poor land led to erosion, which in turn jeopardized downstream water-storage projects. In order to restore the land, the government during the 1930s ordered a drastic reduction in livestock numbers. By 1940 it was clear that the Navajo would have to consider irrigation in order to survive. This would require a large upstream water storage project along the San Juan River. By the end of WWII there were only 12,000 acres under irrigation in the New Mexico portion of the reservation, but this total included floodwater irrigation projects, which were inoperable during times of drought.

The Bureau of Indian Affair's (BIA's) proposed Shiprock Project was to use 585,000 ac-ft of water to irrigate 117,000 acres of Navajo land by gravity flow below an impoundment. In the early 1950s the Navajo had asserted their prior Winter doctrine rights to San Juan River water but recognized they would be unable to finance such a huge project on their own to use all this water. In 1955 the tribe agreed to limit their Winter rights to 508,000 ac-ft of water to irrigate 115,000 acres. They later revised this area down to 110,630 acres. In 1957 the tribe relinquished additional Winters rights by agreeing to share water with other users during drought (Clark 1987).

The Colorado River Basin Compact of 1948 had accelerated the planning of a number of water projects and laid the foundation for the Colorado River Storage Act of 1956. The upper-basin states of Colorado and New Mexico were acutely aware that it was in their best interests to complete their projects as soon as possible to avoid downsteam users establishing a claim on the unused upper-basin water. The act called for the construction of a number of dams and reservoirs to regulate flow, assure delivery of water required by statute to Lee's Ferry, and to generate electric power, the sale of which would be applied to offset construction costs (Clark 1987). At the time much of the water in the San Juan River was rushing downslope to users in Arizona and California. The act authorized the construction of Navajo Dam on the San Juan River as part of the Colorado River Storage Project (CRSP). The so-called Navajo Unit, one of four units of the CRSP (Flaming Gorge, Glen Canyon, and Wayne Aspinall), would provide municipal and industrial water for the surrounding region, regulate a continuous flow of water for power generation at Glen Canyon Dam, and support the irrigation needs of the Navajo Nation (Figure 210).

Navajo Indian Irrigation Project (NIIP)

The Colorado River Storage Act of 1956 also prioritized the planning for the San Juan-Chama Diversion Project and the Navajo Indian Irrigation Project, or "NIIP." The Tribe would receive 508,000 ac-ft of water to irrigate 110,000 acres. The government would build the dam and the hydraulic system to deliver water the 25 miles (as the crow flies) from the reservoir to the farmland. The Navajo Tribe successfully argued that the dam and reservoir would go a long way to satisfy the terms of the 1868 treaty with the U.S. (Price 1985).

Navajo Dam was constructed in 1958–1962 by the U.S. Bureau of Reclamation (BR). Within the area of the reservoir, five cemeteries, four miles of Colorado state highway, and 6.5 miles of railroad were relocated. In 1962 the BR's San Juan-Chama Project and the BIA's NIIP were authorized (Figure 211).

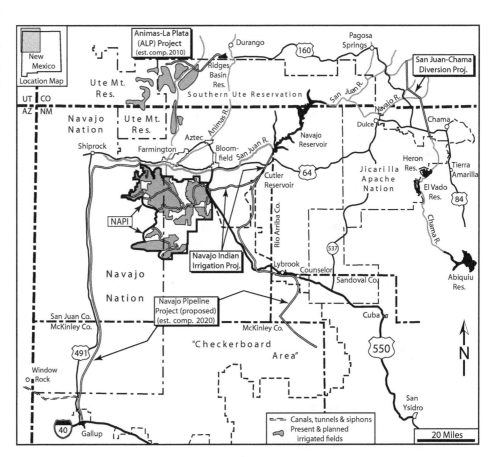

Figure 211. Present and proposed major water projects in the Upper Colorado River basin.

The BIA and the BR were direct competitors with regard to their projects. The BR had greater political influence than the BIA and therefore the former project moved ahead while the NIIP lagged behind. During this time a joint BR/BIA task force re-analyzed the NIIP and recommended that the acreage west of the Chaco River be eliminated due to its questionable value. Instead the task force recommended that the project be fully located on the high mesa south of the San Juan River and east of the Chaco River, mainly lying outside the Navajo Reservation. In 1970 the Interior Department was authorized to acquire such lands east of the reservation for incorporation into the NIIP (Clark 1987).

The NIIP's delivery system consists of a series of canals, pipelines, and the associated lift stations and siphons to transport the water to the perimeter of the farm (Figure 211). A series of gravity-feed canals then delivers water into the farm itself (see NAPI below), which is located on a pair of flat to gently-rolling mesas on the windblown remnants of the Chaco Dune Field. In March 1976 the water was turned on for the first time and delivered to the 9,000-ac Block 1, and the next month some 8,000 people turned out for the formal dedication.

The San Juan-Chama Diversion and the NIIP together used up most of New Mexico's upper-basin water allotment. These water projects coincided with explosive economic growth in the San Juan basin in general due to the development of huge deposits of coal, oil and gas. As the value of water for such development was recognized, the fundamental question of how much real, "wet" water was actually available begged for an answer. As mentioned earlier, the Bureau of Reclamation in the 1920s had estimated that the upper basin had a flow of 7.5 M ac-ft/year, and New Mexico used that figure to plan its projects. Since that time the estimates were revised downward a number of times. In 1966 the BR had announced that by switching NIIP from a gravity-flow to a sprinkler system the necessary diversion could be decreased from 508,000 ac-ft to 370,000 ac-ft, later revised to 330,000 ac-ft. The decision was therefore made to install a sprinkler system to water a huge swath of the Chaco Dune Field. This decision had been made possible by a remarkable invention by a Colorado tenant farmer.

The miracle of center-pivot irrigation

In the late 1940s a tenant farmer and tinkerer living on the plains of eastern Colorado, named Frank Zybach, constructed a device that would revolutionize irrigated agriculture and make the Navajo Indian Irrigation Project possible. Until that time the word "irrigation" was synonymous with human drudgery. Epic quantities of labor were required to dig the ditches, to open and close them as needed, and to repair the damage caused by erosion and rodents. The development of siphon tubes in the 1940s eliminated some of the labor, but the heavy tubes still had to be deployed by hand. The post-WWII invention of "gated" pipe and the availability of cheap aluminum tubing set the stage for the next big step (Splinter 1976).

In 1947, Zybach had just viewed a demonstration using hand-moved irrigation pipe and was convinced that there had to be a better way to distribute water. He went home and began to cobble together the first center-pivot irrigator. Over the next few years he refined the design to maximize its efficiency and in 1952 patented his invention. The center-pivot system was the first to automatically, efficiently, and uniformly irrigate a field, even one with sloping terrain. In 1954 Zybach sold the manufacturing rights and launched a world-wide industry.

The hallmark of center-pivot irrigation is the pattern of circular fields strikingly discernable from an airplane and even from space (Figure 212A). The field size is typically one quarter of a section, or 160 acres, the conventional agricultural unit in the U.S. Because the field is round, only about 133 of the 160 acres are actually irrigated, and the amount of "wastage" is reduced by close-spacing of the circular plots. The irrigation tube is conveyed by a series of triangular frames with wheels propelled by a mechanical linkage (Figure 212B). The water is supplied by a pump at the pivot point, and is distributed by sprinklers arrayed along the length of the pipe. The first center-pivots used high pressure to spray water into the air. Loss to blowing winds and evaporation was about 35%. Newer, less wasteful and more energy-efficient designs employ low-pressure nozzles that hang down from the main pipe and spray water gently onto the crops.

Center-pivot irrigation is ideal for sandy, porous soils, such as those of the Chaco Dune Field. Water delivered to such soils by the traditional, difficult-to-finesse ditch system tends to pass through the soil and bypass the plant roots. In contrast, the center-pivot system provides a controlled dose of water in just the right amounts needed by the crops. Without Zybach's invention the Chaco Dune Field would have remained a wind-swept, barren expanse of sandy wasteland.

Navajo Agricultural Products Industry (NAPI)

Gradually, along US-550 from about MP-24 north of the Blanco Trading Post to about MP-29 just past Huerfano, the landscape becomes higher and smoother as we step up onto the expanses of the Chaco Dune Field (Figure 213). After a few miles a number of water tanks, perched on tall legs, pop into view off to the west, and by about MP-140 we meet the first center-pivot irrigated fields on the west side of the highway. This is the eastern edge of the Navajo Nation irrigated farm administered by the Navajo Agricultural Product Industries, known as "NAPI." This 70,000-ac farm, eventually to encompass about 110,000 acres, is the fruit of the Navajo Indian Irrigation Project (NIIP), mentioned above. The farm is the largest contiguous one in the nation.

In 1967 the Navajo Tribal Council had formed NAPI to operate the farm supplied with water by NIIP. The organization was inactive for the next three years while it engaged in staffing and planning activities. Despite years of under-appropriations, the first water reached Block I in 1976. Each year additional blocks were brought into production. In the beginning the main crops were corn, pinto beans, and alfalfa for cattle feed. NAPI operates a fresh-pack potato plant and markets the product under the Navajo Pride label. Most of the potatoes are sold to Frito-Lay, but much of the crop is destined to the Campbell Soup Company and the Del Monte Foods Company (Larrañaga 2000). The public's recent concern about low-carbohydrate foods has depressed potato prices and NAPI has cut back on the potato acreage.

At the same time the drought of recent years has made alfalfa a high-priced commodity and acreage in alfalfa was accordingly increased. In 2005 total acreage in alfalfa was 16,500 acres. Eighty percent of the hay is sold in New Mexico (*Crop Quest Perspectives* 2005). Thus the common sight of flatbed trucks, loaded to the gunnels with hay, barreling south on US-550 to dairy markets throughout the state. As of the end of 2005, under center-pivot irrigation were 14,000 acres of corn, 7,000 acres of small grains (barley, wheat, oats), 5,000 acres of pinto beans, and 2,000 acres of potatoes. In addition, 10,000 acres are leased out. The farm also includes a 5,000-head feedlot and 200 acres of fruit crops (2005; Roberson 2006). During the two months of fall harvest NAPI employs about 1,200 people, and about 150 work year round. The farm clearly is, and is destined to remain, a major economic entity in San Juan County.

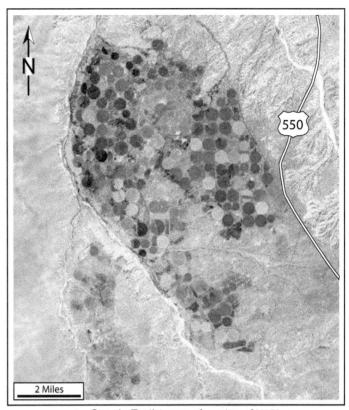

A. *Google Earth* image of portion of NAPI.

B. Center-pivot irrigation booms at NAPI. (Photo by author 2004.)

Figure 212. Center-pivot irrigation at Navajo Agricultural Products Industry, NAPI.

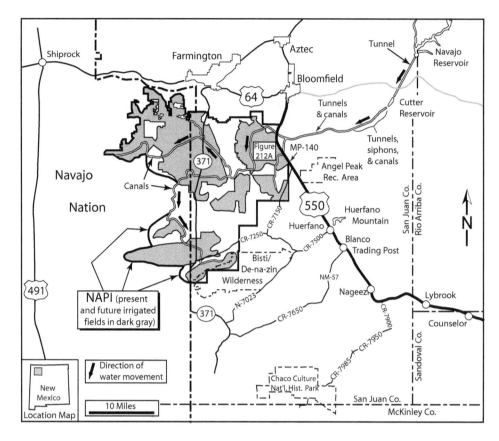

Figure 213. Navajo Indian Irrigation Project (NIIP) water–delivery system, and Navajo Agricultural Products Industry (NAPI) farm. (Modified from Goodman 1982.)

Navajo-Gallup Water Supply Project

In April 2005 the Navajo Nation reached an agreement with the state of New Mexico to settle the Nation's long-standing legal claim to most of the San Juan River's water. As the price for accepting only 56% of the water the Nation claims, it will receive almost $1 billion in federally-funded projects over 20 years. The main project will consist of a pair of buried pipelines to transport water from the river to the Checkerboard Area, the New Mexico side of the reservation, and the City of Gallup (Figure 211). Part of the project will be to fund completion of NIIP and increase the area under irrigation from 70,000 acres to 110,000 acres.

The end result will be that the Navajo Nation will control about 56% of the water from the San Juan River drainage basin, or about 600,000 acre-ft. Of the remainder, 16% will go to non-Indian water users in the basin, 5% to the Jicarilla Apache Nation, 17% to the San Juan-Chama Project users, and about 6% to area power plants (Linthicum 2004). This festering issue lingered between the slow-moving millstones of the bureaucratic machine until March 25, 2009, when the U.S. Congress formally ratified the tribe's rights to the 600,000 ac-ft. The congressional action provides a funding mechanism for the water pipeline project in particular and a stimulus for economic development in the entire San Juan basin (Fleck 2009).

31

Development of Landscape West of the "Lybrook Gate"

Now we'll talk about the landscape as a whole west of the Lybrook Gate to the San Juan River. The elevation profile of US-550 from the Lybrook Gate to the valley of the San Juan River gradually drops some 3,800 feet, from about 7,200 to 3,450 feet (Figure 181). This amount of topographic relief seems impressive, but the grade of the highway is today actually located far below the top surface of a much higher landscape formed about 26 Ma. For a fuller understanding of what we see here we need to therefore go back to 26 Ma before there was a San Juan River and ponder how profoundly the landscape has changed from that time in the Oligocene Epoch to the present. Geology forces us to think four-dimensionally, inserting time into our three-dimensional world. Now is the time to exercise our mental muscles.

By the close of the Laramide mountain-building event at about 40 Ma, all the early Tertiary and older formations had been warped downward to form the San Juan basin. The upturned rims of the basin had by then been beveled down to a gentle surface that rose in elevation from the basin center toward a rim of highlands to the northwest—a belt of uplifted, folded mountains, and to the southwest toward the Mogollon highlands (Figure 214A). Streams draining the highlands found their way southeastward to the Gulf of Mexico (e.g., Figure 151). Roughly between 35 and 26 Ma a number of enormous volcanic fields erupted around three sides of the basin and built themselves up to mountainous heights. These topographic barriers effectively blocked the streams from flowing to the Gulf of Mexico, and the basin thus became topographically isolated, or closed.

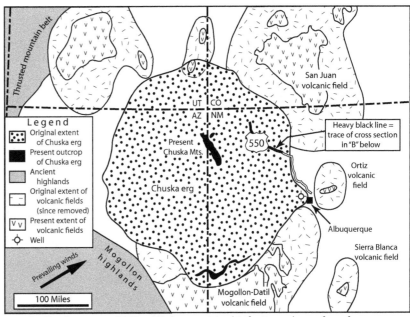

A. Chuska erg, ca. 26 Ma. Modern geographic features shown for reference. (Modified from Cather et al. 2008.)

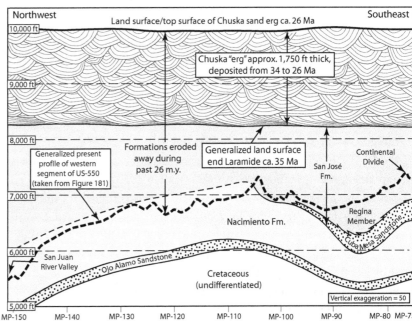

B. Restored schematic profile and geologic cross section along western grade of US-550 (heavy black-line segment in "A" above) between San Juan River at Bloomfield and Continental Divide.

Figure 214. Construction of ancient land surface, ca. 26 Ma.

The time datum of 35 Ma marks a global climate change, from warm and wet (greenhouse) to cool and dry (icehouse) conditions. The volcanic fields and the climate change together caused a "perfect storm." Streams stopped flowing and their beds of loose sand, as well as the up-lifted bedrock of the Mogollon highlands to the southwest, fell prey to the prevailing southwesterly winds. From 35 to 26 Ma several thousand feet of sand was piled up in the closed basin. The area became a huge expanse of wind-blown sand, termed an *erg* (from a Hamitic word used in the Sahara), which sprawled across more than 50,000 square miles and reached a ground-level elevation of some 10,000 feet (Figure 214B).

Enormous degradation of landscape during past 26 Ma

From about 26 to 16 Ma erosion cut about 4,000 feet down into the sandy landscape and removed most of the erg. Today the erg's very existence has been almost obliterated. Its original extent has been only recently established via the efforts of geologists from the New Mexico Bureau of Geology, led by master geologist Steve Cather, in an exquisite piece of forensic geology (Cather et al. 2007). Since the middle 1950s it had been known that the top of the Chuska Mountains of westernmost New Mexico is capped by a thick deposit of wind-blown sandstone, later found to be about 1,750 feet thick (Cather et al. 2008; Figure 214A). It was clearly the remnant of an erg, but where was the rest of it? Then, about 1,000 feet of the erg was found to crop out in the northern edge of the Mogollon-Datil volcanic field. Finally, in 1995 a thickness of almost 1,600-feet of wind-blown sand was encountered about 4,000 feet below the ground in a well drilled just west of Albuquerque in the Rio Grande rift. These big deposits of sand must be explained. How did they once connect? Cather et al. (2008) pulled the pieces together and successfully reconstructed the vast extent of the what they dubbed the Chuska erg (Figure 214A).

The time datum 26 Ma is very significant because it marks the end of aggradation on the Colorado Plateau and the beginning of a prolonged period of degradation initiated by the inception of the Rio Grande rift. It's been all "downhill" from that time. In short, the landscape we drive over along US-550 is only the latest of a progression of landscapes scraped progressively downward from the ultimate elevation of 10,000 feet. It's important to remember that the present landscape hasn't subsided from 10,000 feet to the present level, but rather that relentless erosion

has removed all those thousands of feet of sediment, mainly during the short ten-million-year interval of 26 to 16 Ma. In sum, our present landscape is developed on the eroded carcass of the old landscape formed 26 million years ago.

San Juan River—the driver of local landscape evolution

The name "San Juan" is the most used (31 times) place name in New Mexico (Julyan 1996), and it has evolved over time. In 1678 a Spanish punitive expedition against the Navajo, led by Juan Dominguez de Mendosa, was dispatched to destroy Indian fields in the area of the Navajo River and the "Río Grande" (Figure 211), as the San Juan River was then known. In 1705, another Spanish expedition, this one led by Roque Madrid (Chapter 21), also referred to the San Juan River as the "Río Grande," in contrast to the name "Río del Norte," as our modern Rio Grande was then called (Hendricks and Wilson 1996). The Dominguez-Escalante expedition of 1776 referred to the "Río Grande de Navajo," and the expedition's intrepid cartographer, Bernardo Miera y Pacheco, labeled the river on his map as the "Río Grande de Nabajoó." Gradually, in the early 19th century the modern name "Río San Juan," or San Juan River came into general usage (Julyan 1996).

The San Juan River and its tributary the Chaco River developed sometime about 10 to 5 Ma. By about 1.2 Ma the ancestral San Juan River flowed across a broad, low-relief landscape at an elevation at least 900 feet higher than it is today (Patton et al. 1991). Then, beginning about 800,000 years ago (800 ka), the cyclicity of climate change in North America changed in a major way and the landscape began to take on its modern form. What happened at this time to cause this very significant change? To answer this question we now have to delve into the origin of the San Juan River and to consider the two overarching processes that are principally responsible for the shaping of the landscape of the San Juan River drainage basin: 1) falling base level, and 2) global cooling leading to the Great Ice Ages of the Pleistocene Epoch.

Falling base level

The driver for landscape evolution along US-550 has been the falling or lowering of the "base level" (defined in Chapter 1) of the area's major rivers: the Rio Grande for the region east of the Continental Divide, and the San Juan River for that west of the divide. We'll deal with the latter here. As a re-introduction to this concept, first let's consider one of the most vivid exemplars of what happens to a river when its base level drops. This of course is the world-famous Goosenecks of the San Juan in southeastern Utah, about 120 miles west of Bloomfield (Figure 215).

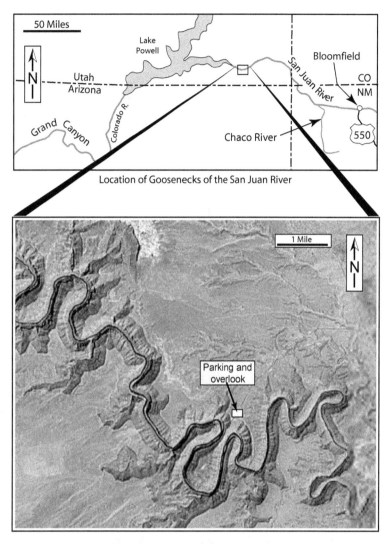

Figure 215. *Google Earth* image of incised meanders at the Goosenecks of the San Juan River, southeast Utah (for better depth perception turn image upside down).

Keep this image in mind (we'll return to it) as we continue with this discussion. Base level is defined as the lowest elevation to which running water can fall. The ultimate base level of course is the sea, which over the short term at least we can assume to be constant. Over the long term though sea level varies considerably. In fact, about 70 - 65 million years ago, when much of the surface of northwestern New Mexico was slightly above sea level, the absolute sea level was somewhere between about 350 feet (Miller 2006) to 750 feet (Haq, et al. 1987) higher than it is today.

To review, there are any number of local base levels upstream from the sea that affect local portions of a drainage basin. For example, the base level for the San Juan River is the elevation of Lake Powell in southern Utah, and base level for the Chaco Wash, a tributary of the San Juan River, is the bed of the San Juan River itself (see location map in Figure 215). Similarly, base level for any one of a number of tributaries of the Chaco Wash is the bed of Chaco Wash. The lower the relevant base level, the greater the vertical fall of running water from upstream. The greater the vertical fall, the more energetic will be the flow and the erosive power of that water on its downward journey to base level. Hypothetically, given enough time a stream will erode and smoothen its course toward base level, thus producing a stream "profile" with a minimum slope. As the profile stabilizes, the stream velocity and erosive power diminish, and by the late stages the stream flows along just barely fast enough to reach its base level.

Now let's return to Figure 215. Base levels for the San Juan River and for its tributary the Chaco River have dropped precipitously during the past million years. The driver of the process was the entrenchment of some 5,000 feet by the Colorado River to create the Grand Canyon far to the west in Arizona. This spectacular gorge was excavated in the space of perhaps only about four million years, from about 5.5 to 1.2 Ma. Up until then, an ancestral Colorado River, again fed by its tributary the San Juan River, lazily flowed off somewhere to the northwest into what is now southern Utah (Lucchitta 1989 and 1990). Whatever the elevation of the local base level of that old river was, it was clearly relatively high such that the river had a low gradient and flowed slowly, lazily meandering back and forth across its flood plain.

Meanwhile, base level of the lower Colorado River—that part today making up the western border of Arizona—was at sea level in the Pacific Ocean. Given the steep gradient of this segment of that old river, beginning about 5.5 Ma, the river vigorously and relentlessly eroded and deepened its channel headward to the east. Eventually the headwaters of this lower river cut far enough eastward to capture the substantial flow of the upper river (with its tributary the San Juan) to merge into a new, expanded Colorado River drainage basin. By 1.2 Ma the newly integrated Colorado River drainage roared through the Grand Canyon on its way down to its new base level in the Gulf of California (Lucchitta 1989, 1990). This spectacular base-level drop sent a pulse of adjustment upstream into our area of interest and lowered the local base levels for every river, creek, or rill upstream from it. When the pulse reached the area of the Goosenecks the river quickly cut some 850 feet down into solid bedrock, entrenching its old meanders. The pulse reached the upper San Juan River and its tributaries, with subsequent entrenchment there as well, although not in such an ostentatious manner.

Global cooling

The second principal process that influences a river's flow is the volume of water available to it. The element of climate now enters the fray. If the climate fluctuates from wet to dry, the amount of flowing water decreases in tandem. The climate has indeed made spectacular swings back and forth between wet and dry periods during the past two million years during what is referred to as the Pleistocene Epoch. The Pleistocene Epoch began 1.8 million years ago (although since 2008 geologists have begun to accept a beginning at 2.6 Ma). For convenience the end of the Pleistocene Epoch is conventionally accepted to be 10,000 years ago. The time since then is called the Holocene (or Recent) Epoch, and the Pleistocene and Holocene Epochs together make up the Quaternary Period (Figure 4). Several Great Ice Ages occupy the middle and latter part of the Pleistocene, ca. 900/800 ka to about 12 ka.

In the last few decades scientists have learned a great deal about the global cooling and intervening warming stages that make up the middle and late Pleistocene. The glacial stages can be plotted in exquisite graphical detail on a time scale due to the extremely fortunate relationship between ice-cap volumes and ocean chemistry (Figure 216). This at-first-intimidating graphical plot can be used as a rough tool to understand the timing of the events that formed the landscape of northwestern New Mexico (as well as that of the Rio Grande basin in central New Mexico). Therefore it is worth while to pause a moment and look at this fascinating relationship.

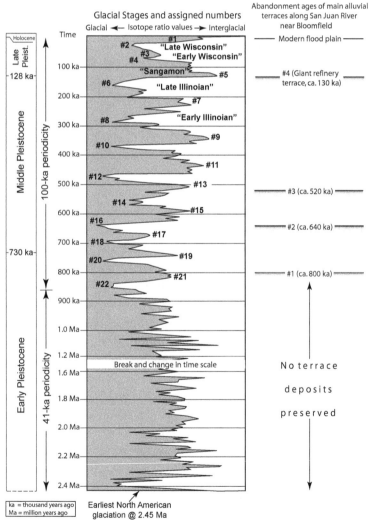

Figure 216. Glacial and interglacial stages (Nos. 1–22), based on oxygen–isotope ratios in oceans (values not shown) and applied to northwestern New Mexico. (Curve modified from Pazzaglia 2005.)

The relevant chemical element is oxygen. Oxygen occurs naturally in two stable isotopic (atomic mass) forms: light, "normal" oxygen with an atomic weight of 16 (written ^{16}O) and the much rarer "heavy" oxygen with an atomic weight of 18 (^{18}O). Both exist naturally in the atmosphere and in the water molecules (H_2O) of the oceans. The ratio of the two types varies slightly, but significantly, due to glaciation. As the climate cools at the onset of a glacial period, the water that evaporates from the ocean precipitates on the land in high latitudes as snow, which then forms ice and ice caps. Because water molecules containing the lighter oxygen evaporate more easily, the ice caps therefore become enriched in light oxygen and the ocean in turn becomes slightly enriched in heavy oxygen. Because the ice caps do not melt and the water locked up in them does not return to the ocean, sea level drops.

Certain types of tiny oceanic animals in shallow water extract both types of oxygen from the water, whatever the ratio, along with dissolved carbon dioxide and calcium to precipitate their shells of calcium carbonate ($CaCO_3$). As these animals die their shells fall to the deep ocean floor and become buried in an accumulating pile of deep-sea sediment. The $^{16}O/^{18}O$ ratios found in the fossilized remains of these animals extracted from deep-sea cores can be measured and plotted as an exquisite proxy for the volume of ice on the land, and for glacial growth and contraction (Figure 216).

The later part of the Pliocene Epoch, ca. 2.6 Ma (more recently considered the beginning of the Pleistocene), saw the first major expansion of ice caps in the Northern Hemisphere. This was the harbinger of the ice ages of the Pleistocene Epoch to follow. The trigger for global cooling was probably a fortuitous change of world geography. At about 2.5 Ma the Himalayan Mountains were greatly uplifted in Asia and the Isthmus of Panama was closed in what would become Central America, severely altering the distribution of global heat by atmospheric and ocean-current flows (Chapin 2008). The stage was thus set for the fateful interaction of the new geography with ongoing fluctuations in the amount of solar heat received and distributed by the oceans.

For the next 1.6 to 1.7 million years after 2.6 Ma the growth and waning of ice sheets, and the fluctuations of the world ocean volumes, occurred in a cyclic fashion having a 41,000-year (41 ka) periodicity. This strange number corresponds to the slight wobble of the earth's axis with respect to the earth's plane of rotation with the sun, ranging from 24.5 to 21.5 degrees. This causes a cyclic variation of solar radiation received in the high latitudes. A high angle leads to hotter summers and colder winters. Then, beginning between 900 ka and 800 ka, glacial episodes became much more energetic with a periodicity of about 100,000 years (100 ka). This fluctuation is in phase with the changing of the eccentricity of the Earth's orbit, from nearly circular to oval (Abreu and Anderson 1998). This causes the distance from the sun to vary by about 11 million miles and again influences the amount of solar radiation received by the earth. Although the two cycles overlap, the 100,000-year fluctuation has dominated from a little before 800 ka onwards. Why the second cycle overtook the first, at what is called the "Mid-Pleistocene Transition" — from 41 ka to 100 ka worlds — is uncertain and is the subject of vigorous current research. This global symphony is nicely illustrated by the marine oxygen isotope ratios (Figure 216).

The Pleistocene Epoch in the San Juan River Valley

The global cycles of net solar radiation have left a legacy in the landscape of the San Juan River Valley. The legacy consists of the the liberal garnishment of the valley walls, especially on the north side, by conspicuous deposits of cobbly, quartzite-rich gravel delivered from source areas to the north in Colorado. During the latter part of the Pleistocene, glaciers formed in the San Juan Mountains of southwestern Colorado, centered on the Silverton area, and flowed outward in all directions. The glaciers transported huge volumes of boulders and cobbles from the mountain heights down to the valleys below (Figure 217).

The interplay of glacial growth and contraction with the development of cobbly-gravel deposits downstream is rather involved, so I will greatly simplify the relationship as follows. The cyclic change of climate — from relatively warm to relatively cold, and then back again to warm — caused the glaciers to expand and contract in tandem. During the warm part of the climatic cycle little snow accumulated in the mountains to form ice, and melting increased at the unreplenished glacial margins in the lower elevations. The large volume of meltwater that was produced at the margins surged downstream with its load of course sediment. The deluge scoured a valley into the soft bedrock of the Nacimiento Formation along the San Juan River and its tributaries, and was quite capable of moving the coarse sediment through and beyond (Figure 218A).

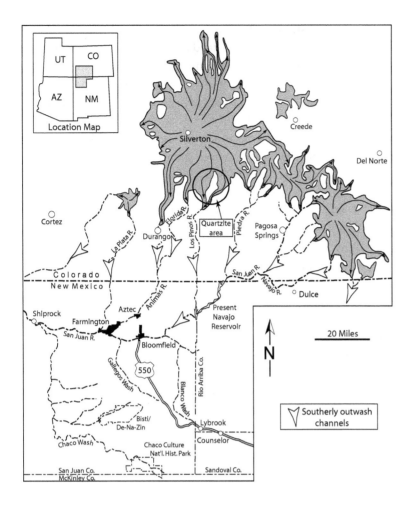

Figure 217. Late Pleistocene mountain glaciers (gray) in southern Rocky Mountains of Colorado. (Modern geographic features shown for reference; modified from Gilliam and Blair 1999.)

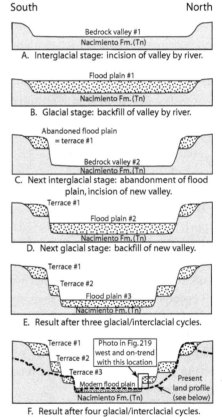

South   North

Bedrock valley #1
Nacimiento Fm. (Tn)

A. Interglacial stage: incision of valley by river.

Flood plain #1
Nacimiento Fm. (Tn)

B. Glacial stage: backfill of valley by river.

Abandoned flood plain
= terrace #1
Bedrock valley #2
Nacimiento Fm. (Tn)

C. Next interglacial stage: abandonment of flood plain, incision of new valley.

Terrace #1
Flood plain #2
Nacimiento Fm. (Tn)

D. Next glacial stage: backfill of new valley.

Terrace #1
Terrace #2
Flood plain #3
Nacimiento Fm. (Tn)

E. Result after three glacial/interclacial cycles.

Terrace #1
Terrace #2
Photo in Fig. 219 west and on-trend with this location
Terrace #3
Modern flood plain
Present land profile (see below)
Nacimiento Fm. (Tn)

F. Result after four glacial/interclacial cycles.

Figure 218. Schematic (not to scale) development of river terraces in the San Juan River Valley during the Pleistocene.

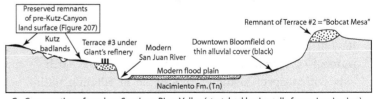

Preserved remnants of pre-Kutz-Canyon land surface (Figure 207)
Remnant of Terrace #2 = "Bobcat Mesa"
Kutz badlands
Terrace #3 under Giant's refinery
Modern San Juan River
Downtown Bloomfield on thin alluvial cover (black)
Modern flood plain
Nacimiento Fm. (Tn)

G. Cross section of modern San Juan River Valley (stretched horizontally for easier viewing).

302

When the climate returned to cold, snow and ice accumulation in the mountains increased and the replenished glacial margins advanced downslope. The lower volume of available meltwater flowing away from the glacial margins was unable to effectively transport all the available sediment, and this gravelly material was accordingly deposited by the river into the old valley, backfilling it and building up a gravelly flood plain at the top (Figure 218B). Another cycle of relatively warm to cold conditions, along with the continually falling base level mentioned earlier, repeated the process: the river cut a new, deeper valley and then nested a new valley-fill into it (Figures 218C and 218D).

Each time the abandoned, previous flood plain became a landform called a "terrace." This cycle occurred several more times (Figures 218E and 218F). The lower and younger terraces are the best preserved, the most extensive, and the easiest to trace. The lowest, most deeply nested level is that of the modern flood plain of the San Juan River (Figure 218G). The relationship of terrace gravels incised into the underlying bedrock of the Nacimiento Formation can be seen in excellent outcrops of the lowest terrace along US-64 between Bloomfield and Farmington (Figures 218G and 219).

Figure 219. Excellent example of cob-blygravel terrace deposit atop bedrock of Nacimiento Formation. (North side of US64 between Bloomfield and Farmington; photo by author 2007.)

In the late 1960s a University of New Mexico graduate student studied the river terraces in the Farmington area. He recognized at least five different levels (Pastuszak 1968), with most of the city constructed on the lower (and younger) two. More recent work in the Farmington area (Gilliam et al. 1999, Love and Connell 2005) has dated most of these terraces. The highest one was abandoned by the downcutting river about 800,000 years ago (Gilliam et al. 1999). This date is significant because it marks the approximate onset of the 100,000-year climate swings mentioned earlier, and, not coincidentally, the beginning of deep entrenchment of the Rio Grande (Chapter 4).

Each terrace has a characteristic "height" above the modern river, and this value provides a tool to trace the terraces upstream to the Bloomfield area. A new and exciting aid for this effort is the satellite imagery available online from *Google Earth*. One such image covering the terrain on the north side of the San Juan River between Farmington and Bloomfield clearly shows the trends of terraces—especially the oldest, highest and widest one—quite nicely (Figure 220). The northern limit of the highest terrace is now a prominent ridge trending east-west. The softer bedrock to the north of it, into which the terrace gravels had been inset, has been partly removed by erosion, producing a classic example of what geologists call "topographic inversion," i.e., where erosion has rendered a feature high that was once low (the inset gravel terrace).

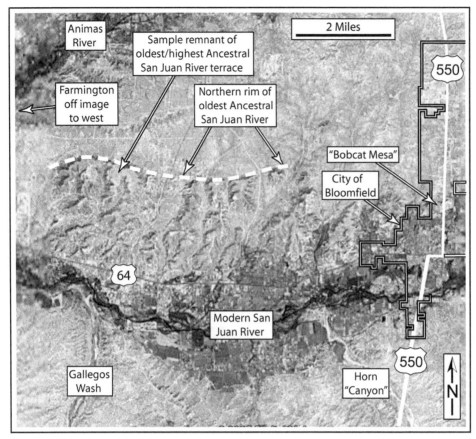

Figure 220. *Google Earth* image of portion of San Juan River Valley west of Bloomfield.

"Cobble Valley"

The San Juan River Valley could have been quite appropriately named "Cobble Valley." Cobbles are everywhere. They tumble down the riverside bluffs onto the roads, and the cobbly deposits frustrate builders who must somehow excavate construction sites on them. Besides being a nuisance, though, they are also fascinating. Truth be told, the cobbles—or rather their composition, shape, and internal constituents—provide a wealth of historical information.

Most of the cobbles are rounded chunks of the metamorphic rock type "quartzite." Quartzite is a rock that started out as sandstone. Due to the compaction and high pressures caused by deep burial beneath many thousands of feet of overburden, the individual sand grains and pebbles of the mineral quartz (SiO_2) have been pushed together and welded into a single, extremely hard massive unit of quartz. Whereas a sandstone will break around the contacts of the individual grains, quartzite—with great effort—will break directly through the grains. A chunk of

quartzite is therefore extremely durable. It can, however, be worn down chip by tiny chip by abrasion in the bed of a roaring stream, but it typically outlasts other rock types in the stream. It is this property of near indestructibility that causes them to be aggressively quarried as aggregate for use in construction and xeroscaping.

But where do the quartzite cobbles come from? Because water runs downhill we must of course look upstream for our source. The nearest source of quartzite is in the southern end of the San Juan Mountains, some 80 miles to the north (Figure 217). This very old material—about 1.7 billion years (1.7 Ga)—has been shoved to the surface from great depths and then, during the past one million years, scoured and plucked loose from its outcrops by glaciers. The ice transported the angular quartzite chunks downslope to the point near the headwaters of the present Los Pinos River between present-day Durango and Pagosa Springs, where the ice front melted. From there the chunks were tumbled along by powerful currents of glacial-outwash water, becoming nicely rounded and polished during their downstream trek. When climate shift reduced the amount of stream flow, the cobbles ceased their downstream movement, piled up atop each other and filled in the river valley. The tops of these old filled-in cobbly stream beds are now abandoned and left behind as terraces (Figure 218), and this is where the gravels reside today.

Being an unrepentant geologist I naturally collect rocks for a rock garden at my Albuquerque home. Some of my favorites specimens are these rounded cobbles from the San Juan River country. Many of them are chock full of interesting colors and textures. A close look reveals a long and involved history. For example, the photo in Figure 221 is of a cobble that I collected at Bloomfield (Figure 218G). The cobble itself is clearly a piece of conglomerate that is composed mainly of large, partly-rounded to angular fragments of quartz (the light-colored particles) with some other rock types (the dark fragments) thrown in. The coarse particles within the cobble therefore had been themselves transported and deposited as a "conglomerate" by an earlier stream sometime before about 1.7 Ga. The newly-formed conglomerate was then deeply buried, metamorphosed to quartzite, uplifted to mountainous heights, ripped loose from its quartzite outcrop during the past 1 Ma by a glacier, rounded in the bed of a glacial-outwash stream, and re-deposited as a new, gravelly sediment. This and other cobbles are clearly time machines that reveal stories within stories.

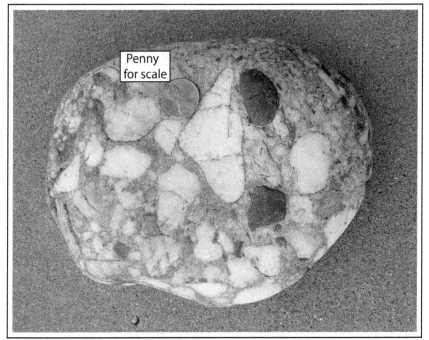

Figure 221. Quartzite cobble from top of "Bobcat Mesa" in Bloomfield. (Photo by author 2005.)

Pleistocene megafauna along the San Juan River Valley

The Bloomfield area has yielded some extremely interesting fossil discoveries of the so-called Pleistocene megafauna. The term refers to a collection of large, ungulate (hoofed) mammals that roamed across parts of North America in a climate that was cooler than today. During the latest Pleistocene the megafauna included elephants, horses, bison, and camels, in that order of frequency (Lucas and Morgan 2005). Fossil remains of elephants and horses have been found in the Bloomfield area. (As mentioned in Chapter 7, large megafaunal fossils have also been recently found near the town of San Ysidro).

The remains of a particular species of an ancient elephant, or mammoth (*Mammuthus columbi*) have been dug out of the old terrace-gravel deposits of the ancestral San Juan River about 4.5 miles east of Bloomfield (Lucas and Morgan 2005). The documentation for this fossil find is somewhat unclear; it may be based on a third-party report of mammoth bones found during construction of a gas-storage facility, but no official report was ever filed (O'Neill 1992). A composite skeleton of this species, but not of this individual, is presently exhibited in the New Mexico Museum of Natural History in Albuquerque.

These beasts evolved in Africa during the Pliocene Epoch (5 to 2.6 Ma). They migrated to North America during the early part of the Pleistocene when sea level fell due to continental glaciation and a land bridge formed across the Bering Strait between Eurasia and North America. By the early Pleistocene the mammoths became extinct in Africa and they began a spectacular evolution in Eurasia and North America. Near the end of the Pleistocene, ca. 11,000 years ago, they became extinct in the Southwest and before ca. 10,000 years ago they were gone from North America (Agenbroad 1984).

These animals probably roamed in small family groups and in small total numbers in the canyon bottoms and on the tops of the river terraces. They were mainly browsers, not grazers like horses or bison. Their diet consisted of vegetation from trees and woody shrubs such as big sagebrush, but also included some grasses (O'Neill 1992). By the end of the Pleistocene the climate was becoming warmer and drier. That sounds good, but for the mammoths it was bad. The woodland savannas were slowly being replaced by grassland savan-

305

nas, to the detriment of the browsing mammoths. But then a more insidious enemy, Man, had arrived.

The existence of Paleo-Indians is well documented elsewhere in New Mexico, especially on the eastern plains, but in the San Juan basin they left no significant traces of their passing (O'Neill 1992). Out on the plains they relentlessly hunted big game such as mammoths and bison. Overhunting in these areas, perhaps combined with the mammoths' poor adaptability to environmental change due to their low genetic diversity, could have reduced the big-game gene pool to the point of unsustainability in the broader range.

There is another explanation for the demise of the megafauna. Recent research has revealed that a comet or comet shower impacted the earth 12,900 years ago (10,900 BCE). The blasts of air would have ignited a conflagration that torched forests and grasslands alike and severely stressed animal life. The megafauna—the mammoths, camels, ground sloths, and camels—abruptly died off. At that time the earth had been steadily warming as it emerged from the last great ice age. Just then the earth abruptly cooled and entered another quasi-ice age that lasted 1,500 years (until 9,400 BCE), a span of time that climatologists call the "Younger Dryas." At numerous sites across North America sediments of this age contain microscopic, "nano-diamonds," caused by some kind of high-pressure/high temperature event, such as an impact, and the showering down of the nano-diamonds from the atmosphere. However, the "smoking gun," i.e., an impact crater, such as the Chicxulub crater in Mexico's Yucatan Peninsula (Chapter 14), has yet to be found (Johnson 2009).

A precise date for the Bloomfield fossil find cannot be determined. The description of the find (Lucas and Morgan 2005) suggests that it came from the gravels under the fourth terrace. This in turn suggests that the animal roamed the ancestral river before about 130,000 years ago. And the residents of Bloomfield thought their history began in the late 1870s!

At about MP-145, US-550 begins a 600-foot bee-line descent down into the valley of the San Juan River (Figure 181). Until 1947, however, the route was far less direct. From mile 147.8 on the modern highway the old road once veered off to the west and careened down to the river through the badlands (Figure 222). The old road has been chopped up and the southern part of this final stretch is no longer usable. A small section however does remain on the north and justifies a short side trip. From mile 149.8 on US-550, take San Juan County Road (CR) 5050 for 0.6 miles to the west until it encounters the old road. Four hundred feet to the south on the old road, on the west side, is a lonely and much-abused 5.5-foot-tall white pillar (inset in Figure 222). This is one of only four such mini-Washington Monuments that I know of along old NM-44 (the first is way back at mile 28.7, inset in Figure 100). Only the faded letters "NRS" remain of the inscription. During the 1930s the Highway Department, under the auspices of the National Recovery effort, had erected these monuments to mark the end/beginning points of major construction efforts and to serve as survey points. From here this northern segment of old NM-44 can be driven all the way down to the valley.

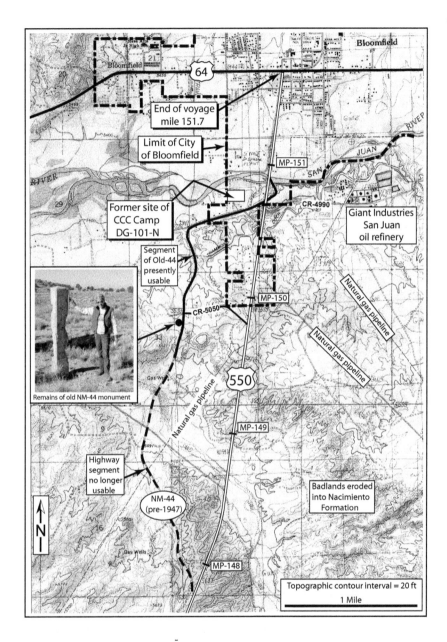

Figure 222. Route of modern US550 vs. old NM44 on final approach to Bloomfield. (From USGS 7.5-minute topographic quadrangle maps: Bloomfield 1985, Horn Canyon 1979.)

The San Juan River Valley cradles the community of Bloomfield. The town's history has been tightly constrained by the valley's geology, the need to overcome transportation hurdles, the vital need to control water, and the production of the fabulous mineral resources that lie deeply buried beneath the ground.

Geology of Bloomfield

After US-550 crosses the San Juan River it first embarks onto the river's flood plain, and then rises a tad and approaches the community of Bloomfield via a wide thoroughfare lined with unpicturesque commercial establishments (Figure 223). The flood plain is a flat surface that very gently slopes to the west at the same rate as the river. The feature is appropriately named because during flood stages the river would spill over its banks and surge across the plain in reckless abandon. After the construction of Navajo Dam those days are now past.

Figure 223. North view of Bloomfield at end of Old NM-44 portion of US-550, with two signature mesas on northern skyline. (Photo by author 2007.)

The low-lying flood plain sharply contrasts with an array of gravel-capped surfaces and mesas that occupy the valley's north and south flanks. The mesas are composite features that consist of a layer of cobbly gravel that was deposited by ancestors of the modern San Juan River, atop the mudstone and sandstone bedrock of the underlying Nacimiento Formation. As mentioned earlier (Chapter 31), such gravels, a.k.a. terrace deposits, are remnants of abandoned valley floors.

Using the *Google Earth* image (Figure 220) and the USGS.7.5-minute topographic maps I have tentatively traced three of the five terrace levels and their dates from near Farmington to the Bloomfield area (Figure 224). The highest terrace level, No. 1, perched some 450 feet above the modern river, has been completely eroded away in the Bloomfield area. The second highest, No. 2, is preserved only on the north side of Bloomfield where it has been eroded into two distinctive mesas. The western of the two, which I have tentatively named "Saddle Mesa," has a characteristic sag in its center. The eastern one, which I call "Bobcat Mesa," has been adorned by the Bloomfield High School (the "Bobcats") class of 2007 with a huge letter "B." This terrace level was probably abandoned about 640,000 years ago (640 ka) and now sits at an elevation some 300 feet above the modern river. The third level, No. 3, lying about 100 feet above the river, is well preserved on the north side of the river west of Bloomfield, and on the south side of the river at Bloomfield where it provides a conveniently-level platform for the big oil refinery there. This terrace was abandoned about 130 ka (Love and Connell 2005).

Rolling terrain that has developed on the Nacimiento Formation occupies the regions of higher elevation where the terrace gravels have been completely stripped away by erosion (Figure 224). These rather barren, impermeable surfaces are not suitable for irrigated agriculture. Gently sloping down from the bedrock hills to the flood plain from both the north and south are sheets of alluvium deposited by minor tributary drainages of the San Juan River. In the Bloomfield area the tributary drainages are well developed on the north side of the valley and their alluvial deposits are therefore widespread. Most of the City of Bloomfield has been constructed on these gentle slopes.

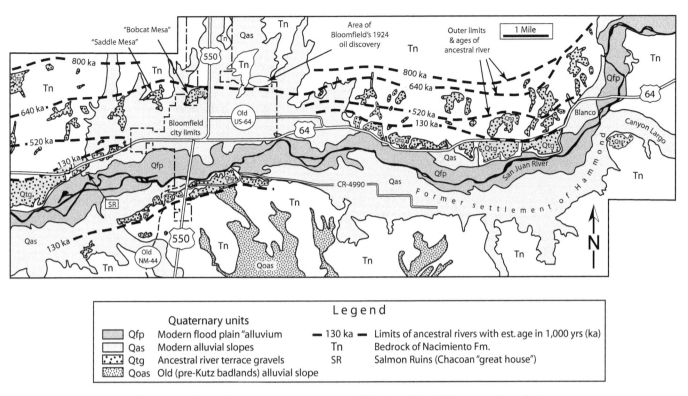

Figure 224. Generalized Quaternary geology of San Juan River Valley near Bloomfield.

Legend

Quaternary units
- Qfp   Modern flood plain "alluvium"
- Qas   Modern alluvial slopes
- Qtg   Ancestral river terrace gravels
- Qoas Old (pre-Kutz badlands) alluvial slope

- — 130 ka — Limits of ancestral rivers with est. age in 1,000 yrs (ka)
- Tn   Bedrock of Nacimiento Fm.
- SR   Salmon Ruins (Chacoan "great house")

Bloomfield's early history

In Chapter 29 we talked about the early settlement of the San Juan River Valley in general. Now we'll focus on this one particular segment of the valley around present-day Bloomfield. To briefly recap, until 1874, when the territory between the river to the New Mexico/Colorado border was proposed as a reservation for the Jicarilla Apaches (Figure 177A), this huge area was virtually unpopulated. Then, in 1874, Hispanic families began to trickle down the Canyon Largo and settle in the valley at what is now Turley (then known as Alcatraz), about four miles east of Blanco (Figure 225). In 1876, the proposed Jicarilla reservation was returned to the public domain and opened for settlement. Stockmen moved down from Colorado into the valley where Farmington would later be located. The first settlers in the area that would become Bloomfield arrived in 1877.

The lush grazing lands were ideal for both sheep and cattle. Back then this was open range and the grass reached up to the horses' stirrups (Cloer 1978). Cattle rustling soon became a problem and area ranchers were forced to organize militias to protect their herds. Then there was the inevitable friction between cattlemen and sheepmen because the animals competed for the same forage. Added to this was the ethnic element because most cattlemen were Anglo and most shepherds were Hispanic. Into this vortex drifted the brothers Ike and Port Stockton.

The Stocktons hailed from a town in central Texas astride a cattle trail pointing north to Kansas. After the Civil War they drifted into northeastern New Mexico. In the late 1870s Port arrived in the San Juan River Valley while brother Ike headquartered in Durango, Colorado where he led a band of cattle rustlers. Port soon built a herd in the Farmington-Bloomfield area with stolen stock, and had a cabin on the north side of the river about four miles west of Bloomfield. He acquired a well-earned reputation as a local thug (MacDonald and Arrington 1970).

In about 1879 or 1880 a group of horses was stolen from a ranch in Mancos, Colorado. The three bandits fled south and camped at the foot of Angel Peak, in Kutz Canyon. A posse, which included Port Stockton, spotted the fugitives and decided to wait until daylight. At dawn the posse opened fire on the three and mortally wounded one, but the other two escaped. The wounded was a boy of about 18, from Texas. He had one boot on, and he asked the posse to remove it because to die wearing his boots would brand him as an outlaw in the eyes of his kin. Someone helped him to pull it off, and then he died. They wrapped him in a blanket and buried him in a shallow grave near the east base of Angel Peak. Later a relative of the boy came to investigate. He identified the boy's remains and found that a finger, supposedly once sporting a diamond ring, had been cut off. It was speculated that Port Stockton had returned and taken the finger—and its ring (MacDonald and Arrington 1970).

Port was gunned down in front of his cabin in 1881 by vigilantes. This event kicked off a series of blood-feud incidents referred to collectively as the 1881 Stockton Cattle War, or the San Juan County War, during which brother Ike led the "Stockton gang" to avenge his brother's death. He vowed to kill a number of Farmington's prominent citizens whom he thought were involved. The convoluted imbroglio finally came to an end later in 1881 with the gunning down of the 29-year-old Ike in Durango. Thus ended the war, and without a leader the Stockton gang fizzled away (MacDonald and Arrington 1970; Stanley 1959). The Stockton boys were not missed.

On the heels of the stockmen came the farmers. They coveted the well-watered flood plain. Dry farming was not an option and irrigation—a vigorous and very challenging enterprise—was essential. The San Juan River formed a capricious barrier to travel and the farmers tended to favor one side or the other. By 1879 a town of sorts began to take shape on the north bank of the river. With the opening of the first post office the town was officially named "Bloomfield" (see below). The early mail came via stage coach from Durango to Farmington, and then to Bloomfield. In 1881 the town was renamed "Porter" to honor Civil War veteran General Horace Porter, who had established trading posts in the area (MacDonald and Arrington 1970). In 1882 it again became Bloomfield, and it remains so today (White 2003).

Namesake—John Bloomfield

In 1878 Mormon colonists began moving into the valley of the lower San Juan River Valley near today's Fruitland (Chapter 29). Soon afterwards the town's future namesake, a Mormon named John Bloomfield, arrived in the valley. The information below comes from Doug Bloomfield and posted on a website (John Bloomfield family 1995). John Bloomfield had been born in 1831 on a farm in eastern England, and at the age of 19 had converted to Mormonism. In response to local persecution he sailed to America in 1856 and settled in New Jersey. The following year he was ordained as a church elder and he married. He made the long trek west to Salt Lake City and helped to colonize the Little Colorado River area of Arizona in 1876, and in an area of New Mexico in 1881 (Figure 203). In 1878 he was ordained a high priest and in 1882 he moved to a farm in Fruitland.

Bloomfield spent considerable time in the upriver communities of Porter (present Bloomfield) and in the mid-1880s at Hammond, near present Blanco, helping the Mormon settlers colonize the area. The Mormon ward at Hammond had made an arrangement with the Mormon church in Salt Lake City to borrow money for construction of an irrigation ditch. A controversy soon developed among the colonists concerning the repayment of the loan. In mid-1882 the colonists of both Hammond and Porter had a public meeting and decided to change the name of the community of Porter back to Bloomfield as an enticement to persuade him to move to the area. He declined the offer because he loathed the idea of becoming involved in a financial dispute between the colonists and the leaders of the Mormon Church. Despite this glitch, the new name was retained. John Bloomfield continued to live in Fruitland while he assisted the Bloomfield area's Mormon colony with its projects.

At about this time many Mormon men were being rounded up and tried for polygamy, and in 1885 Bloomfield and others moved down into old Mexico. He returned to his old home in Ramah in 1894, returned to Kirtland in 1900, and finally returned one last time to Ramah in 1908, where in 1916 he died at the age of 84 (www.bloomfield-family.org, undated). Today John Bloomfield's legacy lives on in the San Juan River town that bears his name.

Hammond

In the mid-1880s the Mormons, led by Francis A. Hammond, president of the San Juan Stake, arrived in the areas east of Bloomfield and aimed to establish a ward. Soon Hammond became a small Mormon community strung out along the south side of the river near and west of the confluence with the Canyon Largo (Figure 224). The people scratched out an irrigation ditch and settled in to farm the alluvial slopes and the flood plain. Soon there were fruit orchards, a few thriving farms, and, by 1900, eight families. The "town" consisted of a church, school, and cemetery.

The ditch was particularly subject to the floods that roared down the Largo and washed out the flumes and siphons, leaving them dry for weeks and dooming the crops already in the ground. Finally, the famously destructive flood of 1911 that surged down the San Juan River following four days of rain in the Durango/Pagosa Springs area, washed out the entire irrigation system. With hopes dashed, the inhabitants abandoned the community. Some moved to the north side of the river at Bloomfield and some returned to the Mormon communities near Fruitland. The flood of 1911 ended irrigation on the south side of the river until the 1960s. Today only the Hammond cemetery remains (Tietjen 1980; Cloer 1978).

Citizens Ditch

Even before the flood of 1911, valley farmers recognized the need to irrigate the north side of the valley, but until then there was no practical way to efficiently do it. Individual farmers had attempted to operate on a small scale, but the endeavor was too much for one or even a few families. In 1909 the Citizens Ditch and Irrigation Company was organized to achieve the task. The river diversion was built at the community of Archuleta, six miles upstream to the east of Blanco, and the ditch stretched for 17 miles along the north side of the river. Tunnels were blasted through countless hillsides and the open areas dug out by men with teams of horses pulling scrapers. By 1910 the project was complete (Figure 225). With the ditch in place people began to migrate to the north side of the river and the way was open for Bloomfield to grow (Cloer 1978).

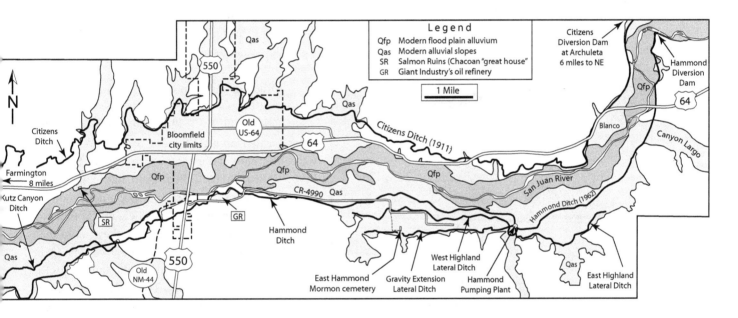

Figure 225. Irrigation efforts in the San Juan River Valley near Bloomfield.

Early transportation woes and early layout of Bloomfield

For many years the only way for people to get around was by foot, wagon, or on horseback over mud tracks. The river presented a formidable barrier for travelers. Before the river was controlled by the Navajo Dam in the early 1960s, the San Juan River was far less predictable than it is today. The first settlements had accordingly favored points where the flood plain was narrow and the river could be more easily forded. One such favorable site was at Blanco, where the alluvium of the Canyon Largo surges onto the flood plain from the south and pinches it to a narrow point (Figure 225).

A second site was at Bloomfield where the lowest Pleistocene-age gravel terrace abuts the river bank on the south and keeps the river channel somewhat in check at that point. This spot became not only a fording place, but in the late 19th century the site of a ferry landing. Commerce funneled across the river at this point and fueled the slow but sure development of the community of Bloomfield. The only connection to a railhead though was far away in Durango. Farmington was more favorably situated and a wagon-freight route and stagecoach line linked Farmington and Durango. Bloomfield folks were very much stuck out in the hinterlands.

In 1905 the Denver and Rio Grande Railroad (D&RG) built a standard-gauge line down from Durango to Aztec and Farmington. Bloomfield area farmers and ranchers were now finally somewhat connected to the outside world and could now receive and ship out goods on the so-called "Red Apple Flier" from the railhead in Aztec (Figure 7). In 1923 the Durango-to-Farmington line was changed over to narrow gauge to be more compatible to the D&RG's extensive narrow-gauge system.

Sometime in the late 1910s or 1920s a one-way bridge was built across the San Juan River at the ferry site (Figure 226). The bridge was a covered structure to enable stockmen to drive their cattle and sheep across. A dirt track, then designated state highway NM-55, led north from the bridge across the flood plain, and up to a slightly elevated area on the alluvial slope where the community of Bloomfield had developed. The town formed at the confluence of NM-55, which continued north to Aztec, and the road to Farmington. Because NM-55 had to outflank "Bobcat Mesa" and had to follow the borders between the quarter-section surveyed fields, the highway took an eastward zag at Bloomfield along what became known as Main Street. Main Street today has been bypassed on the north by US-64 and is now a quiet residential lane.

In 1944 highway NM-55 was redesignated NM-44 and paved between Aztec and Cuba. Then in 1947 the old bridge was replaced by the present structure and the highway realigned to its present course. The paving and the new bridge together provided a hefty impetus to Bloomfield's growth (Cloer 1978).

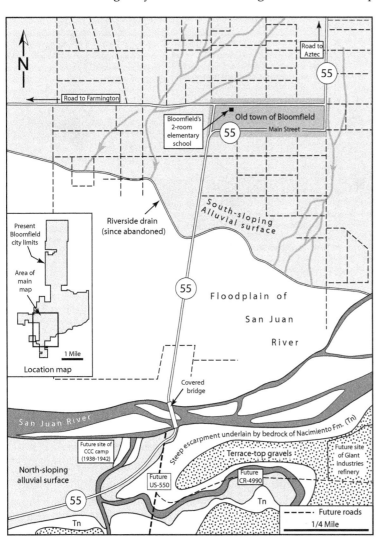

Figure 226. Town of Bloomfield in 1935.
(From 1935 Soil Conservation Service aerial photo.)

Origin of San Juan County

During the 1880s Bloomfield and Turley (the latter located about 12 miles upriver to the east) were the main settlements in the upper San Juan River Valley. In 1881 Bloomfield had a store, and a few years later a saloon, and Bloomfield soon became a gathering place for ranchers and cowhands. As more and more settlers arrived, they raised their collective voice to lobby for a new county to be lopped off from the west side of the then-huge Rio Arriba County (Figure 6B). Their plea was heeded and in 1887 the Territorial Legislature formed San Juan County. Unfortunately the act did not specify the location of the new county seat. This omission launched a protracted and bitter struggle between Farmington and Aztec for the honor. Aztec felt that it had a leg up because it was more at the county's center of population, but Farmington would have none of that. The snit continued until 1892, when the Territorial Supreme Court named Aztec as the county seat of San Juan County. The result of this kerfuffle was an antipathy that endured for years (Duke 1999). Even today its legacy is evident in the intense rivalry between the two cities. Little Bloomfield remained on the sidelines of the dustup.

Even with new county status, the economy of

the San Juan River Valley was mainly based on supplying food for the bustling mining centers to the north in Colorado. The valley communities remained economically joined at the hip with the commercial center of Durango.

First gas boom, 1921–1929

The San Juan basin occupies one end of New Mexico's bipolar distribution of energy resources. The Permian basin on the southeast pole is host for most of the state's oil. The San Juan basin on the northwest pole produces oil too, but it's importance lies chiefly in the basin's prolific deposits of coal, "conventional" natural gas, and "unconventional" coalbed methane gas. The San Juan basin is the second largest producer of natural gas in the U.S. Lower 48 and provides 10% of the nation's supply.

The City of Bloomfield is located near the center of this fabulous resource. Bloomfield is gas country! During the past 85 years the town has experienced five great up-and-down economic cycles, or "booms," so typical of the extractive industries. We are today in the fifth boom. Because natural gas is the bone and sinew of the town's economy, I'll say a few words below about each of the cycles in chronological order, but I'll weave them in and out of the narrative.

It all began in 1921, when a well was drilled at an apparently randomly-selected spot two miles south of Aztec. The well "blew in" from a depth of about 1,000 feet. Oil had been the objective, but the town provided a small and ready gas market and a two-mile pipeline was built. According to Dugan and Arnold (2001), "The gas system was rather primitive with no pressure regulators installed. Since the pressure on the gas line and in the houses was regulated only by how much gas was being used by those consumers connected at a given time, there were several houses burned during that winter and the following spring in Aztec. For instance, a number of people would go to a church social, some would turn the fire off, some would only turn it down. The reduced flow caused the pressure to rise and the fire which had been turned down became larger than the fire boxes and the houses burned."

In the early 1920s the county had only a few oil wells. These were located west and north of Farmington and were remote from markets. In 1924 Art Kittell built a small refinery for this oil in Cortez, Colorado. He soon had supply difficulties with the owners of the wells. In the nick of time an oil discovery was made in Bloomfield, near where the natural gas plants are located today (Figure 224). Kittell promptly moved to Bloomfield and in 1925 built

San Juan County's first refinery, north of Bobcat Mesa. The oil pool, producing from depths of about 600 to 900 feet, soon covered about 100 acres (Dugan and Arnold 2002).

At this time there were only three kinds of cars: Model T Fords, Dodges, and Buicks. Kittell's refinery produced fuel for these, as well as a special aviation fuel for private planes in Farmington (Cloer 1978). Finally in 1952 Kittell sold his refinery and settled down to a comfortable retirement. A street in Bloomfield has been named for him—ignominiously misspelled "Kittle."

Two events in 1927 would herald the San Juan basin becoming a giant gas supplier. A well was drilled at an apparently random location in Kutz Canyon south of Bloomfield and discovered natural gas. Then a second well, also at a random location a few miles north of the little town of Blanco about ten miles northeast of Kutz, discovered gas too. No one then could have known that these two gas fields would eventually connect and become part of the huge Blanco Gas Field, measuring some 70 by 35 miles with more than 5,000 wells (more below). In 1929 the owners of the Kutz pool organized the Southern Union Pipeline Company and built a pipeline to Farmington. In 1930 the company built a 12-inch (big for the time) pipeline to Albuquerque and a 10-inch extension to Santa Fe (these lines are now operated by Public Service Co. of New Mexico, or PNM). A few wells were drilled during the 1930s and 1940s for gas just to keep the pipelines full (Dugan and Arnold 2002), but bigger things would come after World War II when the Blanco Gas Field emerged as the backbone of the San Juan basin. But first a little detour.

Civilian Conservation Corps (CCC) at Bloomfield, 1938–1942

In the summer of 1938 a CCC camp, DG-101-N (New Mexico camp 101 for the Department of Grazing), was established in Bloomfield (Figure 227A). On the evening of July 29, the 200-member company of CCC boys arrived at the new camp. This first group of "boys," aged 17 to 28 (most on the younger end), hailed from the coal fields of Pennsylvania. This group had taken a special train from Philadelphia to Alamosa, Colorado, from which they transferred onto the D&RG narrow-gauge line for Aztec, New Mexico. Because the enlistment period was for six months, with extensions for up to two years, there was a continual supply of new faces. Sometimes entire companies were changed out. For example, by 1941 the boys were mostly from Oklahoma and had been trucked to Bloomfield from the railhead at Gallup.

A. North view of central part of Camp DG-101-N, ca. 1938.

B. East view of west edge of Bloomfield, with CCC-boy Charlie Wilburn, 1941.

Figure 227. The Civilian Conservation Corps (CCC) in Bloomfield, 1938–1942. (Photos courtesy of San Juan County Archaeological Research Center and Library.)

Once at the camp the boys were immediately put to work on grazing projects. One type of project was the building of stock-watering reservoirs to enable a more equitable distribution of livestock, thus preventing over-grazing near natural water sources. They built truck roads to allow stockmen to minimize their impact on the range. They built fences, removed prairie dogs, and rip-rapped sections of the Animas and San Juan Rivers to minimize erosion (*Farmington Times Hustler* 1941).

As one would expect, such companies of young men, chock full of testosterone, had their moments with the locals. The CCC boys had their weekends free, so occasionally Army trucks would haul them to the dance hall in Bloomfield. The local boys had little empathy with these invaders who were competing for the local belles. One such incident in 1939 made the newspapers. Differences between the locals and the CCC boys had resulted in a big fight. The first skirmish was won by the local boys. The CCCers returned to their camp on the south side of the river for reinforcements, and came back to the dance hall and cleaned it out. Law enforcement authorities and the CCC officers agreed that there had been a riot. With the discharge of two of the CCC ringleaders the authorities agreed to not pursue the matter (*Farmington Times Hustler* 1939). Henceforth the CCC made an effort to improve their image with the community by hosting an annual open house to acquaint the general public with the work done by the CCC, and holding baseball games between local and CCC teams. The boys thus became quasi members of the community (Figure 227B). Relations greatly improved.

With the ramping up of preparations for the war, the mission of the CCC ended. The Bloomfield camp was officially shut down in May 1942 and the boys were transferred to southern New Mexico where the Army was fencing a huge area to be used as a bombing range for the Army Air Corps (*Farmington Times Hustler* 1942). Despite efforts to retain and maintain the camp, it was eventually bulldozed over. Today the site is occupied by the Rio Vista Mobil Home Community (Figure 222).

Second gas boom—the big one, 1949–1958

Only in the 1930s, with the development of the new technique of electric welding, was it possible to manufacture pipelines capable of long-distance transmission of natural gas. Later, with the development during WWII of hydraulic pipe-bending machines it became possible to manufacture large-diameter pipelines. These innovations were fortuitously in place when the next event occurred.

In the late 1940s a group of wells were drilled on a block of leases next to the then-small Blanco Field, and they enlarged the productive area by several thousand acres. This caught the attention of El Paso Natural Gas Co., which promptly applied for a permit to construct a large-diameter pipeline to California. Large volumes of proven natural gas plus a pipeline connection to a large market spelled good economics. The 24-inch pipeline was completed in 1951 and the boom was on. In 1950 the Blanco field contained 30 wells, but by the end of the decade the field had expanded to about 1,800 wells covering about 900 square miles and extending into southern Colorado.

The need to keep the pipeline full led to a mighty drilling program around Bloomfield. The utter remoteness of the San Juan basin from supply centers became immediately apparent. The drilling program placed enormous demands on equipment and material. The D&RG narrow gauge railroad to Farmington proved to be not up to the task. At Alamosa, Colorado, huge, difficult transhipments had to be made from standard gauge lines onto the D&RG's narrow-gauge line for the long twisting haul to Farmington. There were no four-lane highways at the time. The two-lane highways, including NM-44 from Albuquerque and US-666 from Gallup to Shiprock, were narrow, crooked, and dangerous. Despite the problems the shipments somehow got through (Dugan and Arnold 2002).

Along with the drilling equipment came people. Job seekers flocked to the Bloomfield and Farmington areas from all over. Trailer parks sprang up faster than Starbucks shops. Also in the late 1950s a new oil refinery, the Bloomfield Refinery, was constructed on the terrace on the south side of the river, and created more jobs. And of course tremendous pressures were placed on Bloomfield's infrastructure, particularly its schools.

Incorporation

In 1955 Bloomfield was incorporated as a village during the middle of the second gas boom. From that point onward the town has grown in great leaps. These leaps were spurred by the construction of several new schools, the construction of the Navajo Dam from 1958 to 1962, the completion of the Hammond Project in 1962, and construction of Navajo Indian Irrigation Project in the 1970s (Chapter 30). In 1966 Bloomfield was incorporated as a city, and, like many communities, it has progressively nibbled up the surrounding region in a series of annexations (inset in Figure 228).

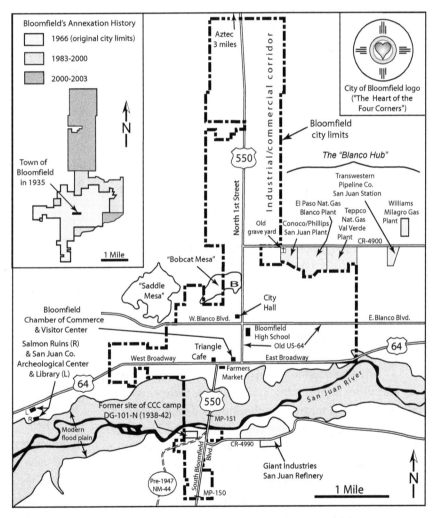

Figure 228. City of Bloomfield and select points of interest. (Taken in part from Bloomfield Chamber of Commerce 2000, 2003.)

315

Schools

The second gas boom of the 1950s was only the latest chapter in the development of Bloomfield's school woes. The San Juan County school system was organized between 1895 and 1897. From the very beginning people took their schools very seriously. Attendance was considered a privilege because no one knew for sure if the school would be there the following year. The school became the community's center of social life. The teacher was often from outside the area and had to board with a local family (Coeler 1978).

By 1902 there was an elementary school (grades one through eight) in Hammond and one in Bloomfield. A single teacher taught for two months in one and then for two months in the other. In 1914 the Bloomfield school district was divided in two, with the Bloomfield District to the west and Rio Vista District to the east. Bloomfield's two-room schoolhouse was built on the site of the Farmers Market parking lot (Figure 228).

The two elementary schools educated Bloomfield kids until the late 1940s. During this time the older kids had to take a bus eight miles to Aztec High School (Coeler 1978). By the late 1940s the two-room Bloomfield school was bursting at the seams. A new school, Central Elementary, was built at a new site in about 1950. As the gas boom got into full swing in the mid-1950s the population exploded. With the 1955 incorporation of the Village of Bloomfield, construction began on Bloomfield High School. The first graduating class was in 1959. Also in the mid-1950s the Rio Vista Elementary School and the Mesa Alta Junior High School were built (Coeler 1978). Now Bloomfield finally had an adequate school infrastructure, which was absolutely vital to the continued growth of the community.

Hammond Project

After the abandonment of the Hammond area in the several years after the big flood of 1911, there was virtually no irrigated farming on the south side of the river for the next 50 years until the completion of the Hammond Project in 1962 (Cloer 1978). The project was intended to divert, store, and distribute water of the San Juan River for irrigation purposes on approximately 3,900 acres of arable land stretching along the south bank of the river from Blanco to near Farmington.

The Hammond Diversion Dam was built across the San Juan River about two miles upstream from Blanco

(Figure 225). The dam diverts water from the river into the 27.4-mile-long Main Gravity Canal. This canal, with an initial capacity of 90 cubic feet per second (cfs), meanders in a southwesterly direction through the project area. About six miles below the dam is the Hammond Pumping Plant, which uses a 30-foot drop to hydroelectrically power a pump to lift 18 cfs to the East Highline Lateral and the West Highline Lateral. In 1968, an auxiliary pumping plant was added about 1,000 feet upstream from the main plant to service the East Highline Lateral. The Gravity Extension Lateral is a branch of the main canal. From the main plant to the lower end of the project, the canal capacity is reduced progressively to five or ten cfs.

In 1974 the project was turned over to the Hammond Conservancy District (HCD), a non-profit Special District for operation and maintenance. The HCD claims rights to 26,705 acre feet of Colorado River Storage Project water.

US-64

In 1962, with the completion of Navajo Dam and the Hammond Project, and the redirection of attention in Colorado mountain towns such as Durango to their undeveloped tourism potential, New Mexico had a tourist-oriented chief executive, Governor Jack Campbell. He realized the tourist value of the new Navajo Lake State Park and authorized the construction of US-64. This "Golden Avenue," completed in 1972, connected the tri-cities of Farmington-Aztec-Bloomfield to Raton in northeastern New Mexico and linked most of New Mexico's major tourist attractions with the Colorado ski resorts (Gómez 1994). Bloomfield thus had become a gateway city.

Third, 1973–1982, and fourth, 1989–1993, gas booms

The so-called third gas boom was the direct result of the Yom Kipper War of 1973, when Egypt attacked America's ally Israel. The Arab world seethed at our support for Israel and retaliated by slapping an oil embargo on sales to the United States. With sudden shortages of crude, prices skyrocketed. Thoroughly alarmed, Congress authorized an assessment of the nation's oil and gas reserves. Based on the new numbers the federal government felt justified to allow the increase in price for natural gas as well. By 1976–1977 the predictable spurt in drilling activity in the Bloomfield area led to another surge in growth. The third boom—like all booms—eventually petered out and by 1983 it was over (Dugan and Arnold 2002).

The fourth gas boom was an entirely new phenomenon. This boom could be called the "CBM boom," or the coal-bed methane boom. What is coal-bed methane and why is it important? Methane (CH$_4$) of course is the "conventional" natural gas produced from the basin that we normally use to heat our homes and cook our food. The "coal-bed" portion of the term requires some elaboration. For years drillers would drill down through the Tertiary-age San José and Nacimiento Formations to get to the gas-bearing Cretaceous sandstones below (Figures 3A and 4). Just below the Tertiary but above the drilling targets they had to penetrate shallow coal seams at depths above 3,000 feet. Oftentimes the coal would give the well a little high-pressure "gas kick" on the way through, but drillers usually considered the zone merely a nuisance. From as early as the early 1950s drillers had known that it was possible to produce gas from the coal itself, but this entailed producing a lot of unwanted water as well. The gas within the coal occurs as a free state in natural fractures ("cleats"), as gas dissolved in the water in the fractures, and as gas absorbed on the surface of the organic matter of the coal.

The issue remained on the proverbial back burner until 1977, when a company got the bright idea to drill and complete a well in the coal about 15 miles northeast of Bloomfield. The well initially produced a modest amount of gas, but after a couple of years production increased by a factor of ten. Whoa! That's not the way "normal" wells behave—production usually declines with time. Although the company invested a considerable sum to deal with the water, the well kept chugging along. Ears perked up everywhere.

Over the next several years a number of important pieces of legislation changed the operating rules in the basin, and restructuring of the tax code made development of this non-conventional gas profitable. But then in 1985, the Kingdom of Saudi Arabia flexed its considerable muscle and changed everything. To greatly simplify, the Saudis had been steadily losing their world-market share of crude-oil sales because their Persian Gulf neighbors were cheating on agreed-on production quotas. The Saudis, now thoroughly exasperated, swung open the valves of their big wells and

flooded the world market with oil. During the first half of 1986 the price of crude plummeted to unimaginable depths (from about $30 to below $10/barrel). Companies saw their operating cash flow dry up. They were forced to slashed their staffs to cut costs and hundreds of thousands of oil industry workers lost their jobs (I was one of them). Today many of these people ruefully refer to this

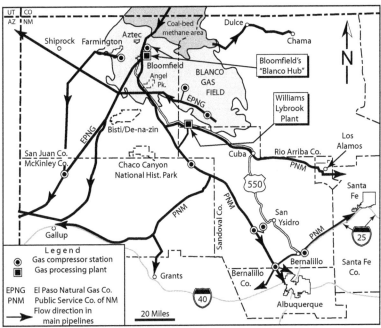

A. Natural gas pipelines.

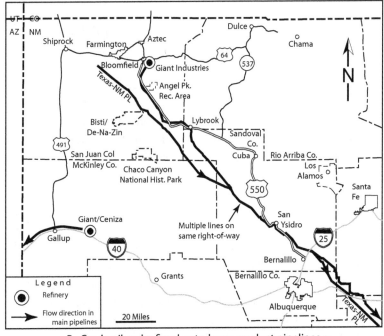

B. Crude oil and refined-petroleum-product pipelines.

Figure 229. Bloomfield and San Juan basin natural gas and oil.

debacle as "the crash." The San Juan basin wasn't as badly hit as some places because it's mainly a gas province, but the ripple effect hurt everyone (Dugan and Arnold 2002).

Recovery was slow, but in a couple of years the fourth boom started in earnest. The center of CBM drilling activity is located to the northeast and east of Bloomfield and extends into southwestern Colorado (Figure 229A). Along with the drilling came the simultaneous need for a webbing of gathering lines to deliver the gas from the wellheads to central treatment and compression facilities. CBM differs from conventional methane gas in that it contains about 10% carbon dioxide (CO_2) that must be removed. Therefore the gas must be transported from the field via different pipelines to its own treatment plant. Once treated, however, it can be mixed with conventional gas.

Five big natural gas treatment and compression plants are aligned along "Plant Row" (CR-4900), just outside the northeastern city limits (Figure 228). One such treatment center is the Williams Field Services' Milagro Plant, which treats CBM. The plant had been in storage in Texas, and had to be disassembled, transported to Bloomfield, and reassembled. By 1991 the plant was in operation. Its tangle of pipes and machinery removes gas liquids (propane and butane), water and carbon dioxide from the gas stream, thus bringing it up to "pipeline quality," The plant adds the rotten-egg-smelling, sulfur-containing "mercaptan" compounds so we can know when there's a gas leak. A second gas-treatment plant, the Conoco/ Phillips San Juan Gas Plant, is located down the street and handles conventional gas. The treated gas from these plants is piped to special compressor plants so that the gas can be pumped into the main, high-pressure lines, such as PNM's line to the Rio Grande Valley and other interstate lines to markets in California and the Pacific Northwest. At the far west end of Plant Row is a little cemetery, snuggled up against the edge of this hulking industrial complex and providing an example of supreme irony.

Fifth gas boom, 2005–?

By 1993 the Fourth Boom too had fizzled (Dugan and Arnold 2002). We are now in the Fifth Boom. Its onset had been delayed by environmental issues, a downturn of the U.S. economy during 2000, and the Al-Qaida attack on the World Trade Center in New York on September 11, 2001 and its aftermath. By 2003 the economy had recovered somewhat and the Fifth Boom was off and running, this time spurred by CBM. During the next decade more than 10,000 new gas wells are expected to be drilled in the CBM area of the San Juan basin.

Because Bloomfield is located near the western edge of the coal-bed methane portion of the basin, it has naturally become the regional hub of the natural-gas industry. The town's oil refinery, gas-treatment plants, and gas compressor stations play a vital role in the area's employment picture. However, because industries such as these handle enormous volumes of fluids via vast networks of pipes, tanks, and towers, and because their extensive array of supply wells spreads relentlessly over the countryside, there is a growing backlash from environmental quarters. There are genuine pollution concerns, and many fear that the anticipated massive expansion of the CBM industry during the next decade will irreparably damage the landscape that most of the valley's citizens treasure. The Fifth Boom is the harbinger of even more growth in the Bloomfield area, but the environmental outcry will likely reach a crescendo. These will indeed be interesting times.

Select points of interest

Although Bloomfield is not known as a tourist town, it does have its points of interest (Figure 228). Some of these are industrial in nature. Many would argue that industrial infrastructure is a blight on the land, while many others (like me) are fascinated by the engineering and problem-solving ingenuity that produced them and the stories behind them.

Giant Industries' San Juan Refinery (recently purchased by Western Refining Inc.)

As US-550 makes it final descent into the San Juan River Valley, virtually the first sight is the oil refinery to the east of the highway, on the terrace overlooking the south bank of the river. Built in the late 1950s and originally called the Bloomfield Refinery, the facility was purchased by Giant Industries in 1996 and renamed the San Juan Refinery. Giant Industries is a refiner and marketer of petroleum products based in Scottsdale, Arizona. Most of the company's operations are in the southwestern United States, centering on the Four Corners area. These include, among other things, two crude-oil refineries in New Mexico and 127 service stations in New Mexico, Arizona, and Colorado. The stations are branded under the company-owned names Giant and Mustang, and under the name Conoco via a licensing agreement. Because this is the only refinery in the Four Corners area, and because it forms such a visible feature in Bloomfield, it is worth a closer look. The account that follows is from James Press (2004).

The rise of Giant Industries is a classic Horatio Alger success story. The founder, James E. Acridge, started his meteoric rise in 1961 when at the tender age of 21 he leased a small gasoline station in Glendale, Arizona. Four years later he leased another station in Phoenix, Arizona and operated it under his own sign, "Giant." Within a year business was booming. When he caught grief from the lessor of the property on which his Giant station was located, Acridge moved into another station, where he again thrived.

In 1968 Acridge plunged into the gas-station big leagues. He purchased a small two-island station and put up the first self-serve sign in the region. He was way ahead of his time. Almost overnight the volume of sales jumped. Next he purchased a plot of land near Mesa, Arizona, borrowed $30,000 from a bank to begin operations, leased all the necessary equipment, and built his own multi-pump gas station. This station too was soon thriving. In 1969 Acridge incorporated his company, and from that time on he was the undisputed king of self-serve gas stations in Arizona.

Acridge built his second self-serve unit in 1970. The design of this unit was the forerunner of all subsequent self-serve gas stations. Acridge aimed to make it a pleasant experience for the customer to pump gas, so he piped in soft, soothing music under the canopies, constructed high retaining walls so that people would not be embarrassed if their friends saw them at a lowly self-service station, and provided ample space between all the gas pumps for efficient filling. Within a short time, the two self-serve stations were successes. By 1973 Giant Industries had expanded to include 12 self-serve gas stations, but was soon forced to close all but four when the 1973 oil embargo by OPEC worsened an already existing supply crisis.

At this time Acridge decided to enter into the refinery business to ensure a steady supply for his stations. He purchased a small gas-processing facility in Texas, had it completely dismantled, and transported the entire plant to Farmington, New Mexico, where it was reassembled and put in working order. As his operation grew, he decided to develop "tie-in" businesses that augmented his self-serve stations. His first idea involved what he called a "C-Store," a huge store situated in back of his gas stations that sold a large line of groceries, sporting goods and automotive parts, and included an on-site dry cleaners. But this time

the customers did not come and Acridge was forced to close the stores within a few months. Evidently he was not invincible.

By the end of 1979, Giant was operating 23 gas stations. During the early 1980s a clash with the major oil companies forced Acridge to close all of his stations in the Phoenix area. He decided to branch out into smaller towns to find a more stable market. In 1982 Giant acquired the Ciniza refinery, located on I-40 about 20 miles east of Gallup (Figure 229B), and upgraded it to increase the facility's ability to produce unleaded gasoline.

Heartened by the success of his C-Stores during the early and mid-1980s, Acridge decided to open what became known as a "highway extravaganza," an enormous combination truck stop/gas station/retail store near the Ciniza refinery on I-40. The sprawling Giant Travel Center opened in 1987 and included just about everything, even a movie theater. Giant began to enjoy a reputation for efficiency and competence. Truckers would sometimes drive 350 miles out of their way just to get their rig washed at the only truck wash between Barstow, California, and Oklahoma City, Oklahoma. Acridge aimed to treat the truckers like royalty, and they responded. A trucker could order a 16-ounce T-bone steak for $10.95, have it cooked to order, and call home from the telephone situated on his or her table!

In 1989 Acridge took Giant Industries public with a listing on the New York Stock Exchange and built a new headquarters building in Scottsdale. In 1995 he acquired a crude-oil gathering operation near Farmington that included approximately 340 miles of pipeline. By 1996, Giant Industries was one of the most successful petroleum products companies in the southwestern United States.

With its niche in the Four Corners area secure, in 1996 Giant purchased its second refinery, the one in Bloomfield, from the Gary-Williams Energy Corporation. This acquisition helped the company to consolidate some of its refining operations, lower transportation costs, and improve production. Also during 1996, Giant purchased stations from Diamond Shamrock in northwestern New Mexico, and in 1997 acquired more stations and convenience stores in Arizona, New Mexico, Colorado, and Utah from Thriftway Marketing Corp.

The next year began with the acquisition of the owner of several Conoco-brand gasoline stations in southwestern Colorado. Giant and Conoco had entered into a branding alliance in 1997, whereby Giant would begin using the oil giant's name at some of its stations. By the end of 1998, 49 of Giant's outlets, along with the Giant Travel

Center, began sporting the Conoco name. Also during 1998, around 50 of the company's service stations adopted a new, company-owned name, "Mustang."

During the late 1990s and early 2000s, Giant went through a number of restructuring moves and dealt with several reversals. Ironically, Acridge, the person who had guided Giant Industries since its founding, was ousted as CEO and president in 2002, although he remained on the board of directors. Even without Acridge, Giant Industries was firmly established on the landscape of northwestern New Mexico. However, in 2007 Western Refining Inc., headquartered in El Paso, TX, purchased Giant's operations and in late 2009, amidst a serious economic recession, announced the closure of the Bloomfield facility (*Albuquerque Journal* 2009).

Navajo Lake State Park

Another, and very appropriate, motto for the City of Bloomfield is "Gateway to Navajo Lake State Park." Navajo Lake, 25 miles east of Bloomfield, is the second largest lake in New Mexico after Elephant Butte and one of state's crown jewels (Figure 230). Sometimes called the "Tur-

quoise of the Four Corners Country," the lake was created in 1962 via the completion of the dam by the U.S. Bureau of Reclamation for flood control as part of the Colorado River Storage Project, and the state park was created the following year. The lake is held back by a 400-foot-high, dirt and rock-filled structure that stretches almost 4,000 feet across the San Juan River (McLemore 2002). It's indeed an impressive piece of work, as is the huge lake behind it. The lake is now a recreational lodestone where there was nothing before, if you could call "nothing" the early 19th century town of Rosa, several old roads, and several historic cemeteries submerged by the rising waters.

The presence of this prime recreational facility had an immediate economic effect on the community of Bloomfield. The park today has four developed recreational sites, three in New Mexico and one in Colorado. Water flowing downstream of the dam provides some of the best fly-fishing river in the United States. Some argue that it rivals the great Tierra del Fuego area of southern Argentina in South America. Trout—some up to two-feet long—thrive in the cold water fed by snow runoff from the San Juan Mountains (Larese 2006).

Salmon Ruins and Heritage Park

This important site is located only three miles west of the intersection of US-550 with US-64 (Figure 228). In 1877 George Salmon was driving his team of horses on the south bank of the San Juan River, searching for a way across to the north side. He found a spot free of quicksand and led his team across. The ford became known as Salmon Crossing. He ascended the low slope above the flood plain and liked what he saw. He also noted a large and very interesting Indian ruin nearby. He decided to homestead on the slope, which naturally became known as Salmon Mesa. Soon he brought his young wife, a girl from the Archuleta family of Archuleta County in southern Colorado.

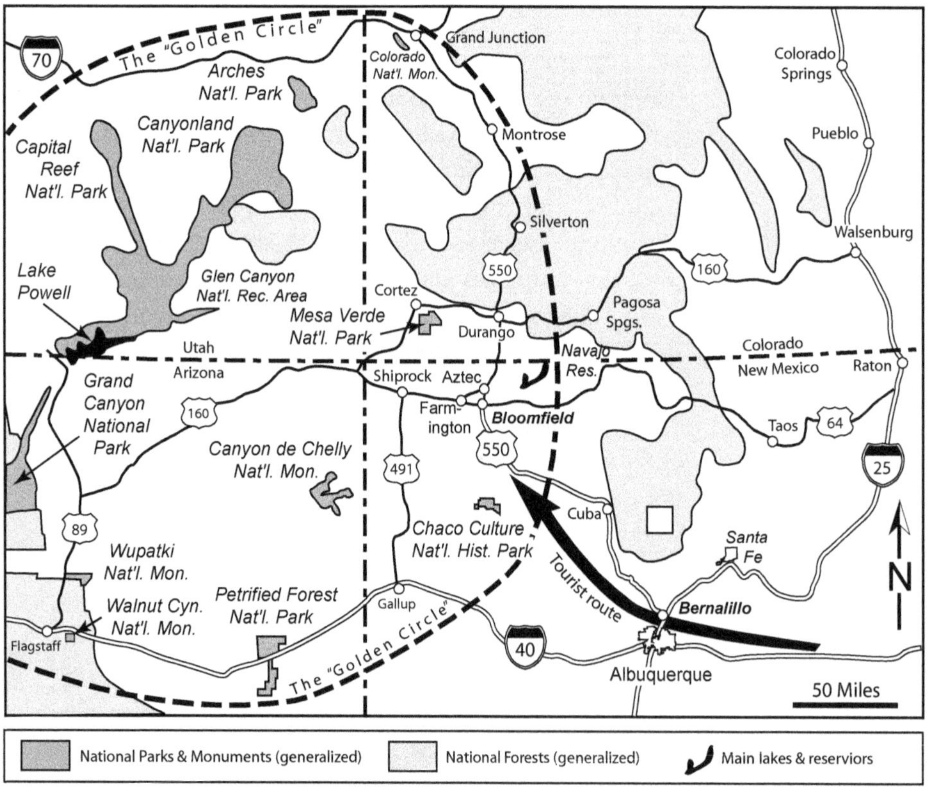

Figure 230. US-550 as gateway route to principal tourist destinations in the Southwest's "Golden Circle."

The Salmons brought young peach trees and soon had the first peach orchard in the county (MacDonald and Arrington 1970).

Salmon and his descendants protected the archeological sight from plunderers and pot hunters for more than 90 years. In 1969 the San Juan County Museum Association purchased the 22-acre tract, containing the Salmon homestead and the ruin. The next year the ruin was nominated to the National Register of Historic Places and to the New Mexico State Register. In 1971 a San Juan County bond was passed to construct a laboratory-museum complex. From 1972 to 1978 professionals from Eastern New Mexico University, under the direction of Dr. Cynthia Irwin-Williams, carried out major excavation work at the ruin. The museum, the San Juan Archaeological Research Center and Library, was opened to the public in 1973 (San Juan County Archaeological Research Center and Library, undated pamphlet).

The huge ancestral-Pueblo ruin, laid out in the shape of a square "C," consists of about 150 ground-floor rooms and more than 67 second-story rooms. It was occupied during two distinct phases: 1068–1130 CE, and 1185–ca. 1285 CE. Today the ruin, the adjacent Salmon homestead, and the wonderful museum (a.k.a. the San Juan County Archaeological Research Center and Library) provide a fascinating day's outing.

Triangle Café

This conveniently-located and very good eating spot is a perfect place to "chow down" and "water up" at the end of old NM-44, and to use as a pit stop on the way to other destinations (Figure 228). The name "Triangle" has been around in Bloomfield since 1957, when the first café was opened where the parking lot of Farmers Market is now located. As early as 1960 it was known as an "oilfield café" and the only 24-hour eatery in Bloomfield. The building burned down in 1966 and was replaced by a new one at the present location across the street. A woman named Wilda Lasater bought the property in 1977 and operated it until her death in 1980. Wilda's daughter, Debbie Mohler, inherited the business. Debbie has run the café ever since, now more than 30 years. The café is one of the oldest businesses in the city and the name Triangle Café has been around now for more than 50 years. As everyone knows, no business—especially a restaurant—survives that length of time unless its owners know what they're doing!

Traffic!

How can a city of fewer than 10,000 people have so much traffic? The intersection of US-550 and US-64 is a non-stop glut of big trucks, all day and much of the night. Bloomfield is an industrial town, but of course not a manufacturing town. Oil and gas operations drive the traffic, as well as everything else. The multitude of natural gas wells to the east are tended by drivers in their fleets of white three-ton pickup trucks, with their little identifying flags attached to their antennas, going to and from the field. And then there is the oil. Crude oil gathered from tanks near the oil wells in the field is hauled to the refinery via big tanker trucks, sporting their Department of Transportation (DOT) placards with the number 1267 (identifying the contents as crude oil). About half of the refined products—mainly gasoline and diesel fuel—is piped to market (Figure 229B). Another half is hauled by tanker trucks to local markets, this time identified by the DOT number 1203 (for gasoline). The traffic will get worse before it gets better.

Bloomfield as Gateway to the southern Rockies and the canyon country

The City of Bloomfield is the end of the line along our segment of interest on US-550 that was once old New Mexico Route 44. The city hails itself as the "Gas Capital of the United States," and also as the "Heart of the Four Corners" (see logo in inset in Figure 228). Bloomfield is the minor member of San Juan County's "tri-cities," that consist of Bloomfield, Aztec, and Farmington. However, Bloomfield has the advantage of being the first of the three seen by visitors from the main population centers to the south and east. Although Bloomfield is not usually considered a destination point, it is very much a decision point: north to Aztec National Monument in Aztec or farther north to Durango and the camping and skiing centers in the San Juan Mountains of southwestern Colorado; west to the fabulous Four Corners and canyon country of the Colorado Plateau; east to Navajo Lake State Park; or a south backtrack to the incomparable Chaco Culture Historical Park (Figure 230). With plans afoot to pave the access road to Chaco Canyon south from US-550 (amidst considerable controversy, Chapter 25), Bloomfield's status as a gateway will certainly increase. And then of course, the southern leg of highway US-550, Old Highway 44, with all its own points of interest—the subject of this book—is the path to Bloomfield.

Epilogue

Finally we arrive at the end of our 151-mile geological and historical excursion from New Mexico's heartland in the Rio Grande Valley to the once-super-remote valley of the San Juan River. The modern highway links two valleys that differ markedly in culture, ethnic makeup, economy, and voting patterns. Some argue that four-laning old NM-44 from San Ysidro to Bloomfield was a political stunt—a Republican governor's (Gary Johnson's) device to buy Republican votes by more efficiently linking the Republican-leaning gas-field population to the center of traditional, Democratic political power in Santa Fe. Perhaps there was a political element underlying the project, but all roads have economic—and therefore political—motivations. Regardless, the modern highway US-550 from the Rio Grande Valley to the San Juan River Valley has effectively integrated the two economic and political blocs.

The long journey between the two valleys is no longer a feat of dogged endurance and questionable survival. Today most travelers make the trek relatively unscathed and they are quite unaware of what it took to traverse this land not that long ago. And, because the four lanes make it possible to stop and take in the intriguing and diverse views along the way, without being converted to road kill by oncoming careening trucks and drunks, we can more fully kick back and appreciate the drive. In this book I have intended to bring this 151-mile drive to life, and to add to its intrinsic interest via some of my idiosyncratic "tangents."

I hope I have been successful.

Abreu, Vitor.S., and John B. Anderson, 1998. Glacial eustasy during the Cenozoic: sequence stratigraphic implications. *American Association of Petroleum Geologists Bulletin* 82 (7): 1385-1400.

Aby, Scott, Allen C. Gellis, and Milan J. Pavich, 1997. The Rio Puerco arroyo cycle and the history of landscape changes. *Impact of Climate Change and Land Use in the Southwestern United States*, U.S. Geological Survey workshop, September 3-5. Tucson AZ: University of Arizona

Agenbroad, Larry D., 1984. New World Mammoth Distribution. *Quaternary Extinctions: a Prehistoric Revolution*, edited by P.S. Martin and R.G. Klein, Tucson AZ: University of Arizona Press, 90-106.

Albuquerque Journal, 1920. Marriage: Miss Marion Watlington and Mr. Sidney Weil. June 6.

Albuquerque Journal, 1934. J.R. Haynes obituary. December 29.

Albuquerque Journal, 1940a. L.B. Putney Jr., taken by death. July 11.

Albuquerque Journal, 1940b. Pioneer woman dies soon after husband taken to hospital. August 13.

Albuquerque Journal, 1940c. Home from hospital, C.A. Watlington. August 25.

Albuquerque Journal, 1940d. M.S. Otero obituary. August 31.

Albuquerque Journal, 1950. Obituary, Mr. Charles Watlington. May 25.

Albuquerque Journal, 1966. J.E. Counselor obituary. Aug. 26, 1966.

Albuquerque Journal, 1970. Cuba Drifts along with Changing Times. June 28.

Albuquerque Journal, 1972a. State of Emergency Requested for Cuba. September 20.

Albuquerque Journal, 1972b. State of Emergency Declared in Cuba. September 21.

Albuquerque Journal, 1979. Water-short town fights for state aid. April 14.

Albuquerque Journal, 1982a. A. Counselor obituary. January 1.

Albuquerque Journal, 1982b. Namesake of town dies at 84. January 2.

Albuquerque Journal, 2000. Cuba gets $5,000 to burnish its welcome mat for tourists, supplement *Business Outlook*. August 28.

Albuquerque Journal, 2001. Indians living on reservation lands in New Mexico. April 29.

Albuquerque Journal, 2003a. Military looking into Sandoval hole mystery. April 5.

Albuquerque Journal, 2003b. Mysterious hole filled with dirt. December 16.

Albuquerque Journal, 2003c. N.M. sawmill to close doors. April 9.

Albuquerque Journal, 2009. Bloomfield refinery closing, Western to shift output to Gallup. November 10.

Albuquerque Journal Democrat, 1903. Article about formation of Sandoval County. February 12.

Albuquerque Tribune, 1940e. M.S. Otero obituary. September 1.

Albuquerque Tribune, 1967. Streets get new names, park renamed. August 9.

Akers, Joshua, 2004. Rio Rancho continues its boom pace, *Albuquerque Journal*, December 20, supplement *Business Outlook*.

American Heritage, 1966. *The American Heritage Pictorial Atlas of United States History*, New York: American Heritage Publishing Co.

Anderholm, Scott K.,1979. Hydrogeology and water resources of the Cuba Quadrangle, Sandoval and Rio Arriba Counties, New Mexico. Unpublished M.S. dissertation. Socorro NM: New Mexico Institute of Mining and Technology.

Anderson, George B.,1907. *History of New Mexico: its Resources and People, v. 2*. Chicago IL and Los Angeles CA: Pacific States Publishing Co.

Arango, Polly, 1980a. Bernalillo Merc echoes eras past. *Rio Rancho Roadrunner*, January 15.

Arango, Polly, 1980b. Cuba: its own place for 200 years. *Rio Rancho Roadrunner*, March 15.

Aragón, Kelly, 2002. Personal communication, December 12.

Armijo, Patrick, 2001a. Drywall plant seeks emissions revision. *Albuquerque Journal Business Week*, May 28.

Armijo, Patrick, 2001b. Ranchers figure promises will erode like land. *Albuquerque Sunday Journal*, August 5.

Arrington, Leonard J., and Davis Bitton, 1992. *The Mormon Experience: a History of the Latter Day Saints*. Urbana and Chicago IL: University of Illinois Press.

Atteberry, Liz, 1990. Lybrook Legacy. *New Mexico Magazine*. 68(10).

Austin, George S., and James M. Barker, 1990. Commercial travertine in New Mexico. *New Mexico Geology* 12 (2): 49-58. Socorro NM: New Mexico Bureau of Geology and Mineral Resources.

Axelrod, Daniel I., 1983. Paleobotanical history of the western deserts. In *Origin and Evolution of Deserts*, eds Stephen G. Wells and D.R. Haragan, 113-129. Albuquerque NM: University of New Mexico Press.

Baca, Jacabo, 2005. The Dixon Case: 1947–1951: End of the Catholic Era in New Mexico Public Education. In *La Crónica de Nuevo México*, issue #65, July. Santa Fe NM: Historical Society of New Mexico.

Baker, Larry L., 1984. The stabilization of Guadalupe Ruin. In *Contributions to Anthropology*, v. 11. Portales NM: Eastern New Mexico University.

Baker, Larry L., 2003. Anasazi settlement distribution and the relationship to environment. In *Prehistory of the Middle Rio Puerco Valley, Sandoval County New Mexico*, Special Publication No. 3, 155-177. Albuquerque NM: Archaeological Society of New Mexico.

Baltz, Elmer H., Jr., 1967. "Stratigraphy and regional tectonic implication for part of Upper Cretaceous and Tertiary rocks, east-central San Juan Basin, New Mexico. U.S. Geological Survey, Professional Paper 552.

Bartlett, Richard A,. 1962. *Great Surveys of the American West*. Norman OK: University of Oklahoma Press.

Bauer, Paul W., Richard P. Lozinsky, Carol J. Condie, and L. Greer Price, 2003. *Albuquerque: a Guide to its Geology and Culture*; Socorro NM:New Mexico Bureau of Geology and Mineral Resources.

Baxter, John O., 1987. *Las Carneradas: Sheep Trade in New Mexico, 1700–1860*. Albuquerque NM: University of New Mexico Press.

Bayer, Laura, 1994. *Santa Ana: the People, the Pueblo, and the History of Tamaya*. Albuquerque NM: University of New Mexico Press.

Beck, Warren A., 1962. *New Mexico – a History of Four Centuries*. Norman OK: University of Oklahoma Press.

Beck, Warren A., and Ynez D. Haase, 1969. *Historical Atlas of New Mexico*. Norman OK: University of Oklahoma Press.

Bernalillo High School 1976, *Viva El Pasado: a History of the Bernalillo Area*. Bernalillo NM: Bernalillo High School Southwest History Class of 1974–1975.

Betancourt, Julio L., and Thomas R. Van Devender, 1981. Holocene vegetation in Chaco Canyon, New Mexico. *Science* 214 (6): 656-658.

BLM (U.S. Bureau of Land Management), 1994. Map of surface management responsibility, state of New Mexico, scale 1:500,000.

BLM (U.S. Bureau of Land Management), 1997. *Surface Management Status*, Chaco Canyon 30 x 60-minute topographic map, scale 1:100,000.

BLM (U.S. Bureau of Land Management), 1999a. *Surface Management Status*, Chaco Mesa 30 x 60-minute topographic map, scale 1:100,000.

BLM (U.S. Bureau of Land Management), 1999b. *Surface Management Status*, Los Alamos 30 x 60 minute topographic map, scale 1:100,000.

BLM (U.S. Bureau of Land Management), 2004. *Surface Management Status*, Abiquiu 30 x 60 minute topographic map, scale 1:100,000.

Bloomfield Chamber of Commerce, 2000. *Map of Bloomfield, New Mexico*.

Bloomfield Chamber of Commerce, 2003. *Map of Bloomfield, New Mexico*.

Bloomfield family, 1995. Website of John Bloomfield family. Bloomfield-family.org/DAD/Family%20Bios/John%20Bloomfield.htm, last visited May 2009.

Boyle, Susan C., 1997. *Los Capitalistas: Hispano Merchants and the Santa Fe Trade*. Albuquerque NM: University of New Mexico Press.

Brayer, Herbert O., 1939. Pueblo Indian Land Grants of the Rio Abajo, New Mexico. *University of New Mexico Bulletin*. Albuquerque NM: University of New Mexico Press.

Brennan, Benita,. no date. The Baca House. Unpublished hand-written note. Bernalillo NM: Sandoval County Historical Society.

Brett, Linda C., 2003. Basic lithic raw material acquisition and reduction patterns. In *Prehistory of the Middle Rio Puerco Valley, Sandoval County New Mexico*, 135-153, Special Publication No. 3. Albuquerque NM: Archaeological Society of New Mexico.

Brokaw, Tom., 1998. *The Greatest Generation*. New York: Random House.

Brugge, David M., 1980. *A History of the Chaco Navajos*; Report #4 of the Albuquerque: National Park Service, Chaco Center, Division of Chaco Research, Report #4.

Brugge, David M., 2004. The Chaco Navajos. In *In Search of Chaco: New Approaches to an Archaeological Enigma*, ed. David G. Noble, 61-69. Santa Fe NM: School of American Research Press.

Bryan, Howard, 1953. Navajo tours to Jémez Springs. *Albuquerque Tribune*, June 15, Off the Beaten Track.

Bryan, Howard, 1964. Public is now fenced off from Cabezón ghost town. *Albuquerque Tribune*, July 7, Off the Beaten Track.

Bryan, Howard, 1966. Off the beaten path. *Albuquerque Tribune*, March 17.

Bryan, Kirk, 1929. Flood-water farming. *Geographic Review* (19): 444-456.

Bryan, Kirk., and Franklin T. McCann. 1936. Successive pediments and terraces of the Upper Rio Puerco in New Mexico. *Journal of Geology* (44) 145-172.

Bunting, Bainbridge, undated. La Casa Vieja. Bainbridge Bunting Papers, Box 2, Folder 11. Albuquerque NM: University of New Mexico, Center of Southwest Research.

Calkins, Hugh G., 1937. *A Report on the Cuba Valley*. U.S. Soil Conservation Service, Region 8, Regional Bulletin No. 36, Conservation Economics Series No. 9, March.

Carlson, Roy L., 1966. Twin Angels Pueblo. *American Antiquity* 31: 676-682.

Carrier, James, 1991. The Colorado: a River Drained Dry. *National Geographic* 179(6, June): 4-35.

Cather, Steven M., 2004. Laramide orogeny in central and northern New Mexico and southern Colorado. In *The Geology of New Mexico – a Geologic History*. Special Publication 11, 203-248. Socorro NM: New Mexico Geological Society.

Cather, Steven M., Richard M. Chamberlin, Sean D. Connell, Richard M. Chamberlin, William C. McIntosh, Glen E. Jones, Andre R. Potochnik, Spencer G. Lucas, and Peggy S. Johnson, 2008. The Chuska erg: paleogeographic and paleoclimatic implications of an Oligocene sand sea on the Colorado Plateau. *Geological Society of America Bulletin* 120 (1-2): 13-33.

Cebada-Córdova, J., 1995. *Natural-ly Wonder-ful Cuba, a Regional Directory*. Cuba NM: Cuba Regional Economic Development Board.

Chafetz, Henry S., and Robert L. Folk, 1984. Travertines – Depositional morphology and the bacterially constructed constituents. *Journal of Sedimentary Petrology*. 54(1): 289-316.

Chappell, Gordon S., 1969. *To Santa Fe by Narrow Gauge: the D&RG's Chili Line*. Golden CO: Colorado Railroad Historical Foundation.

Chávez, Angelico, 1957. A Brief History of Bernalillo. Unpublished essay produced for 100th anniversary of Our Lady of Sorrows Church, Bernalillo NM.

Chávez, Angelico, 1992. Origins of New Mexico Families: a Genealogy of the Spanish Colonial Period. Santa Fe NM: Museum of New Mexico Press.

Chilton, Lance, 1984. *New Mexico: a Guide to the Colorful State*. Albuquerque NM: University of New Mexico Press.

Church, Fermor S., and John T. Hack, 1939. An exhumed erosion surface in the Jémez Mountains, New Mexico. *Journal of Geology* 47, 613-629.

Civilian Conservation Corps, 1936. *Official Annual–1936, Albuquerque District, 8th Corps Area.* Baton Rouge LA: Direct Advertising Company.

Civilian Conservation Corps Commemoration Committee, 2003. *Civilian Conservation Corps 1933–1942, 70th Anniversary*. Albuquerque, NM: Chapter 141 of National Association of Civilian Conservation Corps Alumni.

Clark, Ira G., 1987. *Water in New Mexico: a History of its Management and Use*. Albuquerque NM: University of New Mexico Press.

Cloer, Carol (editor), 1978. *Bloomfield's History as Told by Those Who Lived It*. Bloomfield NM: Bloomfield High School Social Studies Project, 10th and 11th grade classes. Originally available at www.bloomfieldhistory.tripod.com (last accessed 2006)

Coen, C.F., 1925. *A History of New Mexico, vol. II, Historical and Biographical*. Chicago IL and New York: The American Historical Society.

Cole, Thomas J., 2003. Santa Ana may get state loan. *Albuquerque Journal*, July 26.

Coleman, Michael, Allen Gellis, David W. Love, and R. Hadley, 1998. Channelization effects on the Rio Puerco above La Ventana, New Mexico. U.S. Geological Survey. Http://climchange.cr.usgs.gov/rio_puerco/papers/channel/html (last accessed 2005).

Connell, Sean D., 1998. *Geology of the Bernalillo quadrangle, Sandoval County, New Mexico*; M Map OF-DM-16 (scale 1:24,000). Socorro NM: New Mexico Bureau of Mines and Natural Resources.

Connell, Sean D., 2004. Geology of the Albuquerque basin and tectonic development of the Rio Grande rift in North Central New Mexico. In *Geology of New Mexico – a Geologic History*, Special Publication No. 11, eds. Greg H. Mack and Katherine A. Giles, 359-388. Socorro NM: New Mexico Geological Society.

Connell, Sean D., Steven M. Cather, Bradley Ilg, Karl E. Karlstrom, Barbara Menne, Mark Picha, Chris Andronicus, Adam S. Read, Paul W. Bauer, and Peggy S. Johnson, 1995. *Geology of the Placitas quadrangle, Sandoval County, New Mexico*. Map OF-DM-2 (scale 1:24,000). Socorro NM: New Mexico Bureau of Mines and Natural Resources.

Córdova-May, Esther V., 1973. Cuba, early 20s–1973. *Cuba News*, April 20. Cuba NM: Com-Press.

Córdova-May, Esther V. 2007. Antes. *Cuba News*, November 17. Cuba NM: Com-Press.

Córdova-May, Esther V., 2009a. Las Casas de Antes. *Cuba News*, January 16. Cuba NM: Com-Press.

Córdova-May, Esther V., 2009b. La Estafeta de Antes (The Post Office from Before). *Cuba News*, April 17. Cuba NM: Com-Press.

Counselor, James, and Ann Counselor, 1954. *Wild, Wooly and Wonderful*. New York: Vantage Press.

Craig, Phillip, 2002. *The Cox-Truby Feud*. Flora Vista NM: San Juan County Historical Society.

Craigg, Steven D., 1992. Water resources on the pueblos of Jémez, Zia, and Santa Ana, Sandoval County, New Mexico. U.S. Geological Survey Water-Resources Investigations Report 89-4091.

Crampton, Charles G., and Steven K. Madsen, 1994. *In Search of the Spanish Trail: Santa Fe to Los Angeles, 1829-1848*. Salt Lake City UT: Gibbs-Smith.

Cuba News,1964. Old Timers. April 15, 2005. Cuba NM: Com-Press.

Cuba News, 2000. History of the Library in Cuba. December 15. Cuba NM: Com-Press.

Cuba News, 2001a. Cuba Village Council meeting. February 16. Cuba NM: Com-Press.

Cuba News, 2001b. School Board meeting. September 21. Cuba NM: Com-Press.

Cuba News, 2001d. Cuba 4-Lane Fiesta: record-setting burrito. December 21. Cuba NM: Com-Press

Cuba News, 2002a School Board meeting, January 18. Cuba NM: Com-Press.

Cuba News, 2002b. Clara May Johnson receives award. January 18. Cuba NM: Com-Press.

Cuba News, 2002c. Note of appreciation. Letter by E. Maharg, Cuba Mayor Pro Tem. February 15. Cuba NM: Com-Press.

Cuba News, 2002d. Cuba Village Council meeting, February 15. Cuba NM: Com-Press.

Cuba News, 2003. Cuba Library. November 21. Cuba NM: Com-Press.

Cuba News, 2004. Cuba Ranger District Horseshoe Springs decommissioning. September 17. Cuba NM: Com-Press.

Cuba News,2005a. First wood chip-fired biomass heating system in New Mexico. April 15. Cuba NM: Com-Press.

Cuba News, 2005b. Obituary for Alice Francis Wiese. March 18. Cuba NM: Com-Press.

Cuba News, 2005c. People in the News: Hazel Herrera. May 20. Cuba NM: Com-Press.

Cuba News, 2005d. Sarah Harris speaks about her weed-eating goats. September 15. Cuba NM: Com-Press.

Cuba News, 2005e. Capitol Holiday tree leaves Cuba in swirl of music, color, and a big, big parade! November 18. Cuba NM: Com-Press.

Cuba News, 2006a. Suspects indicted on murder and other charges. February 17. Cuba NM: Com-Press.

Cuba News, 2006b. More changes in the air for Cuba News, outcome uncertain. January 21. Cuba NM: Com-Press.

Cuba News, 2006c. El Bruno's, a New Mexico landmark in Cuba, destroyed by fire. June 16. Cuba NM: Com-Press.

Cuba News, 2006d. El Bruno's is back! September 21. Cuba NM: Com-Press.

Cuba News, 2008a. Obituary for Eleanor Law Hernández. April 18. Cuba NM: Com-Press.

Cuba News, 2008b. Cuba Library under Construction at 14 East Cordova Ave. May 16. Cuba NM: Com-Press.

Cuba News, 2008c. Cuba landmark gone. June 20. Cuba NM: Com-Press.

Cuba News, 2008d. Cuba Visitor Center to close. November 21. Cuba NM: Com-Press.

Cuba Visitor Center. Undated. *Cuba*.

Cunliffe, B., 1978. *Rome and Her Empire*, New York, San Francisco CA, St. Louis MO: McGraw-Hill Book Co.

Cutler, Alan, 2003. *The Seashell on the Mountaintop: How Nicolaus Steno Solved an Ancient Mystery and Created a Science of the Earth*. New York: Penguin Group.

Davis, Michael, 2003. Zia Pueblo has its eye on 14,000 acres. *Albuquerque Journal*, February 13.

Dick-Peddie, William A., 1993. *New Mexico Vegetation: past, present, and future*. Albuquerque NM: University of New Mexico Press.

Diven, Bill, 2004. Bernalillo plans one, maybe two train stations. *Sandoval Sentinel*, August.

Donovan, William, 2003. Family feud: McDonald family split over operation of Counselor trading post. *Navajo Times Online*, September 25. Www.thenavajotimes.com. Window Rock AZ.

Dortignac, E.J., 1962. An 1890 irrigation venture in the Rio Puerco. *Professional Engineer*, March.

Douglas, Michael W., Robert A. Maddox, Kenneth Howard, and Sergio Reyes, 1993. The Mexican Monsoon. *Journal of Climate* 1665-1677. American Meteorological Society.

Dugan, Thomas., and Emery Arnold, 2002. *Gas: adventures into the history of one of the world's largest gas fields – the San Juan Basin of New Mexico*. Farmington NM: Dugan Production Corp.

Duke, Robert W.,1999. *San Juan County: the Early Years*. Flora Vista NM: San Juan County Historical Society.

Dane, Carle H., and George O. Bachman, 1965. *Geologic map of New Mexico*. U.S. Geological Survey (scale 1:500,000).

Durand, Stephen R., and Larry L. Baker, 2003. Population, settlement patterns, and paleoenvironment: culture change in the Middle Rio Puerco Valley. In Prehistory of the Middle Rio Puerco Valley, Sandoval County New Mexico 179-189. Special Publication No. 3, p. 179-189. Albuquerque NM: Archaeological Society of New Mexico.

Dykeman, Willilam, and James Stokely, 1968. *The Border States: Kentucky, North Carolina, Tennessee, Virginia, West Virginia*. New York: Time-Life Books.

Eddington, Patrick, and Susan Makav, 1995. *Trading Post Guidebook: Where to find the Trading Posts, Galleries, Auctions, Artists, and Museums of the Four Corners Region*. Flagstaff AZ: Northland Publishing.

Elias, Scott A., 1997. *The Ice-Age History of Southwestern National Parks*. Washington DC: Smithsonian Institution Press.

Elmore, Francis H., 1976. *Shrubs and Trees of the Southwest Uplands*. Tucson AZ: Southwest Parks and Monuments Association.

Farmington Times Hustler, 1939. Riot at Dance in Bloomfield Saturday Night. October 13.

Farmington Times Hustler,1941. CCC Camp at Bloomfield Has Record of Service to Stockmen. April 11. Farmington NM.

Farmington Times Hustler, 1942. CCC Camp at Bloomfield Abandoned, May 15. Farmington NM.

Fassett, James E., 1989. Coal resources of the San Juan basin. In *Southeastern Colorado Plateau*. 40th annual field conference guidebook 303-307. Socorro NM: New Mexico Geological Society.

Fassett, James E., Robert A. Zielinski, and James R. Budahn, 2002. Dinosaurs that did not die: Evidence for Paleocene dinosaurs in the Ojo Alamo Sandstone, San Juan Basin, New Mexico. In *Catastrophic Events and Mass Extinctions: Impacts and Beyond*, eds. Christian Koeberl and Kenneth G. MacLeod 307-336. Geological Society of America Special Paper 356.

Federal Works Agency, 1946. New Mexico Transportation Map. Sheet 3 (scale 1:250,000). Santa Fe NM: Public Roads Administration

Fenton, Carroll L., and Mildred A. Fenton, 1952. *Giants of Geology*. Garden City NY: Doubleday and Company

Fisher, Betty. undated. Hand-written recollections of teaching in Cuba area. Bernalillo NM: Sandoval County Historical Society.

Fischer, Edward E., and John P. Borland, 1983. Estimation of natural streamflow in the Jémez River at the boundaries of Indian lands, central New Mexico. *Water-Resources Investigations Report 82-4113*. U.S. Geological Survey.

Fitzpatrick, George, 1980. New Mexico Magazine—Mirror of a State. *New Mexico Magazine*, July.

Fleck, John, 2001. Retracing a pioneer's trail: scientist studies the life and work of New Mexico's first fossil-hunter. *Albuquerque Sunday Journal* 79(11).

Fleck, John, 2009. Navajos stand to get water windfall. *Albuquerque Journal*, March 26.

Force, Eric R., R. Gwinn Vivian, Thomas C. Windes, and Jeffrey S. Dean, 2002. *Relation of Bonito paleo-channels and base-level variations to Anasazi occupation, Chaco Canyon, New Mexico*. Archaeological Series 194. Tucson AZ: Arizona State Museum.

Formento-Trigilio, Merri L., 1997. The tectonic geomorphology and long-term landscape evolution of the southern Sierra Nacimiento, northern New Mexico. Unpublished M.S. dissertation. Albuquerque NM: University of New Mexico.

Formento-Trigilio, Merri L., and Frank J. Pazzaglia, 1998. Tectonic geomorphology of the Sierra Nacimiento: traditional and new techniques in assessing long-term landscape evolution in the southern Rocky Mountains. *Journal of Geology* 106: 433-453.

Forrest, Suzanne, 1989. *The Preservation of the Village: New Mexico's Hispanics and the New Deal*. New Mexico Land Grant Series. Albuquerque NM: University of New Mexico Press.

Frankel, Charles, 1999. *The End of the Dinosaurs – Chicxulub Crater and Mass Extinction*. New York: Cambridge University Press.

Frazier, Kendrick, 2005. *People of Chaco: a Canyon and its Culture*. New York: W.W. Norton and Co.

Gabriel, Kathryn, 1991. *Roads to Center Place: a Cultural Atlas of Chaco Canyon and the Anasazi*. Boulder CO: Johnson Books.

Gabriel, Kathryn, 1992. *Marietta Wetherill: Reflections on Life with the Navajos in Chaco Canyon*. Boulder CO: Johnson Books.

Gambling Magazine, 1999. Gambling News: Casino's poorly run loans, spending, nepotism cited in Jicarilla deficit. Www.gamblingmagazine. com/articles/14/14-243.htm.

Gantner, B., 1973. Highway 44: New Mexico's primary road is criticized. Albuquerque NM: *Albuquerque Sunday Journal*, October 14.

García, Nasario, 1987, *Recuerdos de los Viejos: Tales of the Rio Puerco*. Albuquerque NM: University of New Mexico Press.

García, Nasario, 1992, *Abuelitos: Stories of the Rio Puerco Valley*. Albuquerque NM: University of New Mexico Press.

García, Nasario, 1994, *Tata: A Voice from the Rio Puerco*. Albuquerque NM: University of New Mexico Press.

García, Nasario, 1997a, *Comadres: Hispanic Women of the Rio Puerco Valley*. Albuquerque NM: University of New Mexico Press.

García, Nasario, 1997b, *Mas Ántes: Hispanic Folklore of the Rio Puerco Valley*. Albuquerque NM: University of New Mexico Press.

Gellis, Allen., Milan Pavich, Paul Bierman, Amy Ellwein, Scott Aby, and Eric Clapp, 2000. Comparison of geomorphic and isotopic measurements for erosion in the Rio Puerco, New Mexico. In *U.S. Geological Survey Middle Rio Grande Basin Study – Proceedings of the Fourth Annual Workshop. Albuquerque, New Mexico, February 15-16, 2000*, 46-47.

Gill, Donald A., 1994. *Stories behind the Street Names of Albuquerque, Santa Fe, and Taos*; Chicago IL: Bonus Books.

Gillette, David D., 1994. *Seismosaurus: the Earth Shaker*. New York: Columbia University Press.

Gilliam, Mary L., Robert W. Blair, Christopher J. Carroll, Mark D. Johnson, and Thomas W. Perry, 1999. *Quaternary and environmental geology of the Animas River Valley, Colorado and New Mexico*; Friends of the Pleistocene/Rocky Mountain Cell 1999 Field Conference Guidebook, September 10-12, Days 2 and 3.

Glischinski, Steve, 1997. *Santa Fe Railway*. Osceola WI: Motorbooks International.

Glover, Vernon, 1990. *Jémez Mountains Railroads, Santa Fe National Forest, New Mexico*; Santa Fe NM: Historical Society of New Mexico.

Goff, Fraser, Jamie N. Gardner, W. Scott Baldridge, Jeffrey B. Hulen, Dennis L. Nielson, David Vaniman, Grant Heiken, Michael A. Dungan, and David Broxton, 1989. Volcanic and hydrothermal evolution of Valles caldera and Jémez volcanic field: in *Field excursions to volcanic terranes in the western United States, Vol. 1 – Southern Rocky Mountain region*. Memoir 46, 381-434. Socorro NM: New Mexico Bureau of Mines and Mineral Resources.

Gómez, Arthur R., 1994. *Quest for the Golden Circle: the Four Corners and the Metropolitan West, 1945-1970*. Albuquerque NM: University of New Mexico Press.

Goodman, James M., 1982. *The Navajo Atlas: Environment, Resources, People, and History of the Diné Bikeyah*; Civilization of the American Indian Series. Norman OK: University of Oklahoma Press.

Goodwin, Jason, 2003. *Greenback: the Almighty Dollar and the Invention of America*. New York: Henry Holt and Company.

Goudsmit, Samuel A., and Robert Claiborne, 1966. *Time*; New York: Time-Life Books.

Gross, Frederick A., and William A. Dick-Peddie, 1979. A map of primeval vegetation in New Mexico. *Southwestern Naturalist* 115-122.

Gutierrez, Alberto A.,1980. Channel and hillslope geomorphology of badlands in the San Juan basin, New Mexico. Unpublished M.S. dissertation. Albuquerque NM: University of New Mexico.

Gutierrez, Alberto A., 1983. Geomorphic processes and sediment transport in badland watersheds, San Juan County, New Mexico. In *Chaco Canyon Country*, eds Stephen G. Wells, David W. Love, and Thomas W. Gardner, 113-120. American Geomorphological Field Group Field Trip Guidebook, 1983 Conference, Northwestern New Mexico.

Hall, G. Emlen, 1987. The Pueblo Grant Labyrinth. In *Land, Water, and Culture: New Perspectives on Hispanic Land Grants*, eds. Charles L. Briggs and John R. Van Ness. Albuquerque NM: University of New Mexico Press.

Hall, Stephen A., 2005. Ice Age vegetation and flora in New Mexico. In *New Mexico's Ice Ages*, eds. Spencer G. Lucas, Gary S. Morgan, and Kate E. Zeigler, 171-183. Bulletin 28. Albuquerque NM: New Mexico Museum of Natural History and Science.

Hallett, R. Bruce, 1992. Volcanic geology of the Rio Puerco Necks. In *San Juan Basin IV*, 43rd fall field conference guidebook 135-144. Socorro NM: New Mexico Geological Society.

Hallett, R. Bruce, Philip R. Kyle, and William C. McIntosh, 1997. Paleomagnetic and ^{40}Ar/^{39}Ar age constraints on the chronologic evolution of the Rio Puerco volcanic necks and Mesa Prieta, west-central New Mexico: implications for transition zone magmatism. *Geological Society of America Bulletin* 109: 95-106.

Hammond, George P., and Agapito Rey, 1966. *The Rediscovery of New Mexico, 1580–1594: the Explorations of Chamuscado, Espejo, Castaño de Sosa, Moriete, and Leyva de Bonilla and Humaña*. Coronado Historical Series III. Albuquerque NM: University of New Mexico Press.

Haq, Bilal U., Jan Hardenbol, and Peter R. Vail, 1987. Chronology of fluctuating sea levels since the Triassic (250 million years ago until the present). *Science* 235: 1156-1167.

Hartranft, Michael, 1986. Prairie Star Restaurant Has No History to Tout. *Albuquerque Journal*, June 19.

Hendricks, Rick., and John P Wilson, 1996. *The Navajos in 1705: Roque Madrid's Campaign Journal*. Albuquerque NM: University of New Mexico Press.

Henry, Susan, 2004. Interview with Susan Henry, Aztec NM, May 11.

Hernández, John S., 2000a. Interview at Young's Hotel, Cuba NM, June 9.

Hernández, John S., 2000b. Interview at Young's Hotel, Cuba NM, June 23.

Hernández, John S., 2004. Interview at Young's Hotel, Cuba, NM, December 28.

Herrera, Mary (compiler), 2008. *New Mexico Blue Book, 2007-2008*. Santa Fe NM: Office of Secretary of State.

Hilton, George W., 1990. *American Narrow Gauge Railroads*. Stanford CA: Stanford University Press.

Hollenshead, Charles T., and Roy L. Pritchard, 1961. Geometry of producing Mesaverde sandstones, San Juan basin. In *Geometry of Sandstone Bodies*, 98-118. Tulsa OK: American Association of Petroleum Geologists.

Hume, Bill, 1971. Santa Ana Pueblo now three villages. *Albuquerque Sunday Journal*, January 24.

Hurst, William B.,2003. Typological analysis of ceramics from the Middle Rio Puerco of the East. In *Prehistory of the Middle Rio Puerco Valley, Sandoval County New Mexico*. Special Publication No. 3: 55-117. Albuquerque NM: Archaeological Society of New Mexico.

Hyer, Sally, 2001. HABS recording in New Mexico. In *Recording a Vanishing Legacy: the Historic American Buildings Survey in New Mexico, 1933–Today*, ed. Sally Hyer. Santa Fe NM: Museum of New Mexico Press.

Ikenson, Ben, 2001. Santa Ana Pueblo works to renew river. *Albuquerque Journal*, April 19.

Irick, John, and Patricia Irick, undated. Hazards of Largo Canyon. Website largocanyon.org/largo/general/quicksand.htm, site no longer active.

Irick, John. and Patricia Irick, 1997. Largo Canyon Station. *The Flora Vista Times*, summer 1997. Website no longer active.

Iverson, Peter, 1981. *The Navajo Nation*. Westport CN: Greenwood Press.

Iverson, Peter, 2002. *Diné: a History of the Navajos*. Albuquerque NM: University of New Mexico Press.

Jackson, David G., 1996. The Rio Puerco Valley: Where the Spanish Dream Failed. *Tradición Revista* 1(1).

Jackson, Hal, 2006. *Following the Royal Road: a Guide to the Historic Camino Real de Tierra Adentro*. Albuquerque NM: University of New Mexico Press.

Jaffe, Mark, 2000. *The Gilded Dinosaur: The Fossil War between E.D. Cope and O.C. Marsh and the Rise of American Science*. New York: Crown Publishers.

James, Harry, undated. *Territorial Trails*. Marceline MO: Walsworth Publishing.

James Press, 2004. *International Directory of Company Histories*, v. 61. Website www.fundinguniverse.com/company-histories/Giant-Industries-Inc-Company-History.html

Jémez Mountains Electrical Cooperative, Inc. Website www.jemezcoop.org.

Jenkinson, Michael, 1965. Coal camp—closed. *New Mexico Magazine* 43(4).

Jersig, Anne S., 2001. Abenicio Sálazar, Master Builder. *Prime Time*, August, Albuquerque, NM.

Jersig, Anne S., 2003. Rio Puerco Villages: Cabezón, Casa Salazar, Guadalupe and San Luís. *El Cronicón*, 143(3). Bernalillo NM: Sandoval County Historical Society.

Jicarilla Apache Nation undated. Website www.jicarillaonline.com.

Johnson, Brian F., 2009. Comet finished off North American big game animals, cooled the planet. Earth 54 (3). Alexandria VA: American Geological Institute.

Johnson, Carla May, 2003. Interview in Cuba NM, September 15.

Johnson, Carla May, 2005. Interview in Cuba NM, August 9.

Johnson, Mae, 1953. Homesteaders in the Cuba Country. *New Mexico Magazine* 31(10).

Jones, Donna, 1988. Sandoval towns invite land dispute decision: talks between Rio Rancho, Bernalillo deadlock. *Albuquerque Journal*, May 26.

Jones, Jeff, 2003. Potties fill in on US-550—Jicarillas shut rest area despite agreement with state. *Albuquerque Journal*, May 7.

Judd, Neil M., 1925. Everyday life in Pueblo Bonito. *National Geographic* 48(3, September): 227-261.

Julyan, Robert, 1996. *The Place Names of New Mexico*. Albuquerque NM: University of New Mexico Press.

Kallas, John, 1998. Euell Gibbons: the father of modern wild foods. *Wild Food Adventurer Newsletter*, November. Website www.wildfoodadventures.com/newsletter.html.

Kaime, Elizabeth, 2004. Interview in Aztec NM, May 12.

Kelley, Klara B., 1977. Commercial networks in the Navajo-Hopi-Zuni region. Unpublished Ph.D. dissertation. Albuquerque NM: University of New Mexico.

Kelley, Klara B., and Harris Francis, 2006. *Navajo Sacred Places*. Bloomington IN: University of Indiana Press.

Kelley, Klara B., and Harris Francis, 2006. Navajoland Trading Posts—preliminary version, June: Navajo Trading Post Encyclopedia in Progress. Website www.navajotradingposts.info/.

Kelley, Vincent C., and Albert M. Kudo, 1978. *Volcanoes and Related Basalts of Albuquerque Basin, New Mexico*. Circular 156. Socorro NM: New Mexico Bureau of Geology and Mineral Resources.

Kelly, Lawrence C., 1968. *The Navajo Indians and Federal Indian Policy, 1900-1935*. Tucson AZ: University of Arizona Press.

Kelly, Lawrence C., 1970. *Navajo Roundup*. Boulder CO: Pruett Publishing Co.

Kessell, John L., 1975. *Nuevo México 1776–1789, the Miera y Pacheco map*. Compiled for the Albuquerque Bicentennial Commission.

Keystone Environmental and Planning, Inc., 1997, *Environmental assessment: NM-44 from NM-528 to NM-4, Sandoval County, New Mexico*. Santa Fe NM: New Mexico State Highway and Transportation Department.

Kline, Doyle, 1978. New Mexico's golden fleeces. *New Mexico Magazine* 56(1).

Knight, Paul J., 1992. Vegetation and plant communities on the San Juan basin. In *San Juan Basin IV*, 43rd annual field conference guidebook 34-36. Socorro NM: New Mexico Geological Society.

Kues, Barry S., 1992. James Hervey Simpson and the first record of San Juan basin geology. In *San Juan Basin IV*, 43rd annual field conference guidebook 83-101. Socorro NM: New Mexico Geological Society.

Kutac, Connie, 2001. The Carlisle Cattle Company. *Elbow Creek Magazine* 1(3). Website www.elbowcreek.com.

Lagasse, Peter F., 1981. Geomorphic response of the Rio Grande to dam construction. In *Environmental Geology and Hydrology in New Mexico*, Special Publication No. 10, 27-46. Socorro NM: New Mexico Geological Society.

Larese, Steve, 2006. Tight Lines. *New Mexico Magazine* 84(9).

Larrañaga, Rene, 2000. New Mexico's Northwest, the connecting point. *Resources Magazine*, winter 2000. Las Cruces NM: New Mexico State University, College of Agriculture and Home Economics.

Law, B.E., and Dickinson, W.W., 1985. Conceptual model for origin of abnormally pressured gas accumulations in low-permeability reservoirs. *American Association of Petroleum Geologists Bulletin* 69(8): 1295-1304.

Leckson, Stephen H., 2004. Architecture: the central matter of Chaco Canyon. In *In Search of Chaco: New Approaches to an Archaeological Enigma*, ed. D.G. Noble, 23-31.Santa Fe NM: School of American Research Press.

Leopold, Leopold B., 1951. Vegetation of the southwestern watersheds in the nineteenth century. *Geographical Review* 41: 295-316.

Liebert, Martha, 1993. Fruit of the Vine—a History of Wine Making in Sandoval County. Bernalillo NM: Sandoval County Historical Society.

Liebert, Martha, 2005. A Brief History of the Bernalillo Merchantile Company, April 3. Bernalillo NM: Sandoval County Historical Society.

Linford, Laurance D., 2000. *Navajo Places: History, Legend, Landscape.* Salt Lake City UT: University of Utah Press.

Linklater, Andro, 2002. *Measuring America: How the United States was Shaped by the Greatest Land Sale in the World.* New York: Penguin Group.

Linthicum, Leslie, 2000. Work improves U.S. 550 safety. *Albuquerque Journal*, December 30.

Linthicum, Leslie. 2001. Smooth ride: rebuilt for safety's sake at a cost of $314 million, U.S. 550 (formerly N.M. 44) is wide open. *Albuquerque Sunday Journal*, December 9.

Linthicum, Leslie, 2002. Wilderness ways: because law prohibits machines in park cowboy builds creek dam by hand. *Albuquerque Sunday Journal*, September 22.

Linthicum, Leslie, 2003. For sale: used town. *Albuquerque Journal*, September 8.

Linthicum, Leslie, 2004. Making waves: an agreement that gives the Navajos much of San Juan River's water is a source of distrust. *Albuquerque Sunday Journal*, August 8.

Linthicum, Leslie, 2007. Navajos Buy Counselor. *Albuquerque Journal*, April 2.

Lister, Robert H., 1984. Chaco Canyon archeology thru time. In *New Light on Chaco Canyon*, ed. D.G. Noble. Santa Fe NM: School of American Research Press.

Lister, Robert H., and Florence C. Lister, 1981. *Chaco Canyon: Archaeology and Archaeologists.* Albuquerque NM: University of New Mexico Press.

Lister, Robert H., and Florence C. Lister, 1985. The Wetherills: Vandals, Pothunters, or Archaeologists. In *Prehistory and History in the Southwest: Collected Papers in Honor of Alden C. Hayes*, ed. N.L. Fox, 147-153. Papers of the Archaeological Society of New Mexico No. 11. Albuquerque NM: Archaeological Society of New Mexico,

Looney, Ralph, 1968. *Haunted Highways: The Ghost Towns of New Mexico.* Albuquerque NM: University of New Mexico Press.

López, Larry S.,1980. The Founding of San Francisco on the Rio Puerco. *New Mexico Historical Review* 55(1): 71-78.

López, Larry S.,1982. The Rio Puerco Irrigation Company. *New Mexico Historical Review* 57(1): 63-79.

Love, David W., 2002. Badlands in the San Juan Basin. *New Mexico's Energy, Present and Future*, 26-27. Socorro NM: New Mexico Bureau of Geology and Mineral Resources.

Love, David W., and Sean D. Connell, 2005. Late Neogene drainage developments on the southeastern Colorado Plateau, New Mexico. In *New Mexico's Ice Ages*. Eds. Spencer G. Lucas, Gary S. Morgan, and Kate E. Zeigler. Bulletin 28, 151-169. Albuquerque NM: New Mexico Museum of Natural History and Science.

Lucas, Spencer G., 2004. The Triassic and Jurassic systems in New Mexico. In *The Geology of New Mexico—a Geologic History*, eds. Greg H. Mack and Katherine A. Giles, 137-152. Special Publication No. 11. Socorro NM: New Mexico Geological Society.

Lucas, Spencer G., and Anderson, O.J., 1997. The Jurassic San Rafael Group, Four Corners region. In *Mesozoic Geology and Paleontology of the Four Corners Region*. 48th annual field conference guidebook, 115-132. Socorro NM: New Mexico Geological Society.

Lucas, Spencer G., and Gary S. Morgan, 2005. Ice Age Proboscideans of New Mexico. In *New Mexico's Ice Ages*, eds Spencer G. Lucas, Gary S. Morgan, and Kate E. Zeigler, 255-261. Bulletin 28. Albuquerque NM: New Mexico Museum of Natural History and Science.

Lucas, Spencer G., Thomas E. Williamson, Lawrence N. Smith, Robyn Wright-Dunbar, R. Bruce Hallett, Barry S. Kues, Gretchen K. Hoffman, Adrian P. Hunt, David W. Love, Virginia T. McLemore, and R.F. Hadley, 1992. First-day road log, from Cuba to La Ventana, San Luis, Babeon, Mesa Portales, Mea de Cuba and return to Cuba. In *San Juan Basin IV.* 43rd annual field conference guidebook, 1-32. Socorro NM: New Mexico Geological Society.

Lucchitta, Ivo, 1989. History of the Grand Canyon and of the Colorado River in Arizona. In *Geologic Evolution of Arizona*. Digest 17, 701-715. Tucson AZ: Arizona Geological Society.

Lucchitta, Ivo, 1990. History of the Grand Canyon and of the Colorado River in Arizona. In *Grand Canyon Geology*, eds Stanley S. Beus and Michael Morales, 311-332. New York: Oxford University Press, and Flagstaff AZ: Museum of Northern Arizona Press.

Luna, Hilario, 1975. *San Joaquin del Nacimiento*. Self-published, limited edition 1,000 copies.

MacDonald, Eleanor D., and John B. Arrington, 1970. *The San Juan Basin: My Kingdom was a County.* Denver CO: Green Mountain Press.

Manley, Kim, Glenn R. Scott, and Reinhard A. Wobus, 1987. *Geologic Map of the Aztec 1 Degree x 2 Degree Quadrangle, Northwestern New Mexico and Southern Colorado*; U.S. Geological Survey Misc. Investigation Series, Map I-1730 (scale 1:250,000).

Mann, Daniel H., and David J. Meltzer, 2007. Millennial-scale dynamics of valley fills over the past 12,000 ^{14}C yr in northeastern New Mexico, USA. *Geological Society of America Bulletin* 119(11/12): 1433-1448.

Marshall, Michael P., and Patrick Hogan, 1991. *Rethinking Navajo Pueblitos*, Bureau of Land Management Cultural Resources Series 8.

Mather, Keith H., 1963. Why do roads corrugate? *Scientific American* 208(1): 128-136.

McCabe, Gregory J., Julio L. Betancourt, and Hugo Hidalgo, 2007. Associations of decadal to multi-decadal sea-surface temperature variability with upper Colorado River flow. *Journal of the American Water Resources Association* 43(1): 183-192.

McCarty, Frank, 1969. Land grant problems in New Mexico. *Albuquerque Journal*.

McFadden, L.D., Wells, S.G., and Jerald D. Schultz, 1983. Development of late Quaternary eolian deposits, San Juan basin, New Mexico. In *Chaco Canyon Country*, eds. Stephen G. Wells, David W. Love, and Thomas W. Gardner, 167-175. Guidebook, American Geomorphological Field Group Field Trip.

McKee, John D., 1981. The Unrelenting Land. In *The Spell of New Mexico*, ed. Tony Hillerman, 71-76. Albuquerque NM: University of New Mexico Press.

McKee, Jennifer, 2001. Cross-country trail hits obstacle; *Albuquerque Sunday Journal*, June 24.

McKenna, Peter J., 2006. *Cultural Resource Survey: Archaeology in the Dragonfly Range Unit Weed Patches, Pueblo of Jémez Holdings in the Espiritu Santo Grant, New Mexico*. Bureau of Indian Affairs, Southwest Regional Office, report JEO4-190, NMCRIS 97659.

McLemore, Virginia T.,1996. Mineral resources in the Jémez and Nacimiento Mountains, Rio Arriba, Sandoval, Santa Fe and Los Alamos Counties, New Mexico. In *Jémez Mountains Region*. 47th annual field conference guidebook, 161-168. Socorro NM: New Mexico Geological Society.

McLemore, Virginia T., 2002. Navajo Lake State Park. *New Mexico Geology* 24(3): 91-96. Socorro NM: New Mexico Bureau of Geology and Mineral Resources.

McNitt, Frank, 1962. *The Indian Traders*. Norman OK: University of Oklahoma Press.

McNitt, Frank, 1966. *Richard Wetherill, Anasazi*. Albuquerque NM: University of New Mexico Press.

McPhee, John, 1962. A remembrance of Euell Gibbons. Preface to *Stalking the Wild Asparagus*, by Euell Gibbons. Newfane VT: Putney Publishing.

McTighe, James, 1984. *Roadside History of Colorado*. Boulder CO: Johnson Books.

Miller, Kenneth G., 2006. New synthesis of global sea-level change. In *The Redbeds*, 10, annual Newsletter of the Department of Geological Sciences. New Brunswick NJ: Rutgers—the State University of New Jersey.

Moore, Sally, 1999. *Country Roads of New Mexico: Drives, Day Trips, and Weekend Excursions*. Chicago IL: Country Roads Press, NTC/Contemporary Publishing Group.

Moorhead, Max L., 1954. *New Mexico's Royal Road: Trade and Travel on the Chihuahua Trail*. Norman OK: University of Oklahoma Press.

Moorland, Max L. (ed.), 1954. *Commerce of the Prairies, Letters of Josiah Gregg*. Norman OK: University of Oklahoma Press.

Morgan, Gary S., and Spencer G. Lucas, 2005. Pleistocene vertebrate faunas in New Mexico from alluvial, fluvial, and lacustrine deposits. In *New Mexico's Ice Ages*, eds. Spencer G. Lucas, Gary S. Morgan, and Kate E. Zeigler, 185-248. Bulletin 28. Albuquerque NM: New Mexico Museum of Natural History and Science.

Morgan, Murray E.,1958. Land Acquisitions. *New Mexico Magazine* 36(11).

Mosk, Sanford A., 1944. Land Tenure Problems in the Santa Fe Railroad Grant Area. Berkeley and Los Angeles CA: University of California Press.

Mraz, David A., 1982. La Hacienda de Baca—a Taste of History. *New Mexico Magazine* 60(9).

MRCOG (Mid-Region Council of Governments of New Mexico), 2004. *Comprehensive Land Use Plan for the Village of Cuba, New Mexico Magazine*, October 26.

Mueller, Claire J., 1990. A synopsis of the activities of the CCCers of Camp NP-2-N, Chaco Canyon, NM. Unpublished newsletter. Albuquerque: Chapter 141 of the National Association of Civilian Conservation Corps Alumni.

Mytton, James W., and Gary B. Schneider, 1987. *Interpretative Geology of the Chaco Area, Northwestern New Mexico*. U.S. Geological Survey, Miscellaneous Investigation Series, Map I-1777 (scale 1:24,000).

National Park Service, 1993. *Pueblo Bonito*. Chaco Culture National Historical Park booklet.

New Mexico Magazine, 1938. Report on highway projects, 16(12).

New Mexico Magazine, 1942. Passing cars hazard at 35, 20(12).

New Mexico Magazine, 1943. Estimated 45 mph speed limit, 21(10).

NMBG&MR (New Mexico Bureau of Geology and Mineral Resources), 2003. *Geologic Map of New Mexico* (scale 1:500,000). Socorro NM: New Mexico Bureau of Geology and Mineral Resources.

New Mexico Department of Transportation. Current *New Mexico Transportation Map*, (scale 1:124,000).

New Mexico Highway Journal, 1926a. Status of construction projects, April.

New Mexico Highway Journal, 1926b. Status of construction projects, August.

New Mexico Highway Journal, 1926c. Status of construction projects, October.

New Mexico Highway Journal, 1931. The origin of the Federal Aid for roads; January.

New Mexico State Highway Commission, 1929. *Highway map of New Mexico* (scale 1:1,650,000).

New Mexico State Highway Department, 1923. *Highway map of the state of New Mexico* (scale 1:1,200,000).

New Mexico State Highway Department, 1939. *Biennial Report of the State Highway Engineer of the State of New Mexico*, January 1, 1937–December 13, 1938.

New Mexico State Highway Department, 1959. *Biennial Report of the Chief Highway Engineer of the State of New Mexico,* January 1, 1957–December 13, 1958.

New Mexico State Highway and Transportation Department, 1998. *Environmental Assessment: New Mexico State Highway 44 from NM 4 to NM 537.* Project #SP-0441(210)23/SP-0442(229)64.

Nickelson, Howard B., 1988. *One hundred years of coal mining in the San Juan Basin, New Mexico.* Bulletin No. 111. Socorro NM: New Mexico Bureau of Mines and Mineral Resources.

Nobis, Dorothy, 2001. A sense of community: family ties bind Nageezi, Counselor. *New Mexico Magazine* 79(9): 62-67.

Nordhaus, Robert J., 1995. *Tipi Rings: a Chronicle of the Jicarilla Apache Land Claim.* Albuquerque NM: Bowarrow Publishing Co.

Oberg, Kalervo, 1940. Cultural Factors and Land-Use Planning in Cuba Valley, New Mexico. *Rural Sociology,* 5(4): December.

Oil and Gas Journal, 2000. Tragedy in New Mexico, August 28. Tulsa OK: The Petroleum Publishing Company.

Ojo del Espiritu Santo Company, undated. Box of company correspondence, MSS 42BC. Albuquerque NM: University of New Mexico, Center of Southwest Research.

O'Neill, F.Michael, 1992. Paleo-Indians in the San Juan basin: a paleontological perspective. In *San Juan Basin IV.* 43rd annual field conference guidebook, 333-339. Socorro NM: New Mexico Geological Society.

Orth, Charles J., James S. Gilmore, and Jere D. Knight, 1987. Iridium anomaly at the Cretaceous boundary in the Raton basin. In *Northeastern New Mexico,* 38th annual field conference guidebook, 265-269. Socorro NM: New Mexico Geological Society

Palace of the Governors Photo Archives, Santa Fe NM. Photo negatives #6153, #51466, #51282, #51283, #130232, and #158158.

Papich, William, 2000. Airline first to pay for Zia use: Southwest, pueblo struck a deal on Zia use. *Albuquerque Journal,* supplement *Business Outlook,* September 28.

Pastuszak, Robert A., 1968. Geomorphology of part of the La Plata and San Juan rivers, San Juan County, New Mexico. Unpublished M.S. dissertation. Albuquerque NM: University of New Mexico.

Patton, Peter C., Norma Biggar, Christopher D. Condit, Mary L. Gilliam, David W. Love, Michael N. Machette, Larry Mayer, Roger B. Morrison, and John N. Rosholt, 1991. Quaternary geology of the Colorado Plateau: Navajo and Acoma-Zuni Sections. In *The Geology of North America, Vol. K-2: Quaternary Non-Glacial Geology, Conterminous U.S.,* 374-397. Boulder CO: Geological Society of America.

Pazzaglia, Frank J., 2005. River responses to Ice Age (Quaternary) climates in New Mexico. In *New Mexico's Ice Ages,* eds. Spencer G. Lucas, Gary S. Morgan, and Kate E. Zeigler, 115-124. Bulletin 28. Albuquerque NM: New Mexico Museum of Natural History and Science.

Pazzaglia, Frank J., Sean D. Connell, John W. Hawley, Richard H. Tedford, Steve Personius, Gary A. Smith, Steven M. Cather, Spencer G. Lucas, Patricia Hester, John Gilmore, and Lee A. Woodward, 1999, Second-day trip 2 road log, from Albuquerque to San Ysidro, Loma Crestón, La Ceja, and San Hill Fault. In *Albuquerque Geology,* eds. Frank J. Pazzaglia, and Spencer G. Lucas, 47-66. 50th Fall Field Conference Guidebook. Socorro NM: New Mexico Geological Society.

Pearce, T.M., 1965. *New Mexico Place Names – a Geographical Dictionary.* Albuquerque NM: University of New Mexico Press.

Perrigo, Lynn I., 1985. *Hispanos: Historic Leaders in New Mexico.* Santa Fe NM: Sunstone Press.

Peterson, C.S., ed., 1912. *Representative New Mexicans: The National Newspaper Reference Book of the New State.* Denver CO: C.S. Peterson.

Pinel, Sandra Lee, 1987. Plant latest chapter in Pueblo Business Plan: Wallboard factory would join restaurant, farm at Santa Ana. *Albuquerque Journal,* sec. Metro Plus, May 27.

Poling-Kempes, L., 1997. *Valley of the Shining Stone: the Story of Abiquiu.* Tucson AZ: University of Arizona Press.

Post, Stephen S., 1994. Cuba North: excavation of a lithic artifact scatter (LA 66471) and an Archaic Period Structure (LA 66472) along State Road 44 near Cuba, Sandoval County, New Mexico. *Archaeology Notes 26.* Albuquerque NM: Museum of New Mexico Office of Archaeological Studies.

Powers, Margaret A., and Byron P. Johnson, 1987. Defensive sites of Dinétah. Bureau of Land Management, Cultural Resources Series #2.

Price, Jess, 1985. NAPI: Navajo agribusiness in the San Juan basin. *New Mexico* 63(3). Santa Fe NM: New Mexico Magazine.

Rae, Steven R., Joseph E. King, and Donald Abbe, 1984. *Phase I Report: Historic Bridge Survey.* Santa Fe NM: New Mexico Highway Department.

Rayburn, Rosalie, 2003. Tamaya among top resort choices. *Albuquerque Journal,* supplement *Business Outlook,* November 17.

Rayburn, Rosalie, 2008. After 5 years, El Zócalo is reborn. *Sunday Rio Rancho Journal,* August 10.

Reagan, A.B., 1903. Geology of the Jémez-Albuquerque region, New Mexico. *American Geologist* 31(2): 67-111 (original not seen, cited in Summers 1976, listed below).

Reager, Nell, 1965. Valley of the Angels. *New Mexico Magazine* 43(10).

Reeve, Agnesa, and Richard Schalk, 2001. The Bainbridge Bunting Years. In *Recording a Vanishing Legacy: The Historic American Buildings Survey in New Mexico, 1933–Today,* ed. Sally Hyer. Santa Fe NM: Museum of New Mexico Press.

Reeve, Frank D., 1946. A Navajo struggle for land. *New Mexico Historical Review* 21: 1-21.

Reeve, Frank D., 1956. Early Navaho Geography. *New Mexico Historical Review* 31: 290-309.

Reeve, Frank D., 1959. "The Navaho-Spanish Peace: 1720s–1770s. *New Mexico Historical Review* 34: 9-40.

Reeve, Frank D., 1961. *History of New Mexico, v. 3: Family and Personal History,* New York: Lewis Historical Publishing Company.

Reid, Kevin D., Fraser Goff, and Dale A. Counce, 2003. Arsenic concentration and mass flow rate in natural waters of the Valles Caldera and Jémez Mountains region, New Mexico. *New Mexico Geology* 25(3): 75-82. Socorro NM: New Mexico Bureau of Geology and Mineral Resources.

Reisner, Marc, 1987. *Cadillac Desert: the American West and its Disappearing Water*. New York: Penguin Books.

Remini, Robert V., 2002. *Joseph Smith*. New York: Viking Press.

Renick, B.Coleman, 1931. Geology and Ground-Water Resources of Western Sandoval County, New Mexico. U.S. Geological Survey, Water Supply Paper 620.

Ricketts, Orval, 1930. Highways to the sunny San Juan. *New Mexico Highway Journal*, May.

Ricketts, Orval, 1932. A road reconnaissance by rowboat. *New Mexico Magazine*, February.

Rinehart, Larry F., Spencer G. Lucas, and Gary S. Morgan, 2006. Tectonic development of late Pleistocene (Rancholabrean) animal-trapping fissures in the Middle Jurassic Todilto Formation (north central New Mexico. *New Mexico Geology* 28(3): 84-87. Socorro NM: New Mexico Bureau of Geology and Mineral Resources.

Riner, Steve, 2004. Details of New Mexico State Routes. Website www.steve-riner.com/nmhighways.

Ripp, Bart, 1980. Rancher tired of government 'messing' *Albuquerque Journal*, March 9.

Ripp, Bart, 1985. Downtown barber remains a cut above the rest. *Albuquerque Tribune*, Nov. 25.

Ritter, Ken, 2007. Colorado River Pact Signed. *Albuquerque Journal*, December 14.

Rittenhouse, Jack D., 1965. *Cabezón: A New Mexico Ghost Town*. Santa Fe NM: Stagecoach Press.

Rivero, Jordi. undated. Virgín de la Caridad de Cobre, Patrona de Cuba. Website www.corazones.org/maria/america/cuba_caridad_cobre.htm.

Roberson, Roy, 2006. Navajo Agricultural Products Industry a farming success in New Mexico. Western Farm Press (E-mail daily newsletter), www.westernfarmpress.com/news/3-14-06-Navajo-Farming-Success-New-Mexico (last accessed April 2, 2009).

Roberts, Willow, 1987. *Stokes Carson: Twentieth-Century Trading on the Navajo Reservation*. Albuquerque NM: University of New Mexico Press.

Roessel, Monty, 1988."The Dinetah—Navajo sacred homeland, holy place reveals Navajo origins, provides strength. *New Mexico*, August.

Rogers, Margaret A., Barry S. Kues, Fraser Goff, Frank J. Pazzaglia, Lee A. Woodward, Spencer G. Lucas, and Jamie N. Gardner, 1996. First-day road log, from Bernalillo to San Ysidro, southern Nacimiento Mountains, Guadalupe Box, Jémez Springs, Valles Caldera, and Los Alamos. In *Jémez Mountains Region*, eds. Fraser Goff, Barry S. Kues, Margaret A. Rogers, McFadden, L.D., and Gardner, J.N., 1-39. 47th annual field conference guidebook. Socorro NM: New Mexico Geological Society.

Romero, Dora B., 1984. Just thinking about my past. Unpublished manuscript. Bernalillo NM: Sandoval County Historical Society.

Romo, Rene, 2000. Logging caps fell another sawmill. *Albuquerque Journal*, August 20.

Roser, Donald R., 1951. New Mexico has only 40 years of modern highway history. *New Mexico*, July. Santa Fe NM: New Mexico Magazine.

Ruíz-Esparza, Robert, 1993. *History of Multi-Cultural Education in Sandoval County, 1540–1993*. Unpublished research project. Bernalillo NM: Sandoval County Historical Society.

Russell, L.R., and Snelson, S., 1994. Structural and tectonics of the Albuquerque Basin segment of the Rio Grande rift: Insights from reflection seismic data. In *Basins of the Rio Grande Rift: Structure, Stratigraphy and Tectonic Setting*, Special Publication No. 291, 83-112. Boulder CO: Geological Society of America.

Ryan, Michael J., and Genie Ryan, 1996. Rio Rancho—City of Vision. In *Albuquerque's Environmental Story: towards a Sustainable Community*; Section II, 33-36. Albuquerque NM: City of Albuquerque, Albuquerque Public Schools, Albuquerque Conservation Association, and others.

Sánchez, Joseph P., 1997. *Explorers, Traders, and Slavers: Forging the Old Spanish Trail*. Salt Lake City UT: University of Utah Press.

Sandoval County Times-Independent, 1972a. Bernalillo water supply. July 21, 1972.

Sandoval County Times-Independent, 1972b. The Old Mill is Gone. July 7.

Sandoval County Times-Independent, 1978. Old roadway story told. Specific date not available.

Sandoval County Historical Society, undated. Putney family history. Unpublished manuscript

Sandoval County Historical Society, 1992. Unpublished transcript of interview of L. Wood by S.B. Doyle, August 26.

San Juan County Archaeological Research Center and Library, undated. *Salmon Ruins and Heritage Park*. Informational pamphlet.

Sauter, J., 1994. Unpublished notes on Cuba churches by Father John Sauter. Bernalillo NM: Sandoval County Historical Society.

Schmedding, Joseph, 1951. *Cowboy and Indian Trader*. Caldwell ID: Caxton Printers, Ltd.

Schultz, Jerald D., 1980. Geomorphology, sedimentology, and Quaternary history of the eolian deposits, west-central San Juan basin, northwest New Mexico. Unpublished M.S. dissertation. Albuquerque NM: University of New Mexico.

Schultz, Jerald D., 1983. Geomorphology and Quaternary history of the southeastern Chaco Dune Field, northwestern San Juan basin. In *Chaco Canyon Country*, eds. Stephen G. Wells, David W. Love, and Thomas W. Gardner, 159-166. American Geomorphological Field Group Field Trip Guidebook,

Schumm, S.A., and Chorley, R.J., 1964. The Fall of Threatening Rock. *American Journal of Science* 262: 1041-1054.

Scurlock, Daniel, 1998. *From the Rio to the Sierra: an Environmental History of the Middle Rio Grande Basin*. U.S. Forest Service, General Technical Report RMRS-GTR-5, Rocky Mountain Research Station, Ft. Collins, Colorado.

Shah, Subhas, 2001. Keeping the Valley Green: the Story of the Middle Rio Grande Conservancy District. *Albuquerque Sunday Journal*, June 10.

Sherman, C.E., 1925. Original Ohio Land Subdivisions, vol. III (of 4 volumes), Final Report. Ohio Cooperative Topographic Survey. Ohio State Reformatory Press.

Sherman, James E., and Barbara H. Sherman, 1975. *Ghost Towns and Mining Camps of New Mexico*. Norman OK: University of New Mexico Press.

Simpson, George G., 1981. History of vertebrate paleontology in the San Juan basin. In *Advances in San Juan Basin Paleontology*, eds. Spencer G. Lucas, J.Keith Rigby, Jr., and Barry S. Kues, 3-25. Albuquerque NM: University of New Mexico Press.

Sinclair, John L., 1943. *Time of Harvest*. Santa Fe NM: Clear Light Publishers.

Sinclair, John L., 1947a. Coronado's Headquarters. *New Mexico Magazine*, March.

Sinclair, John L., 1947b. *Death in the Claimshack*. Denver CO: Sage Books.

Sinclair, John L., 1976. The Place Where De Vargas Died. *New Mexico Magazine*, August.

Sinclair, John L., 1980a. *New Mexico – the Shining Land*. Albuquerque NM: University of New Mexico Press.

Sinclair, John L., 1980b. *Cousin Drewey and the Holy Twister*. Frenchtown NJ: Columbia Publishing Co.

Sinclair, John L., 1982. *Cowboy Riding Country*. Albuquerque NM: University of New Mexico Press.

Sinclair, John L., 1996. *A Cowboy Writer: the Memoirs*. Albuquerque NM: University of New Mexico Press.

Smith, Lawrence N., 1983. Late Cenozoic fluvial evolution in the northern Chaco River drainage basin, northwestern New Mexico. Unpublished M.S. dissertation. Albuquerque NM: University of New Mexico.

Smith, Lawrence N., 1992. Stratigraphy, sediment dispersal and paleogeography of the lower Eocene San Jose Formation, San Juan basin, New Mexico and Colorado. In *San Juan Basin IV*. 43rd annual fall field conference guidebook, 297309. Socorro NM: New Mexico Geological Society.

Smith, Lawrence N., and Spencer G. Lucas, 1991. *Stratigraphy, sedimentology, and paleontology of the lower Eocene San José Formation in the central portion of the San Juan Basin, northwestern New Mexico*; Resources Bulletin No. 126. Socorro NM: New Mexico Bureau of Mines and Mineral Resources.

Smith, Robert J., 1965. The Cuba High School experimental program in non-graded English. Unpublished M.A. thesis. Albuquerque NM: University of New Mexico.

Snead, James E., 2001. *Ruins and Rivals: The Making of Southwestern Archaeology*. Tucson AZ: University of Arizona Press.

Snider, Harold, 2007. An interview with Joe Liebert. In *El Cronicón* 18(1): 12-15. Bernalillo NM: Sandoval County Historical Society.

Snow, David H., 1976. Santiago to Guache: Notes for a Tale of Two (or more) Bernalillos. In *Collected Papers in Honor of Marjorie Ferguson Lambert*, Papers of the Archeological Society of New Mexico, No. 3, Albuquerque NM: Albuquerque Archeological Society Press.

Sofaer, Anna, Michael P. Marshall, and Rolf M. Sinclair, 1986. The Great North Road: a cosmographic expression of the Chaco Culture of New Mexico. In *World Archaeoastraonomy: selected papers from the 2nd Oxford International Conference on Archaeoastronomy held in Mérida, Yucatan, Mexico*, January 13-17, ed. A.F. Aveni.

Soil Conservation Service (SCS), 1935. Aerial photographs on file at Earth Data Analysis Center (EDAC). Albuquerque NM: University of New Mexico.

Soussan, Tania, 2000. Cows pound life into worn land. *Albuquerque Journal*, August 8.

Soussan, Tania, 2001. Watershed future looking up. *Albuquerque Journal*, August 18.

Soussan, Tania, 2005. 40 years of river work can be wrong: river fought man-made path. *Albuquerque Journal*, January 18.

Splinter, William E., 1976. Center-pivot irrigation. *Scientific American* 234(6): 90-99.

Stamm, Roy A., 1999. *For Me the Sun: the Autobiography of Roy A. Stamm, and early Albuquerque Business Leader*, eds. James S. and Ann L. Carson. Albuquerque NM: The Albuquerque Museum.

Stanley, Francis, 1959. *The Private War of Ike Stockton*. Denver CO: World Press.

Stanley, Francis, 1964. *The Bernalillo (New Mexico) Story*; self-published.

Stanley, Steven M., 1986. *Earth and Life through Time*. New York: W.H. Freeman and Company.

Stone, William J., Forest P. Lyford, Peter F. Frenzel, Nancy H. Mizell, and Elizabeth T. Padgett, 1983. *Hydrogeology and water resources of San Juan Basin, New Mexico*. Hydrologic Report 6. Socorro NM: New Mexico Bureau of Mines and Mineral Resources.

Stout, Joseph A., Jr., 1970. Cattlemen, conservationists, and the Taylor Grazing Act. *New Mexico Historical Review* 45(4): 311-332.

Strong, R.L., 1957. Rio Puerco area was once New Mexico's breadbasket. *Albuquerque Tribune*, September 11, p. 23.

Stuart, David E., 2000. *Anasazi America*. Albuquerque NM: University of New Mexico Press.

Stuart, David E., and Rory P. Gauthier, 1981. *Prehistoric New Mexico: Background for Survey*. Originally published by State of New Mexico, Department. of Cultural Affairs, Historic Preservation Division; reprinted 1988 and 1996, Albuquerque NM: University of New Mexico Press.

Summers, W. Kelly, 1976. *Catalog of Thermal Waters in New Mexico*. Hydrologic Report 4. Socorro NM: New Mexico Bureau of Mines and Mineral Resources.

Talbott, Lyle W., 1984. Nacimiento pit, a Triassic strata-bound copper deposit. In *Ghost Ranch*. 25th annual field conference guidebook, 301-303. Socorro NM: New Mexico Geological Society.

Taylor, Louis H., 1981. The Kutz Canyon local fauna, Torrejonian (middle Paleocene) of the San Juan basin, New Mexico. In *Advances in San Juan Basin Paleontology*, eds. Spencer G. Lucas, J.K. Rigby, Jr., and Barry S. Kues, 242-263. Albuquerque NM: University of New Mexico Press.

Tessler, Denise, 1992. Sandoval History Tangled; *The New Mexico Lawyer*, April 6.

Texas State Historical Association. 2002. Biography of Euell Theophilus Gibbons, Handbook of Texas Online. Website www.tsha.utexas.edu/handbook/online.

Thompson, Fritz, 1994. Dreams turned to sawdust: Cuba's economy struggles to survive sawmill shutdown. *Albuquerque Journal*, July 31.

Tietjen, Gary, 1980. *Mormon Pioneers in New Mexico: a History of Ramah, Fruitland, Luna, Beulah, Bluewater, Virden, and Carson*. Los Alamos NM: G. Tietjen.

Tiller, Veronica E.V., 1983. *The Jicarilla Apache Tribe, a History, 1946-1970*. Lincoln NB: University of Nebraska Press.

Time, 1986. In New Mexico: a Very Special Point of View. October 20.

Time Life Books, 1993. *People of the Desert*, The American Indian series. Alexandria VA:Time-Life Books.

Time Life Books, 1995. Tribes of the Southern Plains, The American Indian series. Alexandria VA:Time-Life Books.

Tobias, Henry J., 1990. *A History of the Jews in New Mexico*. Albuquerque NM: University of New Mexico Press.

Torrez, Robert J., 1988. The San Juan gold rush of 1860 and its effect on the development of northern New Mexico. *New Mexico Historical Review* 63(3): 257-272.

Towner, Ronald H., and Jeffrey S. Dean, 1992. LA 2298: the oldest Pueblito revisited. *Kiva* 57(4): 315-329.

University of New Mexico, 1985. Handbook of issues, opinions, choices for the future of the Village of San Ysidro de los Dolores, Sandoval County, New Mexico. Rural Planning Studio Community and Regional Planning Program, June 1985.

U.S. Court of Private Land Claims, 1860. New Mexico Private Land Claim #44, Surveyor General's Office, Santa Fe NM, December 15.

U.S. Department of the Interior, 1874. *Letter from the Secretary of the Interior, Reports of the surveyor general of New Mexico on the private land-claims under grants of Felipe Gutierres and Juan José Gallegos, Nos. 83 and 84*, Ex. Document No. 38, pg. 687-695, March 25.

Van Devender, T.R., and Spaulding, W.G., 1983. Development of vegetation and climate in the southwestern United States. In *Origin and Evolution of Deserts*, eds. Steven G. Wells and D.R. Haragan, 131-160. Albuquerque NM: University of New Mexico Press.

Van Zandt, Franklin K., 1976. *Boundaries of the United States and the Several States*, U.S. Geological Survey Professional Paper 909.

Velarde, Victor, 2006. Personal communication, Cuba NM, February.

Velasco, Diane, 2003. Mayor predicts enormous Rio Rancho. *Albuquerque Journal*, supplement *Business Outlook*, November 6.

Vierra, Bradley J., 1987. The Tiguex Province: a Tale of Two Cities, In *Secrets of a City: Papers on Albuquerque Area Archaeology*, eds. Anne V. Poore and John Montgomery 70-86. Archaeological Society of New Mexico No. 13. Albuquerque NM: Archaeological Society of New Mexico.

Vogel, Chris, 2003. Military looking into Sandoval hole mystery. *Albuquerque Journal*, April 5.

Walker, Henry P., and Donald Bufkin, 1979. *Historical Atlas of Arizona*. Norman OK: University of Oklahoma Press.

Wallace, Joseph, 1987. *The Rise and Fall of the Dinosaur*. New York: Gallery Books.

Watkins, Thomas H., 1969. *The Grand Colorado: the Story of a River and its Canyons*; Palo Alto CA: American West Publishing Co.

Wattles, Ruth J., and Hilda F. Wetherill, 1977. To Chaco Canyon in horse-and-buggy days. *New Mexico Magazine*, October.

Watts, Susan, 2000. Biologic responses and restoration challenges associated with the altered flow pattern of the Rio Grand/Rio Bravo. In *Finding the Balance: Growth and Water in the Rio Grande*, Rio Grande/Rio Bravo Basin Coalition Final Report of the 2000 Uniting-the-Basin Congress, November 9-11.

Waybourn, Marilu, and Paul Horn, 1999. *Gobernador*. Farmington NM: Paul B. and Dorothy M. Horn Living Trust.

Weingroff, Richard F., 2003. U.S. 666: 'Beast of a Highway?' Federal Highway Administration, Department of Transportation. Website, www.fhwa.dot.gov/infrastructure/us666.htm.

Weingroff, Richard F., 2004. Federal Aid Road Act of 1916: Building the Foundation. Website: www.fhwa.dot.gov/infrastructure/rw96a.htm.

Wells, Stephen G., 1983. Regional badland development and a model of late Quaternary evolution of badland watersheds, San Juan basin, New Mexico. In *Chaco Canyon Country*, eds. Stephen G. Wells, David W. Love, and Thomas W. Gardner, 121-132. American Geomorphological Field Group Field Trip Guidebook, 1983 Conference, Northwestern New Mexico.

Wells, Stephen G., Thomas F. Bullard, Lawrence N. Smith, Thomas W. Gardner, 1983. Chronology, rates, and magnitudes of late Quaternary landscape changes in the southwestern Colorado Plateau. In *Chaco Canyon Country*, eds. Stephen G. Wells, David W. Love, and Thomas W. Gardner, 177-185. American Geomorphological Field Group Field Trip Guidebook, 1983 Conference, Northwestern New Mexico.

Westphall, Victor, 1958. The Public Domain in New Mexico, 1854-1891. *New Mexico Historical Review* 33(1): 24-52.

Westphall, Victor, 1973. *Thomas Benton Catron and His Era*. Tucson AZ: University of Arizona Press.

Wetherill, Marietta, 1953. *Pioneer's Foundation Oral History*. Transcriptions of taped conversations of Marietta Wetherill by L.B. Blachly, Tape #488. Albuquerque NM: University of New Mexico, Center of Southwest Research.

Wetherill, Richard, Jr., 1978. Early History of Cuba. *Sandoval County Review*, April.

White, James W., 2003. *The History of San Juan County Post Offices*; privately published by James W. White.

White, Peter, and Mary Ann White, 1988. *Along the Rio Grande: Cowboy Jack Thorp's New Mexico*. Santa Fe NM: Ancient City Press.

Whitehead, Neil.H., III.,1997. Fractures at the surface, San Juan basin, New Mexico and Colorado. In *Fractured Reservoirs: Characterization and Modeling*, 27-42. Denver CO: Rocky Mountain Association of Geologists.

Widdison, Jerold G., 1959. Historical geography of the Middle Rio Puerco Valley, New Mexico. *New Mexico Historical Review* 34(4): 248-284.

Wiese, Alice, 1978. Cuba Putney Family. *Sandoval County Review*, April.

Wiese, H. Louis, III, 2001. Interview at Wiese ranch, Cuba, NM, June 22.

Wiese, H. Louis, III, 2005. Interview at Wiese ranch, Cuba, NM, August 17.

Wiewandt, Thomas, and Maureen Wilks, 2001. *The Southwest Inside Out: an Illustrated Guide to the Land and Its History*. Tucson AZ: Wild Horizon Publishing.

Williams, Gerald W., 2000. *The USDA Forest Service – The First Century*; USDA Forest Service, FS 650. Washington DC: USDA Forest Service.

Williams, Jerry L. (ed.), 1986. *New Mexico in Maps*, 2nd edition. Albuquerque NM: University of New Mexico Press.

Wood, Lovena, 1992. Transcript of interview by S.B. Doyle, in *History of Multicultural Education in Sandoval County*. Bernalillo NM: Sandoval County Historical Society.

Wood, Lovena, 2003. Interview at Wood ranch, Cuba NM, September 15.

Woods, Betty, 1962. Enchanted Desert. *New Mexico Magazine*, June-July.

Woodward, Lee A., 1987. *Geology and mineral resources of Sierra Nacimiento and vicinity, New Mexico*. Memoir 42. Socorro NM: New Mexico Bureau of Mines and Mineral Resources.

Woodward, Lee A., and Ruben Martínez, 1974. *Geologic Map of Holy Ghost Spring Quadrangle, New Mexico*. Geologic Map 33 (scale 1:24,000). Socorro NM: New Mexico Bureau of Mines and Mineral Resources.

Woodward, Lee A., J.B. Anderson, W.H. Kaufman, W.H., and R.K. Reed, 1970a. *Geologic Map of San Pablo Quadrangle, New Mexico*. Geologic Map 26 (scale 1:24,000). Socorro NM: New Mexico Bureau of Mines and Mineral Resources.

Woodward, Lee A., D. McLelland, J.B. Anderson, J.B., and W.H. Kaufman, 1970b. *Geologic Map of Cuba Quadrangle, New Mexico*. Geologic Map 25 (scale 1:24,000). Socorro NM: New Mexico Bureau of Mines and Mineral Resources.

Woodward, Lee A., and Otto Schumacher, 1973. *Geologic Map of La Ventana Quadrangle, New Mexico*. Geologic Map 28 (scale 1:24,000). Socorro NM: New Mexico Bureau of Mines and Mineral Resources.

Woodward, Lee A., and Richard L. Ruetschilling, 1976. *Geology of San Ysidro Quadrangle, New Mexico*. Geologic Map 37 (scale 1:24,000). Socorro NM: New Mexico Bureau of Mines and Mineral Resources.

Woodward, Lee A., Leonard, M.L., Bruce A. Black, Spencer G. Lucas, and James R. Connolly, 1989. Road log from Albuquerque to San Ysidro, San Juan basin and return to Albuquerque. In *Energy Frontiers in the Rockies*, eds. John C. Lorenz and Spencer G. Lucas, 13-22. Companion volume for the 1989 meeting of the Rocky Mountain Section of the American Association of Petroleum Geologists. Albuquerque NM: Albuquerque Geological Society.

Workers of the Writers' Program of the Work Projects Administration, 1989. *The WPA Guide to 1930s New Mexico*, Tucson AZ: University of Arizona Press.

Young, Robert, 2004. Largo Canyon homesteads. *New Mexico Magazine*, May.

Index

CPSIA information can be obtained
at www.ICGtesting.com
Printed in the USA
BVHW011413201121
622137BV00011B/312